Theory and Application of Graphs

Network Theory and Applications

Volume 10

Theory and Application of Graphs

by

Junming Xu

Department of Mathematics
University of Science and Technology of China
Hefei, Anhui, China

Springer Science+Business Media, LLC

Library of Congress Cataloging-in-Publication

CIP info or:

Title: Theory and Application of Graphs
Author: Xu

ISBN 978-1-4613-4670-8 ISBN 978-1-4419-8698-6 (eBook)
DOI 10.1007/978-1-4419-8698-6

Originally published by Kluwer Academic Publishers in 2003
Softcover reprint of the hardcover 1st edition 2003

Printed on acid-free paper.

Contents

Preface

In the spectrum of mathematics, graph theory which studies a mathematical structure on a set of elements with a binary relation, as a recognized discipline, is a relative newcomer. In recent three decades the exciting and rapidly growing area of the subject abounds with new mathematical developments and significant applications to real-world problems. More and more colleges and universities have made it a required course for the senior or the beginning postgraduate students who are majoring in mathematics, computer science, electronics, scientific management and others. This book provides an introduction to graph theory for these students.

The richness of theory and the wideness of applications make it impossible to include all topics in graph theory in a textbook for one semester. All materials presented in this book, however, I believe, are the most classical, fundamental, interesting and important. The method we deal with the materials is to particularly lay stress on digraphs, regarding undirected graphs as their special cases. My own experience from teaching out of the subject more than ten years at University of Science and Technology of China (USTC) shows that this treatment makes hardly the course difficult, but much more accords with the essence and the development trend of the subject.

The book consists of seven chapters. Excepting that the first chapter contains the most basic concepts and results, each chapter presents one special topic, including trees and graphic spaces, plane and planar graphs, flows and connectivity, matchings and independent sets, coloring theory, graphs and groups. These topics are treated in some depth, both theoretical and applied, with some suggestions for further reading. Every effort will be made to strengthen the mutual connections among these topics, with an aim to make the materials more systematic and cohesive. All theorems will be stated clearly, together with full and concise proofs, some of them are new. A number of examples and more than 350 figures are given to help the reader to understand the given materials. To explore the mathematical nature of graph theory better, we will specially stress the equivalence of some classical results, such as the max-flow min-cut theorem of Ford and Fulkerson, Menger's theorem, Hall's theorem, Tutte's theorem and König's theorem.

Throughout this book the reader will see that graph theory has closed connection with other branches of mathematics, including linear algebra, matrix theory, group theory, combinatorics, combinatorial optimization and operation research, and wide applications to other subjects, including computer science, electronics, scientific management and so on. The applications carefully selected are arranged in the latter section(s) of each chapter. The aim of such arrangements is to conveniently choose these materials for some readers according to their interesting and available periods.

Exercises at the end of each section, more than 500, from routine practice to challenging, are supplements to the text. Some of them are very important results in graph theory. It is advisable for the reader to be familiar with the new definitions introduced in the exercises since they are useful for further study. The reader is also advised to do the exercises as many as he (or she) can. The harder ones are indicated by bold type.

The style of writing and of presentation of this book have be, to a great extent, influenced by Graph Theory with Applications, a popular textbook written by J. A. Bondy and U. S. R. Murty whom I am grateful to, from which some typical materials have been directly selected in this book.

The book is developed from the text for a senior and first-year postgraduate course in one semester at USTC. I thank the participants of the course for their great interest and stimulating comments. I would like to thank Teaching Affairs Division, Graduate School and Department of Mathematics at USTC for their support and encouragement.

Many people have contributed, either directly or indirectly, to this book. I avail myself of this opportunity to particularly express my heartfelt gratitude to Qiao Li, Feng Tian, Yanpei Liu, Genghua Fan, Yongchuan Chen, Dingzhu Du and Shenggui Zhang for their continuous help and valuable suggestions.

Finally, I would like to express my appreciation to my son, Keli Xu, for his very concrete help, and my wife, Jingxia Qiu, for her support, understanding and love, without which this work would have been impossible.

Jun-Ming Xu
(xujm@ustc.edu.cn)
December 2002

Chapter 1

Basic Concepts of Graphs

In many real-world situations, it is particularly convenient to describe the specified relationship between pairs of certain given objects by means of a diagram, in which points presents the objects and (directed or undirected) lines the relationship between pairs of the objects. For example, a national traffic map describes a condition of the communication lines among cities in the country, where the points represent cities and the lines represent the highways or the railways joining pairs of cities. Notice that in such diagrams one is mainly interested in whether or not two given points are joined by a line; the manner in which they are joined is immaterial. A mathematical abstraction of situations of this type gives rise to the concept of a graph.

In fact, a graph provides the natural structures from which to construct mathematical models that are appropriate to almost all fields of scientific (natural and social) inquiry. The underlying subject of study in these fields is some set of "objects" and one or more "relations" between the objects.

In this chapter, we will introduce basic concepts of a graph used in the remaining parts of the book, including several special graphs, subgraphs of various types, walk, path, cycle, diameter, connection, Euler circuit, Hamilton cycle, adjacency and incidence matrices, as well as the basic results closely related these concepts. At the end of this chapter we will present an application of graphs to matrix theory.

It should, for the beginner specially, be worth noting that most graph theorists use personalized terminology in their books, papers and lectures. Even the meaning of the word "graph" varies with different authors. We will adopt the most standard terminology and notation extensively used by most authors, with a subject index and a list of symbols in the end of the book.

1.1 Graph and Graphical Presentation

Mathematically, a *graph* is a mathematical structure on a set of elements with a binary relation. Concretely speaking, a graph is an ordered triple (V, E, ψ), where V and E are two disjoint sets, ψ is a mapping from $E \to V \times V$. The set V is nonempty and called the *vertex-set* of the graph, an element in V called a *vertex*. The set E is called the *edge-set* of the graph, an element in E called an *edge*. The mapping ψ is called an *incidence function*, which maps an edge into a pair of vertices called *end-vertices* of the edge.

If $V \times V$ is considered as a set of ordered pair (x, y)'s, then the graph is called a *directed graph*, or *digraph* for short. For an edge e of a digraph, sometimes, called a *directed edge* or *arc*, if $\psi(e) = (x, y)$, then the end-vertices x and y are called the *tail* and the *head* of the edge e, respectively; and the edge e is sometimes called an *out-going edge* of x or an *in-coming edge* of y.

If $V \times V$ is considered as a set of unordered pair $\{x, y\}$'s, then the graph is called an *undirected graph*. Usually, it is customary to henceforth denote an unordered pair of vertices by either xy or yx instead of $\{x, y\}$. Edges of an undirected graph are sometimes called *undirected edges*.

From definition, it is possible that two end-vertices of an edge are identical, such an edge is called a *loop*. It is also possible that more than one edges are mapped into the same element in $V \times V$ under the mapping ψ, these edges are called *parallel edges* or *multi-edges*. For $x, y \in V(G)$, set $E_G(x, y) = \{e \in E(G) : \psi(e) = (x, y)\}$ and $\mu_G(x, y) = |E(x, y)|$. The parameter $\mu(G) = \max\{\mu_G(x, y) : \forall\, x, y \in V(G)\}$ is called the *multiplicity* of G.

Example 1.1.1 $D = (V(D), E(D), \psi_D)$ is a digraph, where

$$V(D) = \{x_1, x_2, x_3, x_4, x_5\},$$
$$E(D) = \{a_1, a_2, a_3, a_4, a_5, a_6, a_7, a_8, a_9\}$$

and ψ_D is defined by

$$\psi_D(a_1) = (x_1, x_2), \quad \psi_D(a_2) = (x_3, x_2), \quad \psi_D(a_3) = (x_3, x_3),$$
$$\psi_D(a_4) = (x_4, x_3), \quad \psi_D(a_5) = (x_4, x_2), \quad \psi_D(a_6) = (x_4, x_2),$$
$$\psi_D(a_7) = (x_5, x_2), \quad \psi_D(a_8) = (x_2, x_5), \quad \psi_D(a_9) = (x_3, x_5).$$

In such a digraph D, two edges a_5 and a_6 are parallel edges, but two edges a_7 and a_8 are not. The edge a_3 is a loop.

Example 1.1.2 $H = (V(H), E(H), \psi_H)$ is a digraph, where

$$V(H) = \{y_1, y_2, y_3, y_4, y_5\}$$
$$E(H) = \{b_1, b_2, b_3, b_4, b_5, b_6, b_7, b_8, b_9\}$$

and ψ_H is defined by

$$\begin{array}{lll} \psi_H(b_1) = (y_1, y_2), & \psi_H(b_2) = (y_3, y_2), & \psi_H(b_3) = (y_3, y_3), \\ \psi_H(b_4) = (y_4, y_3), & \psi_H(b_5) = (y_4, y_2), & \psi_H(b_6) = (y_4, y_2), \\ \psi_H(b_7) = (y_5, y_2), & \psi_H(b_8) = (y_2, y_5), & \psi_H(b_9) = (y_3, y_5). \end{array}$$

Example 1.1.3 $G = (V(G), E(G), \psi_G)$ is an undirected graph, where

$$V(G) = \{z_1, z_2, z_3, z_4, z_5, z_6\}$$
$$E(G) = \{e_1, e_2, e_3, e_4, e_5, e_6, e_7, e_8, e_9\}$$

and ψ_G is defined by

$$\begin{array}{lll} \psi_G(e_1) = z_1 z_2, & \psi_D(e_2) = z_1 z_4, & \psi_G(e_3) = z_1 z_6, \\ \psi_G(e_4) = z_2 z_3, & \psi_G(e_5) = z_3 z_4, & \psi_G(e_6) = z_3 z_6, \\ \psi_G(e_7) = z_2 z_5, & \psi_G(e_8) = z_4 z_5, & \psi_G(e_9) = z_5 z_6. \end{array}$$

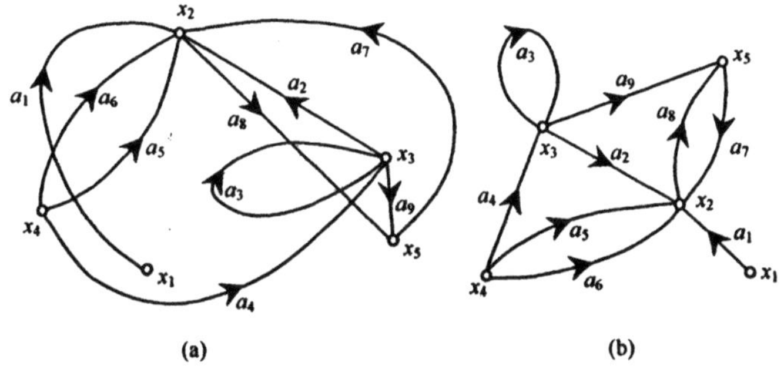

Figure 1.1: Two graphical presentations of the digraph D

A graph can be drawn on the plane. Each vertex x of the graph is indicated by a point. For clarity, such a point is often depicted as a small circle. If the graph is a directed, then each edge with tail x and head y is indicated by a directed line segment or curve joining from x to y. If the graph is undirected, then each edge with end-vertices x and y is indicated by an undirected line segment or curve joining x and y. Such a geometric diagram is called a *graphical presentation* of the graph. It depicts the incidence relationship holding between its vertices and edges intuitively.

For instance, the diagrams shown in Figure 1.1 are two graphical presentations of the digraph D defined in Example 1.1.1. The diagrams shown in Figure 1.2 are graphical presentations of the digraph H and the undirected graph G defined in Example 1.1.2 and Example 1.1.3, respectively. It is this representation that gives graph its name and much of its appeal.

The end-vertices of an edge are said to be *incident* with the edge, and vice versa. Two vertices which are incident with a common edge are *adjacent*, as are two edges which are incident with a common vertex.

A graph is said to be *loopless* if it contains no loop. A graph is said to be *simple* if it contains neither parallel edges nor loops. For a graph without parallel edges, the mapping ψ is injective. In other words, for each edge e there exists a unique pair of vertices corresponding to the edge. Thus it is convenience to directly use a subset of $V \times V$ instead of the edge-set E. In this case, we may simply write (V, E) for (V, E, ψ). For instance, the graph G defined in Example 1.1.3 is simple, which can be written as $G = (V(G), E(G))$, where $E(G) = \{z_1z_2, z_1z_4, z_1z_6, z_2z_3, z_3z_4, z_3z_6, z_2z_5, z_4z_5, z_5z_6\}$.

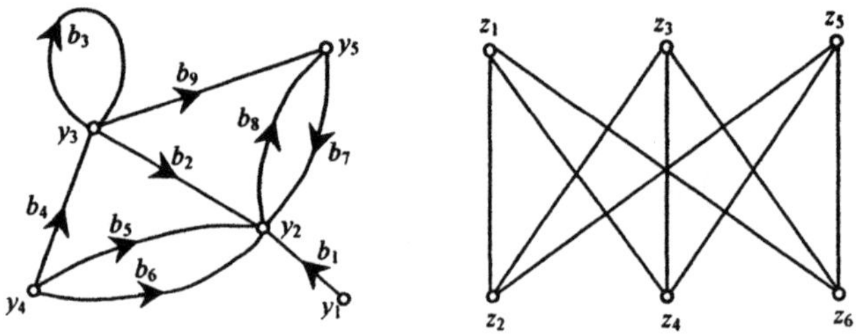

(a) H (b) G

Figure 1.2: Graphical presentations of graphs H and G

An undirected graph can be thought of as a particular digraph, a *symmetric digraph*, in which there are two directed edges called *symmetric edges*, one in each direction, corresponding to each undirected edge. Thus, to study structural properties of graphs for digraphs is more general than for undirected graphs.

There are many topics in graph theory that have no relations with direction of edges. The undirected graph obtained from a digraph D by removing the orientation of all edges is called an *underlying graph* of D. Conversely, the digraph obtained from an undirected graph G by specifying an orientation of each edge of G is called an *oriented graph* of G.

Figure 1.3 shows such graphs, where (a) is an undirected graph, (b) and (c) are its symmetric digraph and an oriented graph, respectively.

Let (V, E, ψ) be a graph. The number of vertices, $v = |V|$, is called *order* of the graph; the number of edges, $\varepsilon = |E|$, is called *size* of the graph. A graph is called to be *empty* if $\varepsilon = 0$. An empty graph is called to be *trivial* if $v = 1$, and all other graphs *non-trivial*. A graph is *finite* if both v and ε are finite. Throughout this book all graphs are always considered to be finite.

The letter G always denotes a graph, which is directed or undirected according to the context if it is not specially noted. Sometimes, to emphasize, we use the letter D to denote a digraph. When just one graph is under discussion, the letters v and ε always denote order and size of the graph, respectively.

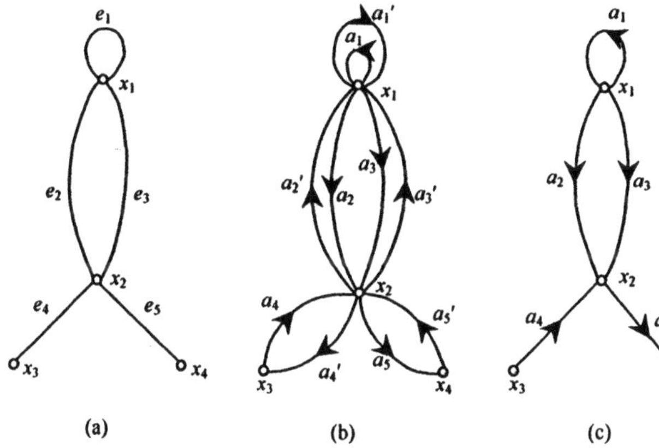

Figure 1.3: The symmetric digraph and oriented graph of an undirected graph

The symbols $\lfloor r \rfloor$ and $\lceil r \rceil$ denote the greatest integer not exceeding the real number r and the smallest integer not less than r, respectively. The symbol

$$\binom{n}{k} = \frac{n(n-1)\cdots(n-k+1)}{k!}$$

denotes the number of k-combinations of n distinct objects ($k \leq n$).

As an application of a graph, we give an example.

Example 1.1.4 In any group of six people, there must be three people who get to either know each other or not.

Proof We use the points A, B, C, D, E, F on the plane to denote these six people, respectively. We draw a red line joining two points if two people have known each other, a blue line otherwise. Use G to denote the resulting diagram. We need to only prove that G certainly contains either a red triangle or a blue triangle. Consider a point, say F. There exist three lines of the same color which are incident with a common point F. Without loss of generality, we can suppose that they are three red lines FA, FB and FC. Consider the triangle ABC. If it has no red line, then it is a blue triangle; if it has a red line, say AB, then the triangle FAB is red. ∎

Exercises

1.1.1 Drawing graphical presentations of the following five graphs without parallel edges B, K, Q, D and G, respectively, where

(a) $V(B) = \{x_1x_2x_3 : x_i \in \{0,1\}\}$ and if $x, y \in V(B)$, $x = x_1x_2x_3$, then $(x, y) \in E(B)$ if and only if $y = x_2x_3\alpha$, $\alpha \in \{0,1\}$;

(b) $V(K) = \{x_1x_2x_3 : x_i \in \{0,1,2\}, x_2 \neq x_1, x_3 \neq x_2\}$ and if $x, y \in V(K), x = x_1x_2x_3$, then $(x, y) \in E(K)$ if and only if $y = x_2x_3\alpha$, $\alpha \in \{0,1,2\}$ and $\alpha \neq x_3$;

(c) $V(Q) = \{x_1x_2x_3 : x_i \in \{0,1\}\}$, if $x, y \in V(Q)$, $x = x_1x_2x_3$ and $y = y_1y_2y_3$, then $xy \in E(Q)$ if and only if $|x_1 - y_1| + |x_2 - y_2| + |x_3 - y_3| = 1$;

(d) $V(D) = \{0, 1, \cdots, 7\}$, and $E(D) = \{(i, j)$: there exists some $s \in \{1,2\}$ such that $j - i \equiv s \pmod 8$ $\}$;

(e) $V(G) = \{0, 1, \cdots, 7\}$, and $E(G) = \{ij$: there exists some $s \in \{1,4\}$ such that $|j - i| \equiv s \pmod 8$ $\}$.

1.1.2 Prove that for any simple graph G,

(a) $\varepsilon \leq v(v-1)$ if G is directed;

(b) $\varepsilon \leq \frac{1}{2}\, v(v-1)$ if G is undirected.

1.1.3 The symbols $\mathcal{D}_v$ and $\mathcal{G}_v$ denote the sets of all simple digraphs and all simple undirected graphs of order v, respectively. Prove that

(a) $|\mathcal{D}_v| = 2^{v(v-1)}$;

(b) $|\mathcal{G}_v| = 2^{v(v-1)/2}$.

1.1.4 Prove that there are 2^{ε} different oriented graphs for any undirected graph.

1.1.5 The symbols $\mathcal{D}(v, \varepsilon)$ and $\mathcal{G}(v, \varepsilon)$ denote the sets of all simple digraph and undirected graphs of order v and size ε, respectively. Prove that

(a) $|\mathcal{D}(v, \varepsilon)| = \binom{v(v-1)}{\varepsilon}$;

(b) $|\mathcal{G}(v, \varepsilon)| = \binom{v(v-1)/2}{\varepsilon}$.

1.2 Graph Isomorphism

Two graphs often have the same structure, differing only in the way their vertices and edges are labelled or the way they are drawn on the plane. To make this idea more exact, we introduce the concept of isomorphism. A graph $G = (V(G), E(G), \psi_G)$ is *isomorphic* to a graph $H = (V(H), E(H), \psi_H)$ if there exist two bijective mappings

$$\theta : V(G) \to V(H) \quad \text{and} \quad \phi : E(G) \to E(H)$$

such that for any $e \in E(G)$,

$$\psi_G(e) = (x, y) \Longleftrightarrow \psi_H(\phi(e)) = (\theta(x), \theta(y)) \in E(H). \tag{1.1}$$

The pair (θ, ϕ) of mappings is called an *isomorphic mapping* from G to H.

Since such two mappings θ and ϕ are bijective, H also isomorphic to G. Thus we often call that G and H are *isomorphic*, write $G \cong H$, the pair (θ, ϕ) of mappings is called an *isomorphism* between G and H.

To show that two graphs are isomorphic, one must indicate an isomorphism between them. For instance, two digraphs D and H defined in Example 1.1.1 and Example 1.1.2, respectively, are isomorphic since the pair of mappings (θ, ψ) is an isomorphism between them, where $\theta : V(D) \to V(H)$ and $\psi : E(D) \to E(H)$ are defined by

$$\begin{array}{ll} \theta(x_i) = y_i, & \text{for each } i = 0, 1, 2, \cdots, 5; \quad \text{and} \\ \psi(a_j) = b_j, & \text{for each } i = 0, 1, 2, \cdots, 9. \end{array}$$

The concept of isomorphism for simple graphs is simple. Two simple graphs G and H are isomorphic if and only if there is a bijection $\theta : V(G) \to V(H)$ such that $(x, y) \in E(G)$ if and only if $(\theta(x), \theta(y)) \in E(H)$. In this case, the condition (1.1) is usually called the *adjacency-preserving condition.*

It is clear that if G and H are isomorphic, then $v(G) = v(H)$ and $\varepsilon(G) = \varepsilon(H)$. But the converse is not always true. Generally speaking, to judge whether or not two graphs are isomorphic is quite difficult.

It is easy to see that "to be isomorphic" is an equivalence relations on graphs; hence, this relation divides the collection of all graphs into equivalence classes. Two graphs in the same equivalence classes have the same structure, and differ only in the labels of vertices and edges. Since we are primarily interested in structural properties of graphs, we will identify two isomorphic graphs, and often write $G = H$ for $G \cong H$. We often omit labels when drawing them on the plane; an unlabelled graph can be thought of as a representative of the equivalence class of isomorphic graphs. We assign labels to vertices and edges in a graph mainly for the purpose of referring to them.

Next, we introduce some special classes of graphs, which frequently occur in our discussion later on.

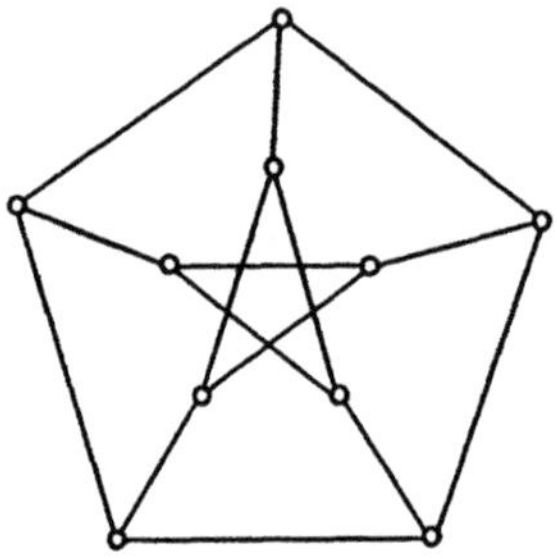

Figure 1.4: Petersen graph

The graph shown in Figure 1.4 is called *Petersen graph*, an interesting graph, which often occurs in the literature and any textbook on graph theory as various counterexamples.

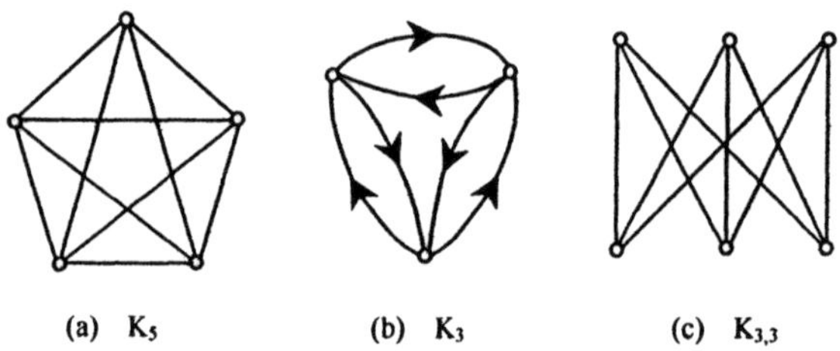

Figure 1.5: (a) K_5, (b) K_3, (c) $K_{3,3}$

A *complete graph* is one in which each ordered pair of distinct vertices is linked by exactly one edge. Up to isomorphism, there is just one complete graph on v vertices, denoted by K_v. The graphs shown in Figure 1.5 (a) and (b) are a complete undirected graph K_5 and a complete digraph K_3, respectively. It is clear that

$$\varepsilon(K_v) = \begin{cases} v(v-1) & \text{if } K_v \text{ is directed;} \\ \frac{1}{2}\, v(v-1) & \text{if } K_v \text{ is undirected.} \end{cases}$$

An oriented graph of a complete undirected graph is called a tournament. The reason why we call it the name is that it can be used to indicate the results of games in a round-robin tournament between v players. A directed edge (x, y) means that the player x has won the player y. Up to isomorphism, the tournament of order one is a trivial graph; there is just one tournament of order two; two tournaments of order three; four tournaments of order four. These not isomorphic tournaments are shown in Figure 1.6.

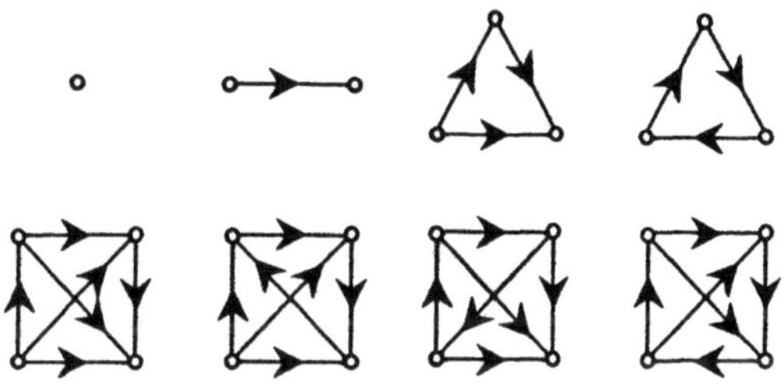

Figure 1.6: Nonisomorphic tournaments of order i for $i = 1, 2, 3, 4$

A *bipartite graph* is one whose vertex-set can be partitioned into two subsets X and Y, so that each edge has one end-vertex in X and another in Y, such a partition $\{X, Y\}$ is called a *bipartition* of the graph. We call a graph to be *equally bipartite* if it is bipartite and has a bipartition with the same number of vertices in each part. We often use the symbol $G(X \cup Y, E)$ to denote a bipartite simple graph $G = (V, E)$ with bipartition $\{X, Y\}$. Similarly, we can define a *k-partite graph* and an equally k-partite graph.

A *complete bipartite graph* is a bipartite simple graph $G(X \cup Y, E)$ in which each vertex of X is joined by exactly one edge to each vertex of Y; if $|X| = m$ and $|Y| = n$, up to isomorphism, such a complete bipartite undirected graph is unique and denoted by $K_{m,n}$. The graph shown in Figure 1.5 (c) is $K_{3,3}$. It is customary to call $K_{1,n}$ a *star.* Usually, write $K_n(2)$ for $K_{n,n}$.

Similarly, we can define *complete k-partite graph* and $K_n(k)$. It is easy to verify that

$$\varepsilon(K_{m,n}) = mn \quad \text{and} \quad \varepsilon(K_n(k)) = \frac{1}{2}\, k(k-1)n^2.$$

It is also easy to verify that for any bipartite simple graph G of order n,

$$\varepsilon(G) \leq \begin{cases} \frac{1}{4}\, n^2 & \text{if } n \text{ is even;} \\ \frac{1}{4}\, (n^2 - 1) & \text{if } n \text{ is odd.} \end{cases}$$

Bipartite graphs are an important class of graphs. In fact, every digraph corresponds a bipartite undirected graph. Let $D = (V, E, \psi)$ be a digraph, where

$$V(D) = \{x_1, x_2, \cdots, x_v\} \quad \text{and} \quad E(D) = \{a_1, a_2, \cdots, a_\varepsilon\}.$$

Construct an equally bipartite undirected graph $G = (X \cup Y, E(G), \psi_G)$ with

$$\begin{aligned} X &= \{x_1', x_2', \cdots, x_v'\}, \quad Y = \{x_1'', x_2'', \cdots, x_v''\}, \\ E(G) &= \{e_1, e_2, \cdots, e_\varepsilon\}, \text{ where } \psi_G(e_l) = x_i'x_j'' \\ &\Longleftrightarrow \text{ there is } a_l \in E(D) \text{ such that } \psi_D(a_l) = (x_i, x_j) \\ &\qquad (l = 1, 2, \cdots, \varepsilon). \end{aligned}$$

So constructing bipartite undirected graph G is called an *associated bipartite graph* with the digraph D. For instance, the graph G shown in Figure 1.7 (b) is an associated bipartite graph with the digraph D shown in (a). It is clear that

$$v(G) = 2\,v(D) \quad \text{and} \quad \varepsilon(G) = \varepsilon(D). \tag{1.2}$$

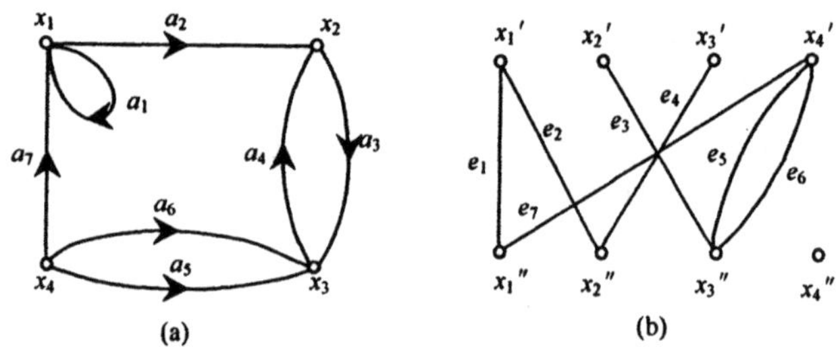

Figure 1.7: A digraph D and its associated bipartite graph G

Example 1.2.1 We construct an equally bipartite simple graph, called *n-cube*, or *hypercube*, denoted by $Q_n = (V(Q_n), E(Q_n))$, where

$$V(Q_n) = \{x_1x_2\cdots x_n : \ x_i \in \{0,1\}, \ i = 1, 2, \cdots, n\},$$

and two vertices $x = x_1x_2\cdots x_n$ and $y = y_1y_2\cdots y_n$ are linked by an undirected edge if and only if they differ in exactly one coordinate, i.e.,

$$xy \in E(Q_n) \Longleftrightarrow \sum_{i=1}^{n} |x_i - y_i| = 1.$$

The graphs shown in Figure 1.8 are Q_1, Q_2, Q_3 and Q_4.

By definition, Q_n is a simple undirected graph, and has 2^n vertices. We show that Q_n is bipartite. To the end, let

$$\begin{aligned} X &= \{x_1x_2\cdots x_n : \ x_1 + x_2 + \cdots + x_n \equiv 0 \pmod 2\}; \\ Y &= \{y_1y_2\cdots y_n : \ y_1 + y_2 + \cdots + y_n \equiv 1 \pmod 2\}. \end{aligned}$$

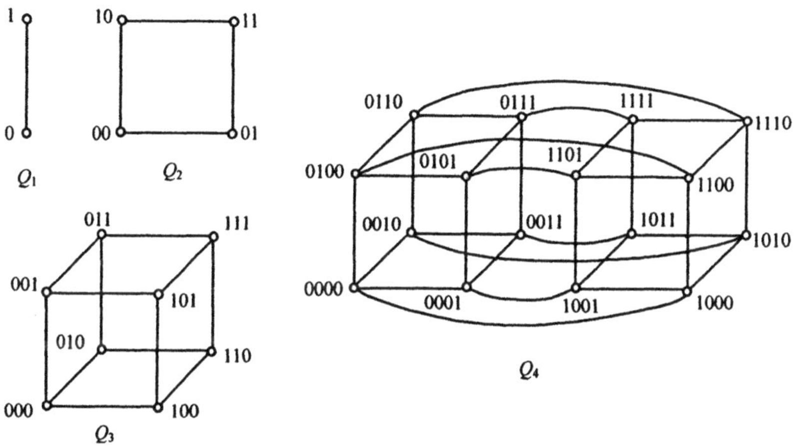

Figure 1.8: n-cubes Q_n for $n = 1, 2, 3, 4$

Then, by definition, $X \cup Y = V(Q_n)$, $X \cap Y = \emptyset$. Therefore, $\{X, Y\}$ is a bipartition of $V(Q_n)$. We can claim that there is no edge between any two vertices in X. Suppose to the contrary that there exist $x = x_1x_2\cdots x_n$, $x' = x'_1x'_2\cdots x'_n \in X$ such that $xx' \in E(Q_n)$. Then $\sum_{i=1}^{n} |x_i - x'_i| = 1$, namely,

$$|(x_1 + x_2 + \cdots + x_n) - (x'_1 + x'_2 + \cdots + x'_n)| = 1.$$

This contradicts the fact that x and x' are in X. There, therefore, is no edge between any two vertices in X.

Similarly, there is no edge between any two vertices in Y. Therefore, Q_n is a bipartite graph with the bipartition $\{X, Y\}$.

Arbitrarily choose $x = x_1x_2\cdots x_n \in V(Q_n)$. For a vertex $y = y_1y_2\cdots y_n$ of Q_n, it is adjacent to x if and only if they differ in exactly one coordinate. This means that vertices adjacent to the vertex x have exactly n, that is, edges incident with x have n since Q_n is simple. Let us use E_X (resp. E_Y) to denote the set of edges incident with vertices in X (resp. Y). Then

$$n|X| = |E_X| = \varepsilon(Q_n) = |E_Y| = n|Y|.$$

As a result, we have that

$$|X| = |Y| = \frac{1}{2}\, v(Q_n) = 2^{n-1} \quad \text{and} \quad \varepsilon(Q_n) = n2^{n-1}.$$

Example 1.2.2 The symbol $T_{k,v}$ denotes a complete k-partite graph of order v in which each part has either $m = \left\lfloor \frac{v}{k} \right\rfloor$ or $n = \left\lceil \frac{v}{k} \right\rceil$ vertices. Prove that

(a) $\varepsilon(T_{k,v}) = \binom{v-m}{2} + (k-1)\binom{m+1}{2}$;

(b) $\varepsilon(G) \leq \varepsilon(T_{k,v})$ for any complete k-partite graph G, and the equality holds if and only if $G \cong T_{k,v}$.

Proof (a) Let $v = km + r$, $0 \leq r < k$. Then $r = v - km$. By the definition of $T_{k,v}$, we have that

$$
\begin{aligned}
T_{k,v} &= \binom{v}{2} - r\binom{m+1}{2} - (k-r)\binom{m}{2} \\
&= \tfrac{1}{2}\left[v(v-1) - rm(m+1) - (k-r)m(m-1)\right] \\
&= \tfrac{1}{2}\left[v(v-1) - 2m(v-km) - km(m-1)\right] \\
&= \tfrac{1}{2}(v-m)(v-m-1) + \tfrac{1}{2}(k-1)m(m+1) \\
&= \binom{v-m}{2} + (k-1)\binom{m+1}{2}.
\end{aligned}
$$

(b) Suppose that $G = K_{n_1,\cdots,n_k}$ is a complete k-partite graph with the largest number of edges. Then

$$\varepsilon(G) = \binom{v}{2} - \sum_{l=1}^{k}\binom{n_l}{2}.$$

If G is not isomorphic to $T_{k,v}$, then there must exist some i and j $(i < j)$ such that $n_i - n_j > 1$. Consider another complete k-partite graph G', the number of vertices in its k-partition are, respectively,

$$n_1, n_2, \cdots, n_{i-1}, (n_i - 1), n_{i+1}, \cdots, n_{j-1}, (n_j + 1), n_{j+1}, \cdots, n_k.$$

Then

$$
\begin{aligned}
\varepsilon(G') &= \binom{v}{2} - \sum_{l=1\neq i,j}^{k}\binom{n_l}{2} - \binom{n_i - 1}{2} - \binom{n_j+1}{2} \\
&= \binom{v}{2} - \sum_{l=1}^{k}\binom{n_l}{2} - (n_i - n_j) - 1 \\
&> \binom{v}{2} - \sum_{l=1}^{k}\binom{n_l}{2} = \varepsilon(G),
\end{aligned}
$$

which contradicts to the choice of G. Thus, $G \cong T_{k,v}$. ∎

Exercises

1.2.1 (a) Prove that if $G \cong H$, then $v(G) = v(H)$ and $\varepsilon(G) = \varepsilon(H)$.

(b) Construct a graph to show that the converse of (a) is not true.

1.2.2 Prove that if G is a bipartite simple graph G of order n, then

$$\varepsilon(G) \leq \begin{cases} \frac{1}{4}\, n^2 & \text{if } n \text{ is even;} \\ \frac{1}{4}\, (n^2 - 1) & \text{if } n \text{ is odd.} \end{cases}$$

In Particular, $\varepsilon(K_{m,n}) = mn$.

1.2.3 Write out definition of k-partite graph and prove that

$$\varepsilon(K_n(k)) = \tfrac{1}{2}\, k(k-1)n^2.$$

1.2.4 Prove that the following three graphs are isomorphic to Petersen graph.

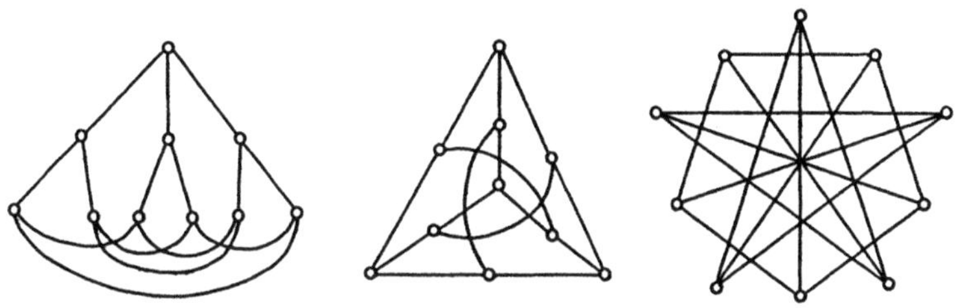

(the exercise 1.2.4)

1.2.5 The *complement* G^c of a simple graph $G = (V, E)$ is the simple graph with the vertex-set V, and $(x, y) \in E(G^c) \Longleftrightarrow (x, y) \notin E(G)$. Prove that

(a) the complement of every tournament is a tournament;

(b) $G^c \cong H^c \Longleftrightarrow G \cong H$ if both G and H are simple.

1.2.6 A simple graph G is *self-complementary* if $G \cong G^c$. Prove that if G is self-complementary, then

(a) $\varepsilon(G) = \frac{1}{2}\, v(v-1)$ if G is directed;

(b) $\varepsilon(G) = \frac{1}{4}\, v(v-1)$ and $v \equiv 0$, or 1 (mod 4) if G is undirected.

1.2.7 Construct that

(a) two self-complementary tournaments of order four;

(a) a self-complementary undirected graph of order five.

1.3 Vertex Degrees

Let G be an undirected graph and $x \in V(G)$. The *degree* of x in G, denoted by $d_G(x)$, is the number of edges incident with x, each loop counting as two edges.

For the graph G shown in Figure 1.9 (a), for instance,

$$d_G(x_1) = d_G(x_3) = 4, \quad d_G(x_2) = d_G(x_4) = 3.$$

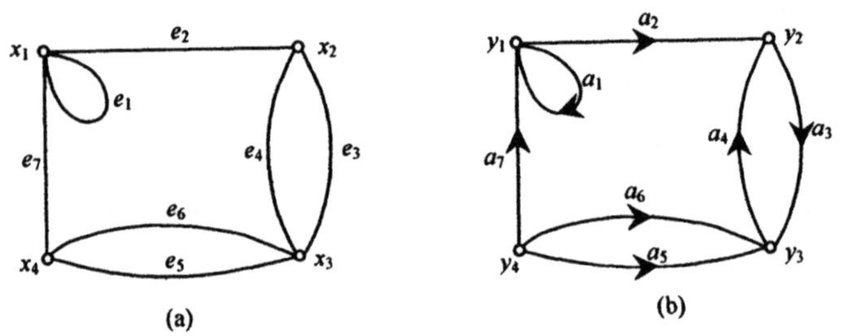

Figure 1.9: (a) an undirected graph G (b) a digraph D

A vertex of degree d is called a *d-degree vertex*. A 0-degree vertex is called an *isolated vertex*. A vertex is called to be odd or even if its degree is odd or even. A graph G is *k-regular* if $d_G(x) = k$ for each $x \in V(G)$, and G is *regular* if it is k-regular for some k, and k is called the *regularity* of G.

For instance, the complete graph K_n is $(n-1)$-regular, the complete bipartite graph $K_{n,n}$ is n-regular; Petersen graph is 3-regular; the n-cube is n-regular. The parameters

$$\begin{aligned} \Delta(G) &= \max\{d_G(x) : x \in V(G)\}, \quad \text{and} \\ \delta(G) &= \min\{d_G(x) : x \in V(G)\} \end{aligned}$$

are the *maximum* and the *minimum degree* of G, respectively. Clearly, $\delta(G) = k = \Delta(G)$ if G is k-regular.

We now give the corresponding terminology and notation for a digraph. Let D be a digraph and $y \in V(D)$. The symbol $E_D^+(y)$ denotes a set of out-going edges of y in D. The cardinality $|E_D^+(y)|$ is the *out-degree* of y, denoted by $d_D^+(y)$. Similarly, the symbol $E_D^-(y)$ denotes a set of in-coming edges of y in D, and $d_D^-(y) = |E_D^-(y)|$ is the *in-degree* of y. For the digraph D shown in Figure 1.9 (b), for instance,

$$\begin{aligned} &d_D^+(y_1) = 2, \quad d_D^+(y_2) = 1, \quad d_D^+(y_3) = 1, \quad d_D^+(y_4) = 3; \\ &d_D^-(y_1) = 2, \quad d_D^-(y_2) = 2, \quad d_D^-(y_3) = 3, \quad d_D^-(y_4) = 0, \end{aligned}$$

A vertex y is called to be *balanced* if $d_D^+(y) = d_D^-(y)$, and D is called to be *balanced* if each of its vertices is balanced. The parameters

$$\Delta^+(D) = \max\{d_D^+(y) : \ y \in V(D)\}, \quad \text{and}$$
$$\Delta^-(D) = \max\{d_D^-(y) : \ y \in V(D)\}$$

are the *maximum out-degree* and *maximum in-degree* of D, respectively. The parameters

$$\delta^+(D) = \min\{d_D^+(y) : \ y \in V(D)\}, \quad \text{and}$$
$$\delta^-(D) = \min\{d_D^-(y) : \ y \in V(D)\}$$

are the *minimum out-degree* and *minimum in-degree* of D, respectively. The parameters

$$\Delta(D) = \max\{\Delta^+(D),\ \Delta^-(D)\}, \quad \text{and}$$
$$\delta(D) = \min\{\delta^+(D),\ \delta^-(D)\}$$

are the *maximum* and the *minimum degree* of a digraph D, respectively. A digraph D is *k-regular* if $\Delta(D) = \delta(D) = k$.

Let G be a bipartite undirected graph with a bipartite $\{X, Y\}$. It is easy to see that the relationship between degree of vertices and the number of edges of G is as follows.

$$\sum_{x \in X} d_G(x) = \varepsilon(G) = \sum_{y \in Y} d_G(y). \tag{1.3}$$

As a result, we have

$$2\,\varepsilon(G) = \sum_{x \in V(G)} d_G(x). \tag{1.4}$$

Generally, for any a digraph D we have the following relationship between degree of vertices and the number of edges of G.

Theorem 1.1 For any digraph D,

$$\varepsilon(D) = \sum_{x \in V} d_D^+(x) = \sum_{x \in V} d_D^-(x).$$

Proof Let G be the associated bipartite graph with D of bipartition $\{X, Y\}$. Note that

$$d_G(x') = d_D^+(x), \quad d_G(x'') = d_D^-(x), \quad \forall\, x \in V(D).$$

By the equality (1.3), we have that

$$\sum_{x \in V} d_D^+(x) = \sum_{x' \in X} d_G(x') = \varepsilon(G) = \sum_{x'' \in Y} d_G(x'') = \sum_{x \in V} d_D^-(x).$$

Since $\varepsilon(D) = \varepsilon(G)$ by (1.2), the theorem follows. ∎

Corollary 1.1 For any undirected graph G,

$$2\varepsilon(G) = \sum_{x \in V} d_G(x)$$

and the number of vertices of odd degree is even.

Proof Let D be the symmetric digraph of G. Then $\varepsilon(D) = 2\varepsilon(G)$. Note that

$$d_G(x) = d_D^+(x) = d_D^-(x), \quad \forall\, x \in V.$$

By Theorem 1.1, we have that

$$\sum_{x \in V} d_G(x) = \sum_{x \in V} d_D^+(x) = \sum_{x \in V} d_D^-(x) = \varepsilon(D) = 2\varepsilon(G).$$

Let V_o and V_e be the sets of vertices of odd and even degree in G, respectively. Then

$$\sum_{x \in V_o} d_G(x) + \sum_{x \in V_e} d_G(x) = \sum_{x \in V} d_G(x) = 2\varepsilon(G).$$

Since $\sum\limits_{x \in V_e} d_G(x)$ is even, it follows that $\sum\limits_{x \in V_o} d_G(x)$ is also even. Since $d_G(x)$ is odd for any $x \in V_o$, thus, $|V_o|$ is even. ∎

The following notation and terminology are useful and convenient to our discussions later on.

Let D be a digraph, S and T are disjoint nonempty subset of $V(D)$. The symbol $E_D(S,T)$ denotes the set of edges of D whose tails are in S and heads are in T, and $\mu_D(S,T) = |E_D(S,T)|$. When just one graph is under discussion, we usually omit the letter D from these symbols and write (S,T) and $\mu(S,T)$ instead of $E_D(S,T)$ and $\mu_D(S,T)$ for short. $[S,T] = (S,T) \cup (T,S)$. If $T = \overline{S} = V(D) \setminus S$, then we write $E_D^+(S)$ (resp. $E_D^-(S)$) instead of $(S, \overline{S})$ (resp. $(\overline{S}, S)$), and $d_D^+(S) = |E_D^+(S)|$ (resp. $d_D^-(S) = |E_D^-(S)|$).

The symbol $N_D^+(S)$ (resp. $N_D^-(S)$) denotes the set of heads (resp. tails) of edges in $E_D[S]$, which is called a set of *out-neighbors* (resp. *in-neighbors*) S in D.

For instance, consider the digraph D shown in Figure 1.9. Let $S = \{y_1, y_2\}$, then

$$\begin{array}{lll} E_D^+(S) = \{a_3\}, & d_D^+(S) = 1, & N_D^+(S) = \{y_3\}, \\ E_D^-(S) = \{a_4, a_7\}, & d_D^-(S) = 2, & N_D^-(S) = \{y_3, y_4\}. \end{array}$$

Similarly, for an undirected graph G and $S \subset V(G)$, the symbols $E_G(S)$ and $N_G(S)$ denote the set of edges incident with vertices in S in G and the set of *neighbors* of S in G, $d_G(S) = |E_G(S)|$.

Example 1.3.1 Prove that $\varepsilon(G) \le \frac{1}{4}\, v^2$ for any simple undirected graph G without triangles.

Proof Arbitrarily choose $xy \in E(G)$. Since G is simple and contains no triangle, it follows that

$$[d_G(x) - 1] + [d_G(y) - 1] \le v - 2,$$

that is,

$$d_G(x) + d_G(y) \le v.$$

Then summing over all edges in G yields

$$\sum_{x \in V} d_G^2(x) \le v\, \varepsilon.$$

By Cauchy's inequality and Corollary 1.1, we have that

$$v\, \varepsilon \ge \sum_{x \in V} d_G^2(x) \ge \frac{1}{v} \left(\sum_{x \in V} d_G(x) \right)^2 = \frac{4}{v}\, \varepsilon^2,$$

that is, $\varepsilon(G) \le \frac{1}{4}\, v^2$. ∎

Example 1.3.2 Let G is a self-complementary simple undirected graph with $v \equiv 1 \pmod 4$. Prove that the number of vertices of degree $\frac{1}{2}\,(v-1)$ in G is odd (the self-complementary graph is defined in the exercise 1.2.6).

Proof Let V_o and V_e be the sets of vertices of odd and even degree in G, respectively. Then $|V_o|$ is even by Corollary 1.1. Since $v \equiv 1 \pmod 4$, v must be odd and, thus, $|V_e|$ is odd and $\frac{1}{2}\,(v-1)$ is even.

Let V_e' be the set of vertices in V_e whose degree are not equal to $\frac{1}{2}\,(v-1)$. To prove the conclusion, we need to only show that $|V_e'|$ is even. To the end, let $x \in V_e'$. Then, since $G \cong G^c$, there must exist $y_x \in V(G)$ such that $d_G(y_x) = d_{G^c}(x)$. Note that

$$d_G(y_x) = d_{G^c}(x) = (v-1) - d_G(x) \tag{1.5}$$

is even. Thus, $y_x \in V_e$. Since $d_G(x) \ne \frac{1}{2}\,(v-1)$, it follows that $d_G(y_x) \ne \frac{1}{2}\,(v-1)$ and $y_x \ne x$ from (1.5). Therefore, $y_x \in V_e'$. Furthermore, $y_x \ne y_z$ if $x, z \in V_e'$ and $x \ne z$. This fact implies that the vertices in V_e' occur in pairs, which shows that $|V_e'|$ is even. ∎

Exercises

1.3.1 Prove that $\delta \leq 2\varepsilon/v \leq \Delta$ for any undirected graph.

1.3.2 Prove that there are always two vertices with exactly the same degree for any simple undirected graph of order at least two.

1.3.3 (a) Prove that if a digraph D is both δ^+-regular and δ^--regular, then $\delta = \delta^+ = \delta^-$, and hence D is δ-regular.

(b) Construct a digraph that is δ^+-regular but not δ^--regular.

1.3.4 Let $v \geq 2$. Prove that

(a) there exists a simple digraph D of order v such that for any two distinct vertices x and y

$$d_D^+(x) \neq d_D^+(y) \quad \text{and } d_D^-(x) \neq d_D^-(y);$$

(b) there exists a simple digraph D of order v such that the number of vertices of odd out-degree and the number of vertices of odd in-degree both are odd;

(c) there exists a r-regular simple digraph for any integer r with $r < v$.

1.3.5 Prove that for any tournament D,

$$\sum_{x\in V} d_D^+(x)^2 = \sum_{x\in V} d_D^-(x)^2 = \sum_{x\in V} (v - d_D^-(x))^2 - v^2.$$

1.3.6 Prove that

(a) any k (> 0)-regular bipartite graph is equally bipartite;

(b) any k-regular tournament has order $v = 2k + 1$.

1.3.7 Let X and Y be two subsets of $V(G)$. Prove that

(a) $d_G^+(X \cap Y) + d_G^+(X \cup Y) \leq d_G^+(X) + d_G^+(Y)$ if G is a digraph;

(b) $d_G^-(X \cap Y) + d_G^-(X \cup Y) \leq d_G^-(X) + d_G^-(Y)$ if G is a digraph;

(c) $d_G(X \cap Y) + d_G(X \cup Y) \leq d_G(X) + d_G(Y)$ if G is an undirected graph.

1.3.8 The symbol $\varepsilon_{\min}$ denotes the minimum number of edges in a simple undirected graph of order v that there is at least one edge among any three vertices. Prove that

$$\varepsilon_{\min} = \begin{cases} k^2 - k, & \text{if } v = 2k; \\ k^2, & \text{if } v = 2k + 1. \end{cases}$$

1.4 Subgraphs and Operations

A subgraph is one of the most basic concepts in graph theory. In this section, we first introduce various subgraphs induced by operations of graphs.

Suppose that $G = (V(G), E(G), \psi_G)$ is a graph. A graph $H = (V(H), E(H), \psi_H)$ is called a *subgraph* of G, denoted by $H \subseteq G$, or G is a *supergraph* of H if $V(H) \subseteq V(G)$, $E(H) \subseteq E(G)$ and ψ_H is the restriction of ψ_G to $E(H)$. A subgraph H of G is called a *spanning subgraph* if $V(H) = V(G)$.

Let S be a nonempty subset of $V(G)$. The *induced subgraph* by S, denoted by $G[S]$, is a subgraph of G whose vertex-set is S and whose edge-set is the set of those edges of G that have both end-vertices in S. The symbol $G - S$ denotes the induced subgraph $G[V \setminus S]$.

Let B be a nonempty subset of $E(G)$, the *edge-induced subgraph* by B, denoted by $G[B]$, is a subgraph of G whose vertex-set is the set of end-vertices of edges in B and whose edge-set is B. The symbol $G - B$ denotes the spanning subgraph $G[E \setminus B]$ of G. Similarly, the graph obtained by adding a set of extra edges F to G is denoted by $G + F$. Subgraphs of these various types are depicted in Figure 1.10.

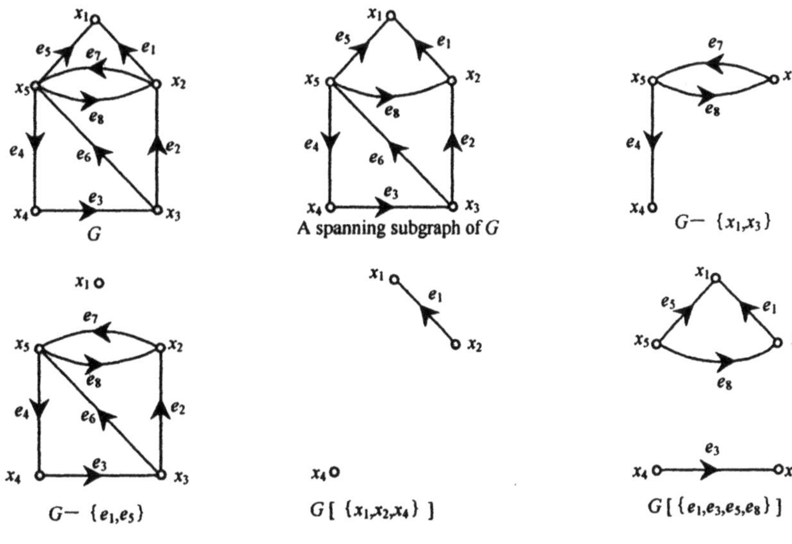

Figure 1.10: A graph and its various types of subgraphs

Let G_1 and G_2 be subgraphs of G. We say that G_1 and G_2 are *disjoint* if they have no vertex in common, and *edge-disjoint* if they have no edge in common. The *union* $G_1 \cup G_2$ of G_1 and G_2 is the subgraph with vertex-set $V(G_1) \cup V(G_2)$ and edge-set $E(G_1) \cup E(G_2)$. We write $G_1 + G_2$ for $G_1 \cup G_2$ if G_1 and G_2 are disjoint, and $G_1 \oplus G_2$ for $G_1 \cup G_2$ if G_1 and G_2 are edge-disjoint. If $G_i \cong H$ for each $i = 1, 2, \cdots, n$, then write nH for $G_1 + G_2 + \cdots + G_n$. The *intersection* $G_1 \cap G_2$ of G_1 and G_2 is defined similarly if $V(G_1) \cap V(G_2) \neq \emptyset$. These operations of graphs are depicted in Figure 1.11.

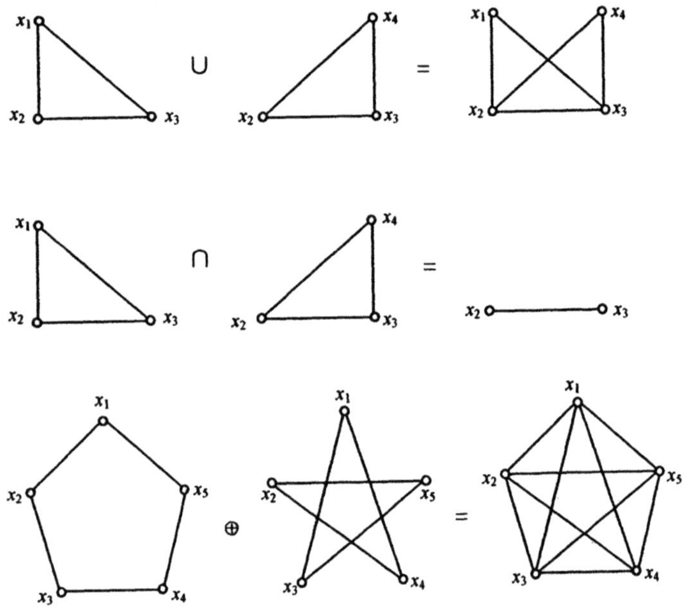

Figure 1.11: Union and intersection of graphs

An edge e of G is said to be *contracted* if it is deleted and its end-vertices are identified; the resulting graph is denoted by $G \cdot e$. This is illustrated in Figure 1.12.

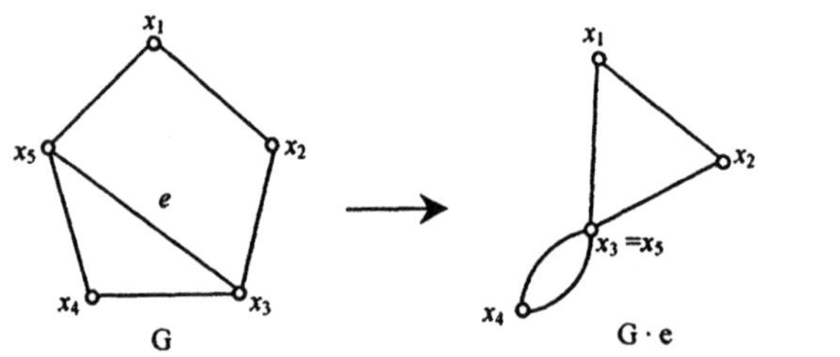

Figure 1.12: A graph $G \cdot e$ by contracting the edge e of G

Example 1.4.1 Let G be a balanced digraph. Then $d^+_G(X) = d^-_G(X)$ for any nonempty proper $X \subset V(G)$.

Proof Let $H = G[X]$. Since G is balanced, $d^+_G(x) = d^-_G(x)$ for each $x \in V(G)$. By Theorem 1.1, we have that $\sum\limits_{x\in X} d^+_H(x) = \sum\limits_{x\in X} d^-_H(x)$. Thus,

$$d^+_G(X) = \sum_{x\in X} d^+_G(x) - \sum_{x\in X} d^+_H(x) = \sum_{x\in X} d^-_G(x) - \sum_{x\in X} d^-_H(x) = d^-_G(X)$$

as required. ∎

Example 1.4.2 Let G be an undirected graph without loops. Then G contains a bipartite spanning subgraph H such that $d_G(x) \leq 2d_H(x)$ for any $x \in V(G)$. Hence $\varepsilon(G) \leq 2\,\varepsilon(H)$.

Proof Let H be a bipartite spanning subgraph of G with edges as many as possible, and let $\{X, Y\}$ be a bipartition. Arbitrarily choose $x \in V(G)$, without loss of generality, say $x \in X$. Let $d = d_G(x) - d_H(x)$.

We claim that $d \leq d_H(x)$. In fact, suppose to the contrary that $d > d_H(x)$. Let $X' = X \setminus \{x\}$ and $Y' = Y \cup \{x\}$. Consider a bipartite spanning subgraph H' of G with the bipartition $\{X', Y'\}$. Then

$$\varepsilon(H) \geq \varepsilon(H') = \varepsilon(H) + d - d_H(x) > \varepsilon(H),$$

a contradiction. Thus, $d_G(x) = d + d_H(x) \leq 2\,d_H(x)$. Summing up all vertices in G yields that $\varepsilon(G) \leq 2\,\varepsilon(H)$ by Corollary 1.1. ∎

The *cartesian product* $G_1 \times G_2$ of two simple graphs G_1 and G_2 is a graph with the vertex-set $V_1 \times V_2$, in which there is an edge from a vertex x_1x_2 to another y_1y_2, where $x_1, y_1 \in V(G_1)$ and $x_2, y_2 \in V(G_2)$, if and only if either $x_1 = y_1$ and $(x_2, y_2) \in E(G_2)$, or $x_2 = y_2$ and $(x_1, y_1) \in E(G_1)$. See Figure 1.8, for example, $Q_2 = K_2 \times K_2$, $Q_3 = K_2 \times Q_2$ and $Q_4 = K_2 \times Q_3$, in general, $Q_n = K_2 \times Q_{n-1}$. Some simple properties are stated in the exercise 1.4.6. Particularly, the cartesian product satisfies commutative and associative laws if we identify isomorphic graphs. It is the two laws that can make us greatly simplify proofs of many properties of the cartesian products.

Let $G_i = (V_i, E_i)$ be a graph for each $i = 1, 2, \cdots, n$. By the commutative and associative laws of the cartesian product, we may write $G_1 \times G_2 \times \cdots \times G_n$ for the cartesian product of $G_1, G_2, \cdots, G_n$, where $V(G_1 \times G_2 \times \cdots \times G_n) = V_1 \times V_2 \times \cdots \times V_n$. Two vertices $x_1x_2\cdots x_n$ and $y_1y_2\cdots y_n$ are linked by an edge in $G_1 \times G_2 \times \cdots \times G_n$ if and only if two vectors $(x_1, x_2, \cdots, x_n)$ and $(y_1, y_2, \cdots, y_n)$ differ exactly in one coordinate, say the ith, and there is an edge $(x_i, y_i) \in E(G_i)$.

Example 1.4.3 An important class of graphs, the well-known hypercube Q_n, defined in Example 1.2.1, can be defined in terms of the cartesian products, that is, $Q_n = K_2 \times K_2 \times \cdots \times K_2$ of n identical complete graph K_2, see Figure 1.8 for Q_1, Q_2, Q_3 and Q_4. The hypercube is an important class of topological structures of interconnection networks, some of whose properties will be further discussed in some sections in this book.

The *line graph* of G, denoted by $L(G)$, is a graph with vertex-set $E(G)$ in which there is an edge (a, b) if and only if there are vertices $x, y, z \in V(G)$ such that $\psi_G(a) = (x, y)$ and $\psi_G(b) = (y, z)$. This is illustrated in Figure 1.13. Some simple properties of line graphs are stated in the exercise 1.4.4.

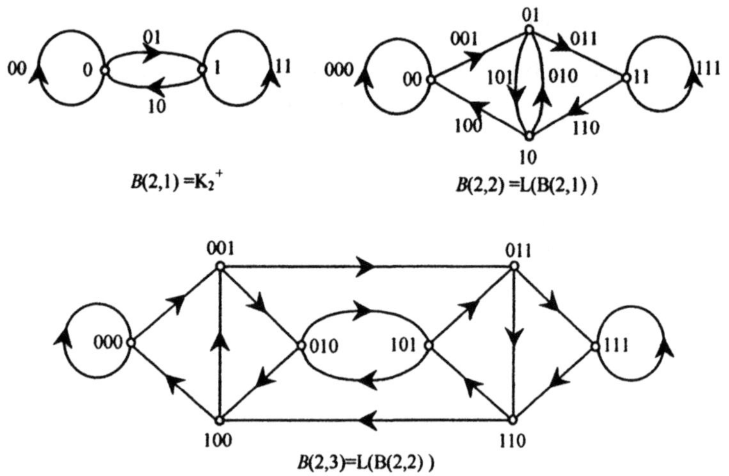

Figure 1.13: Graphs and their line graphs

Assume that $L(G)$ is the line graph of a graph G. If $L(G)$ is non-empty and has no isolated vertices, then its line graph $L(L(G))$ exists. For integers $n \geq 1$, $L^n(G)$ can be recursively defined as $L(L^{n-1}(G))$, where $L^0(G)$ and $L^1(G)$ denote G and $L(G)$, respectively, and $L^{n-1}(G)$ is assumed to be non-empty and has no isolated vertex. The graph $L^n(G)$ is called the nth *iterated line graph* of a graph G.

Example 1.4.4 Two important classes of graphs, the well-known the n-dimensional d-ary *Kautz digraph* and *de Bruijn digraphs*, denoted by $K(d, n)$ and $B(d, n)$, respectively, whose original definitions will be given in Section 1.8, can be defined as $L^{n-1}(K_{d+1})$ and $L^{n-1}(K_d^+)$, where $K_d^+ (d \geq 2)$ denotes a digraph obtained from a complete digraph K_d by appending one loop at each vertex. The digraphs in Figure 1.13 are $B(2, 1)$, $B(2, 2)$ and $B(2, 3)$.

Exercises

1.4.1 Prove that for any bipartite undirected graph G with $\Delta(G) = \Delta$,

(a) there exists a Δ-regular bipartite graph H such that $G \subseteq H$;

(b) there exists a Δ-regular bipartite simple graph F such that $G \subseteq F$ if G is simple.

1.4.2 Prove that any loopless undirected graph G contains a k-partite spanning subgraph H such that $(1 - \frac{1}{k})d_G(x) \leq d_H(x)$ for any $x \in V(G)$.

1.4.3 (a) Let G be an undirected graph of order v, and n be an integer with $2 \leq n < v - 1$. Prove that if $v \geq 4$ and all induced subgraphs by n vertices in G have the same numbers of edges, then G is either complete or empty.

(b) Give an example to show that the conclusion in (a) is false for digraphs.

(c) Let G be a digraph of order v, and n be an integer with $2 \leq n < v-1$. Prove or disprove that if all induced subgraphs by n vertices in G are regular, then G is either complete or empty.

1.4.4 Let $L = L(G)$ be the line graph of G. Prove that

(a) L contains no parallel edges and contains a loop at vertex a if and only if a is a loop in G;

(b) $d_L^+(a) = d_G^+(y)$ and $d_L^-(a) = d_G^-(x)$ for any $a \in E(G)$ with $\psi_G(a) = (x, y)$, in particular, if G is d-regular, then so is L;

(c) if G is undirected then $d_L(e) = d_G(x) + d_G(y) - 2$ for any $e \in E(G)$ with $\psi_G(e) = xy$, particularly, L is $(2d-2)$-regular if G is d-regular;

(d) $\varepsilon(L) = \sum\limits_{x \in V(G)} d_G^+(x) d_G^-(x)$ if G is directed, and

(e) $\varepsilon(L) = \frac{1}{2} \sum\limits_{x \in V(G)} (d_G(x))^2 - \varepsilon(G)$ if G is undirected.

1.4.5 The *join* $G_1 \vee G_2$ of disjoint undirected graphs G_1 and G_2 is the undirected graph obtained from $G_1 + G_2$ by joining each vertex of G_1 to each vertex of G_2. Prove that

(a) $K_{m,n} \cong K_m^c \vee K_n^c$;

(b) $\varepsilon(G_1 \vee G_2) = \varepsilon(G_1) + \varepsilon(G_2) + v(G_1)\, v(G_2)$.

1.4.6 Prove that the cartesian product $G_1 \times G_2$ of two simple graphs G_1 and G_2 satisfies the following properties.

(a) $v(G_1 \times G_2) = \nu(G_1)v(G_2)$.

(b) For any $xy \in V(G_1 \times G_2)$, where $x \in V(G_1)$ and $y \in V(G_2)$,

$$d^+_{G_1\times G_2}(xy) = d^+_{G_1}(x) + d^+_{G_2}(y), \quad d^-_{G_1\times G_2}(xy) = d^-_{G_1}(x) + d^-_{G_2}(y)$$

if G is directed, and

$$d^-_{G_1\times G_2}(xy) = d^-_{G_1}(x) + d^-_{G_2}(y)$$

if G is undirected. In particular, $G_1 \times G_2$ is $r_1 + r_2$ regular if G_1 and G_2 are r_1- and r_2-regular, respectively.

(c) $\varepsilon(G_1 \times G_2) = v(G_1)\varepsilon(G_2) + v(G_2)\varepsilon(G_1)$.

(d) The cartesian product satisfies commutative and associative laws if we identify isomorphic graphs, that is, $G_1 \times G_2 = G_2 \times G_1$ and $(G_1 \times G_2) \times G_3 = G_1 \times (G_2 \times G_3)$.

(e) $Q_n = K_2 \times K_2 \times \cdots \times K_2$ of n identical complete graph K_2.

1.4.7 F.R.Ramsey [149] in 1930 proved the well-known *Ramsey's Theorem*: For given positive integers k and l, there exists a smallest integer $r = r(k,l)$ such that every simple undirected graph of order r contains either K_k or K_l^c as its subgraph. The number $r(k,l)$ is known as the *Ramsey number*. Prove that

(a) $r(k,l) = r(l,k)$, $\quad r(1,k) = 1$, $\quad r(2,k) = k$ and $r(3,3) = 6$;

(b) $r(k,l) \leq r(k,l-1) + r(k-1,l)$, and the strict inequality holds if $r(k,l-1)$ and $r(k-1,l)$ are both even for $k \geq 3$ and $l \geq 3$;

(c) $r(3,4) = 9$, $R(3,5) = 14$, $r(4,4) = 18$.

(Other Ramsey numbers known to date are $r(3,6) = 18$ [102], $r(3,7) = 23$ [77], $r(3,8) = 28$ [126] and $r(3,9) = 36$ [79].)

1.4.8 Prove that if an undirected graph G with vertex set V contains no K_{k+1} as its subgraph, then there exists a complete k-partite graph H with vertex-set V such that $d_G(x) \leq d_H(x)$ for every $x \in V(G)$. Moreover, the equality holds if and only if $G \cong H$. (P.Erdös [54])

1.4.8 Prove that (*Turán's theorem*) if an undirected graph G contains no K_{k+1} as its subgraph, then $\varepsilon(G) \leq \varepsilon(T_{k,v})$. Moreover, the equality holds if and only if $G \cong T_{k,v}$. (P.Turán [161])

1.5 Walks, Paths and Connection

Let x and y be two vertices of a graph G. An *xy-walk* of *length* k in G is a sequence

$$W = x_0e_1x_1e_2\cdots e_kx_k,$$

where $x_0 = x, x_k = y$, whose terms are alternately vertices and edges, such that the end-vertices of the edge e_i are x_{i-1} and x_i for each $i = 1, 2, \cdots, k$.

If G is simple, an edge e_i of W is determined by two vertices x_{i-1} and x_i, we may write $W = (x, x_1, \cdots, x_{k-1}, y)$ for short. The vertices x and y are called the *origin* and the *terminus* of W, respectively, and other vertices its *internal vertices.* A subsequence $(x_i, e_{i+1}, \cdots, e_j, x_j)$ of W is called a *subwalk* of W, denoted by $W(x_i, x_j)$.

If the edges $e_1, e_2, \cdots, e_k$ of W are distinct, W is called a *trail.* If, in addition, the vertices $x_0, x_1, \cdots, x_k$ are distinct, W is called a *path.* A walk (trail) is called to be *closed* if its origin and terminus are identical. A closed trail is called a *circuit*; a circuit is called a *cycle* if its vertices are distinct.

Note that in definition of a walk, all edges are direction-free. An xy-walk W of a digraph G is called a *directed xy-walk*, denoted by (x, y)-walk, if for every edge e_i of W, $\psi_G(e_i) = (x_{i-1}, x_i)$, that is, the direction of all edges in W is accordance with one of W from x to y. Analogous definition can be given for a *directed trail*, *directed path*, *directed circuit* and *directed cycle.*

Figure 1.14 illustrates an x_1x_3-walk W, -trail T, -path P, an (x_1, x_3)-walk W', -trail T', -path P', a directed circuit C and a directed cycle C' in the digraph.

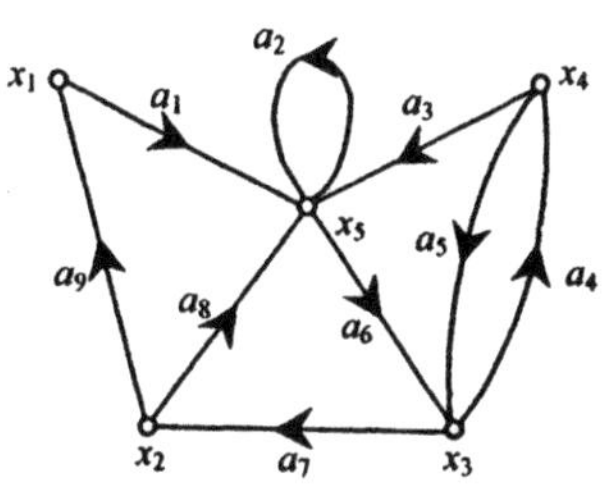

$W = x_1a_1x_5a_2x_5a_3x_4a_3x_5a_8x_2a_7x_3$

$T = x_1a_1x_5a_2x_5a_8x_2a_7x_3$

$P = x_1a_1x_5a_8x_2a_7x_3$

$W' = x_1a_1x_5a_2x_5a_6x_3a_4x_4a_3x_5a_6x_3$

$T' = x_1a_1x_5a_2x_5a_6x_3$

$P' = x_1a_1x_5a_6x_3$

$C = x_1a_1x_5a_2x_5a_6x_3a_7x_2a_9x_1$

$C' = x_1a_1x_5a_6x_3a_7x_2a_9x_1$

Figure 1.14: A digraph G

It should be noted from definition that a path is always supposed that its origin and terminus are different. A path is called to be *longest* if it has the maximum length over all paths in the graph. A path is call a *Hamilton path* if it contains all vertices of the graph.

Example 1.5.1 Every simple graph G must contain a path of length at least δ, where $\delta = \delta(G)$.

Proof We can, without loss of generality, suppose that G is an undirected graph. Let $P = (x_0, x_1, \cdots, x_k)$ be a longest path in G. Then all neighbors of x_0 must lie on P, that is, $N_G(x_0) \subseteq \{x_1, x_2, \cdots, x_k\}$. Since $d_G(x_0) \geq \delta(G) = \delta$, it follows that

$$k \geq |N_G(x_0)| = d_G(x_0) \geq \delta.$$ ∎

Theorem 1.2 (Rédei, 1934 [150]) Every tournament contains Hamilton directed path.

Proof Let T be a tournament of order v and $P = (x_1, x_2, \cdots, x_n)$ be a longest directed path in T. The conclusion holds for $v \leq 2$ clearly. Suppose $v \geq 3$ below, and suppose to the contrary that P is not a Hamilton directed path. Then $n < v$, and there exists some $x \in V(T) \setminus V(P)$ such that (x, x_n), $(x_1, x) \in E(T)$. Thus, there must be some i $(1 < i \leq n)$ such that (x_{i-1}, x) $(x, x_i) \in E(T)$ (see Figure 1.15). It follows that $(x_1, x_2, \cdots, x_{i-1}, x, x_i, x_{i+1}, \cdots, x_n)$ is a directed path whose length is longer than P's. This contradicts to the hypothesis that P is a longest directed path in T. Thus P is a Hamilton directed path in T. ∎

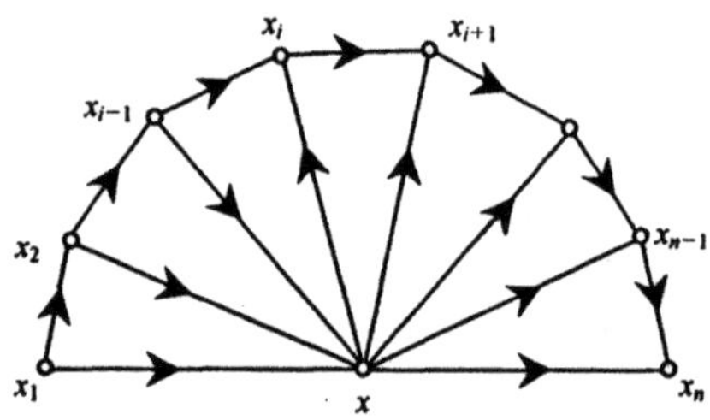

Figure 1.15: An illustration of the proof of Theorem 1.2

Two vertices x and y of G are said to be *connected* if there is an xy-path in G. It is easy to see that "to be connected" is an equivalence relation on $V(G)$. Thus, there exists an equivalence partition $\{V_1, \cdots, V_\omega\}$ of $V(G)$, two vertices are in the same equivalent class V_i if and only if they are connected in G. The subgraphs $G[V_i]$ is called a *connected component* of G. $\omega = \omega(G)$ is called the *number of connected components* of G. If $\omega = 1$, then G is called to be *connected*; otherwise *disconnected.*

For instance, the digraph shown in Figure 1.16 (a) is connected, but the digraph shown in (b) is not for it has three connected components, where one component is a single vertex.

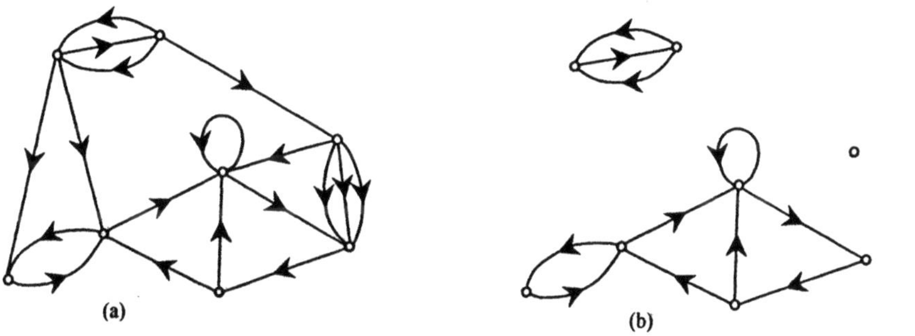

Figure 1.16: a connected digraph and its three strongly connected components

It is easy to prove that a graph is connected if and only if $[S, \overline{S}] \neq \emptyset$ for any nonempty proper subset S of V (see the exercise 1.5.4 (a)).

Example 1.5.2 Let G be a simple undirected graph with vertex-set $V = \{x_1, x_2, \cdots, x_v\}$ satisfying $d_G(x_1) \leq d_G(x_2) \leq \cdots \leq d_G(x_v)$. If $d_G(x_k) \geq k$ for any integer k with $1 \leq k \leq v - d_G(x_v) - 1$, then G is connected.

Proof Suppose to the contrary that G is disconnected. Then there is a nonempty $S \subset V$ such that $[S, \overline{S}] = \emptyset$. Without loss of generality, let $x_v \in \overline{S}$. Then $|\overline{S}| \geq d_G(x_v) + 1$ since G is simple. Let $k = |S|$, then

$$k = |S| = v - |\overline{S}| \leq v - d_G(x_v) - 1.$$

Thus, $d_G(x_k) \geq k$ by our hypothesis, which means that $x_i \in \overline{S}$ for every $i = k, k+1, \cdots, v$. If follows that $k = |S| = v - |\overline{S}| \leq v - (v - k + 1) = k - 1$, a contradiction. Therefore, G is connected. ∎

Let G be a loopless graph, $x \in V(G)$ and $e \in E(G)$. If $\omega(G - x) > \omega(G)$, then x is called a *cut-vertex*; if $\omega(G - e) > \omega(G)$, then e is called a *cut-edge.* For instance, in the graph G shown in Figure 1.17, both x_2 and x_4 are cut-vertices; $x_1 x_2$ is a cut-edge. It is clear that if G contains a cut-edge then it must contain a cut-vertex if the order is at least three, but the converse is not always true in general. A graph is called a *block* if it contains neither cut-vertex nor cut-edge. Every graph can be expressed as the union of several blocks. In Figure 1.17, (a) shows a graph G and (b) shows all blocks of G.

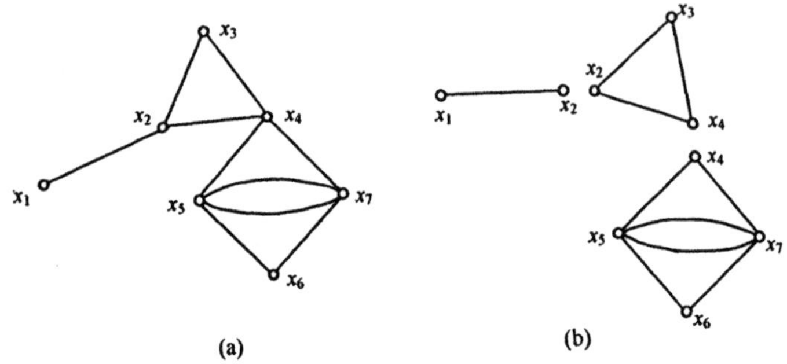

Figure 1.17: a graph G and all of its blocks

Example 1.5.3 Every graph of order at least two contains at least two vertices that are not cut-vertices.

Proof Without loss of generality, suppose that G is a nontrivial connected graph. Let $P = x_0e_1x_1e_2x_2\cdots x_{k-1}e_kx_k$ be a longest path in G. Then $k \geq 1$.

We now prove that each of two end-vertices x_0 and x_k of P is not a cut-vertex. Suppose to the contrary that x_0 is a cut-vertex. Then $\omega(G - x_0) > \omega(G)$. Let G_0 and G_1 be two connected components of $G - x_0$, where G_1 contains x_1. Choose $y \in N_G(x_0) \cap V(G_0)$. Thus, there exists an edge e of G with end-vertices x_0 and y. Since P is not contained in G_0, the path

$$Q = yex_0e_1x_1e_2x_2\cdots x_{k-1}e_kx_k$$

is longer than P, which contradicts to the choice of P. It follows that x_0 is not a cut-vertex of G.

Similarly, we can prove that x_k is not a cut-vertex of G too. ∎

The connection is a concept that has no relations with direction of edges. Next, we introduce a concept of strong connection that has relation with direction of edges.

Let G be a digraph, two vertices x and y of G are said to be *strongly connected* if there are both (x, y)-path and (y, x)-path in G. It is also easy to see that "to be strongly connected" is an equivalence relation on $V(G)$. The subgraph induced by an equivalence class is called a *strongly connected component* of G. A digraph is called to be *strongly connected* if it has exactly one strongly connected component, that is, any two vertices of G are strongly connected.

Two concepts of connectedness are the same for undirected graphs. For a digraph, if it is strongly connected, then it is necessarily connected. The converse is not always true. For instance, the digraph shown in Figure 1.16 (a) is connected, but not strongly connected, its three strongly connected components are shown in (b).

It is also easy to prove that a digraph is strongly connected if and only if both $(S, \overline{S}) \neq \emptyset$ and $(\overline{S}, S) \neq \emptyset$ for any nonempty proper subset S of V (the exercise 1.5.4 (b)).

Example 1.5.4 A simple digraph G with $\varepsilon > (v-1)^2$ is strongly connected.

Proof Suppose to the contrary that G is not strongly connected. Then there exists a nonempty proper subset S of $V(G)$ such that $(S, \overline{S}) = \emptyset$. Let $k = |S|$. Note that $|(\overline{S}, S)| \leq k(v-k)$. It follows that

$$\begin{aligned} \varepsilon &\leq 2\binom{k}{2} + 2\binom{v-k}{2} + k(v-k) \\ &= (v-1)^2 - (k-1)(v-k-1) \leq (v-1)^2, \end{aligned}$$

which contradicts to the hypothesis. Thus, G is strongly connected. ∎

Example 1.5.5 A digraph G is called to be *unilateral connected* if there exists either (x, y)-path or (y, x)-path for any two vertices x and y of G. Show that G is unilateral connected if and only if G contains a directed walk going through all vertices of G.

Proof The condition is sufficient clearly. We now show the condition is necessary. To the end, we construct a simple digraph $G' = (V(G'), E(G'))$ as follows. $V(G') = V(G)$ and $(x, y) \in E(G')$ if and only if there is an (x, y)-path P_{xy} in G. Then G' contains a tournament as its spanning subgraph since G is unilateral connected. By Theorem 1.2, G' contains a Hamilton directed path P'. We can obtain a directed walk going through all vertices of G by replacing an edge (x, y) of P' with the directed path P_{xy} in G. ∎

Exercises

1.5.1 (a) Prove that any xy-walk (resp. (x, y)-walk) necessarily contain an xy-trail (resp. (x, y)-trail).

(b) Prove that any xy-trail (resp. (x, y)-trail) necessarily contain an xy-path (resp. (x, y)-path).

(c) Prove that any directed closed walk can be expressed as the union of several edge-disjoint closed trails, and construct an example to show that the term "directed" can not be deleted.

(d) Prove that any (directed) circuit can be expressed as the union of several edge-disjoint (directed) cycles.

1.5.2 Prove that any simple digraph contains a directed path of length at least $\max\{\delta^+, \delta^-\}$.

1.5.3 Prove that if G is a strongly connected digraph and $x, y \in V(G)$, then there exists an (x, y)-walk going through every vertex in G.

1.5.4 Prove that

(a) a graph is connected if and only if $[S, \overline{S}] \neq \emptyset$ for any nonempty proper subset S of V;

(b) a digraph is strongly connected if and only if both $(S, \overline{S}) \neq \emptyset$ and $(\overline{S}, S) \neq \emptyset$ for any nonempty proper subset S of V;

(c) a balanced digraph is strongly connected if and only if it is connected;

(d) a digraph contains a directed path from a vertex x_0 to any other vertex if and only if $(S, \overline{S}) \neq \emptyset$ for any nonempty proper subset S of V containing x_0.

1.5.5 Prove that a graph G of order at least three is connected if and only if there exist two vertices x and y in G such that $G - x$ and $G - y$ both are connected.

1.5.6 Let G be a simple undirected graph and $\omega = \omega(G)$. Prove that

(a) $\varepsilon(G) \leq \frac{1}{2}\,(v - \omega)(v - \omega + 1)$;

(b) G is connected if $\varepsilon(G) > \frac{1}{2}(v - 1)(v - 2)$;

(c) G is connected if $d_G(x) + d_G(y) \geq v - 1$ for any two nonadjacent vertices x and y.

1.5.7 Let G be a simple digraph with ω strongly connected components. Prove that

(a) $\varepsilon(G) \leq (v - \omega)(v - \omega + 1) + \frac{1}{2}\,(\omega - 1)(2v - \omega)$;

(b) $\omega = 1$, that is, G is strongly connected if $\varepsilon(G) > (v - 1)^2$.

1.5.8 Let G be a simple digraph of order $v > 1$. Prove that

(a) G is strongly connected if $d_G^+(x) + d_G^-(y) \geq v - 1$ for any two vertices x and y satisfying $(x, y) \notin E(G)$;

(b) G is strongly connected if $\varepsilon > v(v-1) - (k+1)(v-k-1)$ and $\delta \geq k$.

1.5.9 Let G be an undirected graph. Prove that

(a) G contains no cut-edge if G contains no vertex of degree odd;

(b) G contains no cut-edge if G is k (≥ 2)-regular and bipartite;

(c) if b_x denotes the number of blocks containing the vertex x in G, then the number of all blocks in G

$$b(G) = \omega(G) + \sum_{x \in V(G)} (b_x - 1).$$

1.5.10 Prove that

(a) any two longest paths in any connected graph must have a vertex in common;

(b) all longest paths in the following graph have no vertex in common.

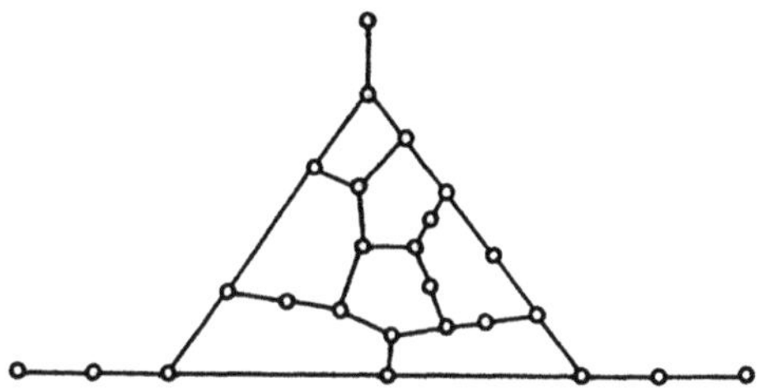

(the exercise 1.5.10)

1.5.11 Let G be a simple undirected digraph. Prove that

(a) if G is disconnected, then G^c is connected;

(b) G and G^c both are connected if and only if G and G^c both contains no complete bipartite graph as their spanning subgraph.

1.6 Distances and Diameters

We in this section introduce the concepts of distance and diameter of graphs which is closely related with paths and connectedness.

Let x and y be two distinct vertices of a graph G. The smallest length over all (x, y)-paths is called the *distance* from x to y in G, denoted by $d_G(x, y)$; if there is no (x, y)-path, we adopt the convention that $d_G(x, y)$ is positive infinite. An (x, y)-path is said to be *shortest* if its length is equal to $d_G(x, y)$.

It is clear that if G is an undirected graph, then the existence of $d_G(x, y)$ means existence of $d_G(y, x)$, and they have the same value, as the reverse of a shortest (x, y)-path is also a shortest (y, x)-path. However, if G is a digraph, then $d_G(x, y)$ and $d_G(y, x)$ exist not always at the same time, even if both exist, they have not necessarily the same value.

The *diameter* of G, denoted by $d(G)$, is defined as the maximum distance between any two vertices of G, that is,

$$d(G) = \max\{d_G(x, y) : \ \forall \ x, y \in V(G)\}.$$

For example, for a path P_n of n vertices, we have $d(P_n) = n - 1$ if P_n is undirected, and $d(P_n) = \infty$ if P_n is directed. For a complete graph K_n, we have $d(K_n) = 1$ no matter whether it is directed or undirected. The diameter of a complete bipartite graph or Petersen graph is 2. For the n-cube Q_n, its diameter $d(Q_n) = n$.

It is clear that $d(G) = 1$ if and only if $K_v \subseteq G$, and the diameter $d(G)$ is well-defined if and only if G is a connected undirected graph or a strongly connected digraph.

Example 1.6.1 Let G be a connected undirected graph with two nonadjacent vertices. Then there exist $x, y \in V(G)$ such that $d_G(x, y) = 2$.

Proof Let x and z be two nonadjacent vertices in G. Since G is connected, there exists a shortest xz-path $P = xe_1x_1e_2x_2 \cdots x_{k-1}e_kz$. Then $k \geq 2$. Let $y = x_2$, then $xe_1x_1e_2y$ is an xy-path in G, and, hence $d_G(x, y) \leq 2$. If $d_G(x, y) = 1$, then there exists some $e \in E(G)$ such that $\psi_G(e) = xy$. The length of the xy-path $P' = xeye_3x_3 \cdots e_{k-1}e_kz$ is shorter than that of P, which contradicts to the choice of P. It follows that $d_G(x, y) = 2$. ∎

Example 1.6.2 Let G be a connected and simple undirected graph of order v and the minimum degree δ, then $d(G) \leq \dfrac{3v}{\delta + 1}$.

Proof Note that the diameter $d(G)$ is well-defined and $d(G) \leq v - 1$ for any connected graph G of order v. Thus we may assume that $\delta \geq 3$. Since $v \geq \delta + 1$, thus, we may also assume that $d(G) \geq 4$.

Let x and y be two vertices of G such that $d_G(x,y) = d(G) = d \geq 4$, and let $P = (x_0, x_1, \cdots, x_{d-1}, x_d)$ be a shortest xy-path in G, where $x_0 = x$ and $x_d = y$. Because of the shortness of P, it is clear that $N_G(x_{3i}) \cap N_G(x_{3j}) = \emptyset$ for any two vertices x_{3i} and x_{3j} $(0 \leq i \neq j \leq \lfloor \frac{d}{3} \rfloor)$. Considering all vertices in P as well as their neighbors, we have

$$v \geq \delta \left(\left\lfloor \frac{d}{3} \right\rfloor + 1 \right) + \left(\left\lfloor \frac{d}{3} \right\rfloor + 1 \right) \geq \delta \cdot \frac{d}{3} + \frac{d}{3} = \frac{d}{3}\,(\delta + 1).$$

This means that the assertion follows. ∎

Example 1.6.3 Let G be a strongly connected digraph of order v and the maximum degree Δ. Then diameter of G

$$d(G) \begin{cases} = v - 1, & \text{for } \Delta = 1; \\ \geq \lceil \log_\Delta (v(\Delta - 1) + 1) \rceil - 1, & \text{for } \Delta \geq 2. \end{cases}$$

Proof Since G is a strongly connected digraph, diameter $d(G)$ is well-defined. Let $d(G) = k$. From any given vertex at most Δ vertices can be reached in a distance of one and, for $i \geq 1$, at most Δ^i vertices can be reached in a distance of i. It follows that

$$\begin{aligned} v &\leq 1 + \Delta + \Delta^2 + \cdots + \Delta^{k-1} + \Delta^k \\ &= \begin{cases} k + 1, & \text{for } \Delta = 1; \\ \dfrac{\Delta^{k+1} - 1}{\Delta - 1}, & \text{for } \Delta \geq 2. \end{cases} \end{aligned} \tag{1.6}$$

For $\Delta = 1$, $v \leq k + 1$ by (1.6), which means that $d(G) = k \geq v - 1$. But $d(G) \leq v - 1$ obviously. Thus, $d(G) = v - 1$ if $\Delta = 1$.

For $\Delta \geq 2$, $(\Delta - 1)v \leq \Delta^{k+1} - 1$ by (1.6). This implies that $d(G) = k \geq \lceil \log_\Delta (v(\Delta - 1) + 1) \rceil - 1$ as required. ∎

The upper bounds of v in the expression (1.6) is called (Δ, k)-*Moore bounds* for digraphs of the maximum degree Δ and diameter k. The digraphs whose order reaches the Moore bounds is called a (Δ, k)-*Moore digraph.* A directed cycle of length $k + 1$ is the unique $(1, k)$-Moore digraph. We can prove there are no (Δ, k)-Moore digraph for $\Delta \geq 2$ and $k \geq 2$ (see Example 1.10.1 in Section 1.10). The Moore bounds for undirected graphs are given in the exercise 1.6.5.

The following two theorems give diameters of the cartesian product and a line graph, respectively.

From definition of $G_1 \times G_2$, we note that if $P = (x_1, v_1, v_2, \cdots, v_m, y_1)$ is an (x_1, y_1)-path in G_1, then for any $b \in V(G_2)$, $(x_1b, v_1b, v_2b, \cdots, v_mb, y_1b)$, denoted by Pb, is an (x_1b, y_1b)-path from the vertex x_1b to the vertex y_1b in $G_1 \times G_2$. Similarly, if $W = (x_2, u_1, u_2, \cdots, u_l, y_2)$ is an (x_2, y_2)-path in G_2, then for any $a \in V(G_1)$, $(ax_2, au_1, au_2, \cdots, au_l, ay_2)$, denoted by aW, is an (ax_2, ay_2)-path from the vertex ax_2 to the vertex ay_2 in $G_1 \times G_2$. Thus, $Q = Px_2 \cup y_1W$ is an (x_1x_2, y_1y_2)-path in $G_1 \times G_2$ with length $\varepsilon(P) + \varepsilon(W)$.

Theorem 1.3 The diameter $d(G_1 \times \cdots \times G_n) = d(G_1) + \cdots + d(G_n)$. Particularly, $d(Q_n) = n$, where Q_n is an n-cube.

Proof Using the associative law and the induction on $n \geq 2$, we need to only prove $d(G_1 \times G_2) = d(G_1) + d(G_2)$. To this purpose, let $x = x_1x_2$, $y = y_1y_2 \in V(G_1 \times G_2)$, where $x_1, y_1 \in V(G_1)$ and $x_2, y_2 \in V(G_2)$. Let P be a shortest (x_1, y_1)-path in G_1 and W be a shortest (x_2, y_2)-path in G_2.

If $x_1 = y_1$, then x_1W is a shortest (x, y)-path in $G_1 \times G_2$ and so

$$d_{G_1 \times G_2}(x, y) = \varepsilon(x_1W) = \varepsilon(W) \leq d(G_2) \leq d(G_1) + d(G_2).$$

If $x_2 = y_2$, then Px_2 is a shortest (x, y)-path in $G_1 \times G_2$, and so

$$d_{G_1 \times G_2}(x, y) = \varepsilon(Px_2) = \varepsilon(P) \leq d(G_1) \leq d(G_1) + d(G_2).$$

If $x_1 \neq y_1$ and $x_2 \neq y_2$, then $Px_2 \cup y_1Q$ is an (x, y)-path in $G_1 \times G_2$, and

$$d_{G_1 \times G_2}(x, y) \leq \varepsilon(Px_2) + \varepsilon(y_1W) \leq d(G_1) + d(G_2).$$

By arbitrariness of x and y, we have $d(G_1 \times G_2) \leq d(G_1) + d(G_2)$.

On the other hand, let $x_1, y_1 \in V(G_1)$ and $x_2, y_2 \in V(G_2)$ such that $d_{G_1}(x_1, y_1) = d(G_1)$ and $d_{G_2}(x_2, y_2) = d(G_2)$. Let P be a shortest (x_1, y_1)-path in G_1 and W a shortest (x_2, y_2)-path in G_2. We now show that $Q = Px_2 \cup y_1W$ be a shortest (x_1x_2, y_1y_2)-path in $G_1 \times G_2$. In fact, if Q' is an (x_1x_2, y_1y_2)-path in $G_1 \times G_2$ whose length is shorter than $\varepsilon(Q)$. Then the path P' determined by distinct vertices in the first coordinates of vertices of Q' in the original order is an (x_1, y_1)-path of G_1. Similarly, the path W' determined by distinct vertices in the second coordinates of vertices of Q in the original order is an (x_2, y_2)-path of G_2. It is clear that $\varepsilon(P') + \varepsilon(W') = \varepsilon(Q') < \varepsilon(Q) = \varepsilon(P) + \varepsilon(W)$. Thus either $\varepsilon(P') < \varepsilon(P)$ or $\varepsilon(W') < \varepsilon(W)$, which contradicts the choice of P or W. It follows that

$$d(G_1 \times G_2) \geq \varepsilon(Q) = \varepsilon(P) + \varepsilon(W) = d(G_1) + d(G_2).$$

Immediately, $d(Q_n) = n$ since $Q_n = K_2 \times \cdots \times K_2$ and $d(K_2) = 1$. ∎

Theorem 1.4 (Aigner [2]) Let G be a strongly connected digraph of order at least two, and L be the line digraph of G. Then

$$d(G) \leq d(L) \leq d(G) + 1.$$

Furthermore, $d(G) = d(L)$ if and only if G is a directed cycle. In particular, $d(K(d,n)) = d(B(d,n)) = n$.

Proof Since G is strongly connected, it is easy to show that $L(G)$ is also strongly connected (the exercise 1.6.3), and so $d(G)$ and $d(L)$ are well-defined. Let $x, y \in V(G)$ such that $d_G(x,y) = d(G)$, P be a shortest (x,y)-path in G. Let $a \in E_G^-(x)$ and $b \in E_G^-(y) \cap E(P)$, and $Q = a + P$. Then $L(Q)$ is a shortest (a,b)-path in L. It follows that

$$d(L) \geq \varepsilon(L(Q)) = \varepsilon(P) = d(G).$$

On the other hand, assume $a, b \in V(L)$ such that $d_L(a,b) = d(L)$. There must be vertices $x, y, z, u \in V(G)$ such that $a = (z,x), b = (y,u)$. Then

$$d(G) \geq d_G(x,y) = d_L(a,b) - 1 = d(L) - 1.$$

Thus we obtain $d(G) \leq d(L) \leq d(G) + 1$. We now prove the second assertion.

If G is a directed cycle, then, by definition of line graph, L is also a directed cycle clearly, and so $d(G) = d(L)$. We need to only prove that $d(L) = d(G) + 1$ if G is not a directed cycle. To this purpose, let $d = d(G)$, $x, y \in V(G)$ such that $d_G(x,y) = d$. Then there exists an (x,y)-path P of length d in G. Since G is strongly connected, $d_G^-(x) \geq 1$ and $d_G^+(y) \geq 1$.

If there are $x', y' \in V(G)$ such that $a = (x', x)$, $b = (y, y')$ and $a \neq b$, then $d_L(a,b) = d+1$. Otherwise, there is $c \in E(G)$ such that $c = (y,x)$. So $P \cup \{c\}$ is a directed cycle in G, denoted by $C = (x_0, x_1, \cdots, x_d, x_0)$, where $x_0 = x$ and $x_d = y$. Since G is strongly connected and not a directed cycle, there are $x_i \in V(C)$ and $z \in V(G)$ such that $(x_i, z) \in E(G)$ (maybe z is some x_j $(0 \leq j < i)$). Choose such a vertex x_i such that $i (0 \leq i \leq d)$ is as large as possible. Then $d_G(x_{i+1}, x_i) = d$. Let $a = (x_i, x_{i+1})$ and $b = (x_i, z)$. Then $d_L(a,b) = d+1$ and so

$$d + 1 \geq d(L) \geq d_L(a,b) = d + 1.$$

This implies $d(L) = d + 1$.

Note that K_{d+1} and K_d^+ $(d \geq 2)$ all are not directed cycles and have diameter 1. It is immediate that $d(K(d,n)) = d(B(d,n)) = n$ since $K(d,n) = L^{n-1}(K_{d+1})$ and $B(d,n) = L^{n-1}(K_d^+)$ (see the end of Section 1.4). ∎

We conclude this section with some other concepts related to distance and diameter.

A vertex is *central* of G if the greatest distance between it and any other vertex is as small as possible. This distance is the *radius* of G, denoted by rad (G). Thus, formally,

$$\text{rad}\,(G) = \min_{x \in V(G)} \{ \max_{y \in V(G)} \{d_G(x,y)\} \}.$$

It is easy to check that rad $(G) \leq d(G) \leq 2\,\text{rad}\,(G)$ (the exercise 1.6.6).

Example 1.6.4 A digraph G of radius at most r has not more than $1 + r\Delta^r$ vertices.

Proof Let x be a central vertex of G, and let J_i denote the set of vertices of G at distance i from x. Then $|J_i| \leq \Delta|J_{i-1}| \leq \Delta^i$ for $i = 1, 2, \cdots, r$. Adding up these inequalities yields

$$v(G) \leq 1 + \Delta + \Delta^2 + \cdots + \Delta^r \leq 1 + r\Delta^r$$

as desired. ∎

Let G be a connected undirected graph or strongly connected digraph with order $v(\geq 2)$. The *mean* or *average distance* of G, denoted by $m(G)$, is defined as

$$m(G) = \frac{1}{v(v-1)} \sum_{x,y \in V} d(G; x, y).$$

We note that $m(G)$ is the arithmetic average of only all non-zero distances. This ensures that $m(G) \geq 1$ and, moreover, the equality holds if and only if G is a complete graph. Of course, the average distance can be calculated by first finding the distances of all pairs of different vertices. For convenience, Let

$$\sigma(G) = \sum_{x,y \in V} d(G; x, y).$$

Consider a directed cycle $C_n (n \geq 3)$. For a fixed vertex x_0 of C_n, the sum of all distances from x_0 to any other vertex is equal to $1 + 2 + \cdots + n - 1 = \frac{1}{2}n(n-1)$. Thus $\sigma(C_n) = \frac{1}{2}n^2(n-1)$ and $m(C_n) = \frac{1}{2}n$. For an undirected cycle C_n, we have

$$m(C_n) = \begin{cases} \dfrac{n+1}{4}, & \text{if } n \text{ is odd;} \\ \dfrac{n^2}{4(n-1)}, & \text{if } n \text{ is even,} \end{cases}$$

the computation is left the reader as an exercise (the exercise 1.6.6).

Exercises

1.6.1 Prove that $d_G(x,z) \le d_G(x,y) + d_G(y,z)$ for any three vertices x, y and z of a strongly connected digraph G.

1.6.2 Prove that if G is a (Δ, k)-Moore digraph, then G is simple and Δ-regular, contains no cycle of length at most k and there is only unique (x, y)-path of length at most k for any pair (x, y) of vertices in G.

1.6.3 Let $L(G)$ be the line digraph of a digraph G of order $v \ge 2$. Prove that

(a) $L(G)$ is strongly connected if and only if G is strongly connected;

(b) $L(G) \cong G$ if and only if G is a directed cycle;

(c) the above conclusions and Theorem 1.4 are true if G is undirected.

1.6.4 Let G be a simple undirected graph. Prove that

(a) if G is disconnected, then $d(G^c) \le 2$;

(b) if $d(G) > 3$, then $d(G^c) < 3$;

(c) if $d(G) = 2$ and $\Delta(G) = v - 2$, then $\varepsilon \ge 2v - 4$.

1.6.5 Prove that if G is a connected undirected graph of the maximum degree Δ and diameter d, then

$$v \le \begin{cases} 2d+1, & \text{for } \Delta = 2; \\ \dfrac{\Delta(\Delta-1)^d - 1}{\Delta - 2}, & \text{for } \Delta \ge 3, \end{cases}$$

and, hence,

$$d \ge \begin{cases} \left\lfloor \dfrac{1}{2} v \right\rfloor, & \text{for } \Delta = 2; \\ \left\lceil \log_{(\Delta-1)} \dfrac{v(\Delta-2)+2}{\Delta} \right\rceil, & \text{for } \Delta \ge 3. \end{cases}$$

1.6.6 Prove that

(a) $\operatorname{rad}(G) \le d(G) \le 2 \operatorname{rad}(G)$;

(b) if C_n is an undirected cycle, then the mean distance of C_n

$$m(C_n) = \begin{cases} \dfrac{n+1}{4}, & \text{if } n \text{ is odd}; \\ \dfrac{n^2}{4(n-1)}, & \text{if } n \text{ is even}. \end{cases}$$

1.7 Circuits and Cycles

A cycle of length k is called a *k-cycle*; a k-cycle is *odd* or *even* according as k is odd or even. A 3-cycle is often called a *triangle*.

We will use the word "cycle" to denote a graph or subgraph whose vertices and edges are the terms of a cycle. Usually, we denote a cycle of order n by C_n, which is undirected or directed according to the context if it is not specified.

Example 1.7.1 Let G be an undirected graph with $\delta \geq 2$. Then G contains a cycle certainly. Moreover, if G is simple, then G contains a cycle of length at least $\delta + 1$.

Proof If G contains loops or parallel edges, then the conclusion holds clearly. Suppose that G is simple below and let $P = (x_0, x_1, \cdots, x_k)$ be a longest path in G. Then all neighbors of x_0 must lie on P, that is

$$N_G(x_0) \subseteq \{x_1, x_2, \cdots, x_k\}.$$

Note that

$$|N_G(x_0)| = d_G(x_0) \geq \delta(G) = \delta \geq 2.$$

Thus, there exists $x_i \in N_G(x_0)$, $\delta \leq i \leq k$. It follows that $(x_0, x_1, \cdots, x_{i-1}, x_i, x_0)$ is a cycle of length at least $\delta + 1$ in G. ∎

The length of a shortest (directed) cycle in a (di)graph G is called the *girth* of G, denoted by $g(G)$.

Example 1.7.2 If G be a k-regular undirected graph with girth g at least three, then

$$v(G) \geq \begin{cases} 1 + k + k(k-1) + \cdots + k(k-1)^{\frac{1}{2}(g-3)}, & \text{if } g \text{ is odd;} \\ 2\left(1 + (k-1) + \cdots + (k-1)^{\frac{1}{2}(g-2)}\right), & \text{if } g \text{ is even.} \end{cases}$$

Proof It is clear that G is simple because of $g \geq 3$. We first consider the case that g is odd. Let $g = 2d + 1$, $d \geq 1$ and, for any vertex x of G, let

$$J_i(x) = \{y \in V(G) : d_G(x, y) = i\}, \ 0 \leq i \leq d.$$

Clearly, $|J_0(x)| = 1$, $|J_1(x)| = k$. Since $g = 2d + 1$, there exists a unique xy-path of length i in G for any $y \in J_i(x)$ $(1 \leq i \leq d)$, and any two vertices of $J_i(x)$ are not adjacent in G for any $i = 1, 2, \cdots, d-1$. This means that

$$|J_i(x)| = k(k-1)^{i-1}, \ i = 1, 2 \cdots, d.$$

It follows that

$$\begin{aligned} v(G) &\geq |J_0(x)| + |J_1(x)| + \cdots + |J_d(x)| \\ &= 1 + k + k(k-1) + \cdots + k(k-1)^{d-1} \\ &= 1 + k + k(k-1) + \cdots + k(k-1)^{\frac{1}{2}(g-3)} \end{aligned}$$

The case that g is even can be similarly proved, and the details are left to the reader an exercise (the exercise 1.7.10 (b)). ∎

We have in Theorem 1.2 seen that every tournament contains a Hamilton directed path. It is perhaps surprising that if a tournament is strongly connected, then it must possess à significantly stronger properties. A (di)graph G of order $v\,(\geq 3)$ is *vertex-pancyclic* if each vertex of G is contained in (directed) cycles of all length between 3 and v.

Theorem 1.5 (Moon [133], 1966) Every strongly connected tournament of order $v\,(\geq 3)$ is vertex-pancyclic.

Proof We prove the theorem by induction on $k \geq 3$. Let G be a strongly connected tournament and u be any vertex of G. For $k = 3$, let $S = N_G^+(u)$, $T = N_G^-(u)$. Then $S \neq \emptyset$ and $T \neq \emptyset$ since G is strongly connected. Moreover, $T \cup \{u\} = \overline{S}$ since G is a tournament. Thus, $(S, T) = (S, \overline{S}) \neq \emptyset$ by the exercise 1.5.4 (b). It follows that there exist $x \in S$ and $y \in T$ such that $(x, y) \in E(G)$ and, hence, (u, x, y, u) is a directed 3-cycle containing u in G (see Figure 1.18 (a)).

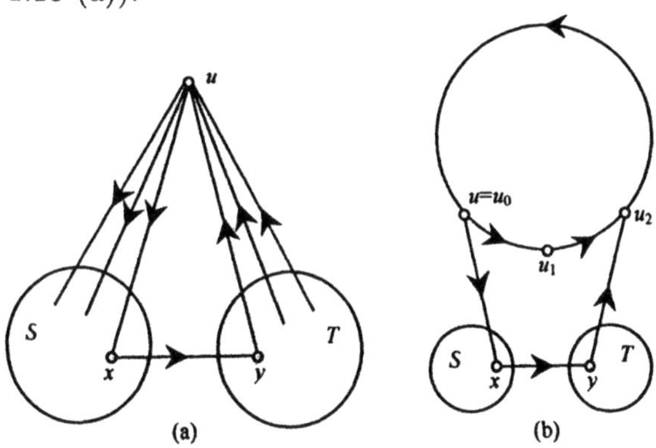

Figure 1.18: Two illustrations of the proof of Theorem 1.3

Suppose that u is contained in directed cycles of all lengths between 3 and n, where $n < v$. We will prove that u is contained in a directed $(n+1)$-cycle.

Let $C = (u_0, u_1, \cdots, u_{n-1}, u_0)$ be a directed n-cycle, where $u_0 = u$. If there is some $x \in V(G) \setminus V(C)$ such that $N_G^+(x) \cap V(C) \neq \emptyset$ and $N_G^-(x) \cap$

$V(C) \neq \emptyset$, then there are adjacent vertices u_i and u_{i+1} on C such that $(u_i, x), (x, u_{i+1}) \in E(G)$. In this case u is contained in the directed $(n+1)$-cycle $(u_0, u_1, \cdots, u_i, x, u_{i+1}, \cdots, u_{n-1}, u_0)$. Otherwise, let

$$S = \{x \in V(G) \setminus V(C) : \ N_G^+(x) \cap V(C) = \emptyset\},$$
$$T = \{y \in V(G) \setminus V(C) : \ N_G^-(x) \cap V(C) = \emptyset\}.$$

It is clear that $S \neq \emptyset$, $T \neq \emptyset$ and $(S, T) \neq \emptyset$. Let $x \in S$ and $y \in T$ such that $(x, y) \in E(G)$ (see Figure 1.18 (b)). Thus u is contained in the directed $(n+1)$-cycle $(u_0, x, y, u_2, \cdots, u_i, u_{i+1}, \cdots, u_{n-1}, u_0)$. ∎

Corollary 1.5 Let x and y be two vertices of a strongly connected tournament G of order $v \geq 5$. Then there exists an (x, y)-walk of length exactly $(d+3)$, where d is diameter of G.

Proof Let P be a shortest (x, y)-path in G. Since $0 \leq d_G(x, y) \leq d \leq v - 1$, it follows that

$$3 \leq d - d_G(x, y) + 3 \leq v + 2.$$

If $m = d - d_G(x, y) + 3 \leq v$ then, by Theorem 1.5, there exists a directed m-cycle C_m containing y. Thus, $P \oplus C_m$ is an (x, y)-walk of length

$$d_G(x, y) + (d - d_G(x, y) + 3) = d + 3.$$

If $d - d_G(x, y) + 3 = v + 1$ then, by Theorem 1.5 and $v \geq 5$, there exist a directed 3-cycle C_3 and a directed $(v-2)$-cycle C_{v-2} containing y. Thus, $P \oplus C_{v-2} \oplus C_3$ is an (x, y)-walk of length

$$d_G(x, y) + (v - 2) + 3 = d_D(x, y) + d - d_D(x, y) + 3 = d + 3.$$

If $d - d_G(x, y) + 3 = v + 2$ then, by Theorem 1.5, there exist a directed 3-cycle C_3 and a directed $(v-1)$-cycle C_{v-1} containing y. Thus, $P \oplus C_{v-1} \oplus C_3$ is an (x, y)-walk of length $d + 3$. ∎

Using the concepts of circuits and cycles, we can give a characterization of a bipartite graph in terms of the parity of cycles (or circuits).

Theorem 1.6 A strongly connected digraph is bipartite if and only if it contains no odd directed circuit.

Proof Suppose that G is a strongly connected bipartite digraph with a bipartition $\{X, Y\}$ and suppose that $C = x_0 e_1 x_1 \cdots x_{k-1} e_k x_0$ is a directed k-circuit in G. We can, without loss of generality, suppose that x_0 is in X. Then $x_1 \in Y$, $x_2 \in X$, $x_3 \in Y, \cdots, x_{k-1} \in Y$. Generally, $x_{2i} \in X$ and $x_{2i+1} \in Y$. Thus, there exists some i such that $k - 1 = 2i + 1$, which implies that $k = 2i + 2$, that is, C is even.

Conversely, let G be a strongly connected digraph that contains no odd directed circuit. Then G has no loops. We choose arbitrarily a vertex u and define a partition $\{X, Y\}$ of $V(G)$ by setting

$$X = \{x \in V(G) : \ d_G(u, x) \text{ is even}\}$$
$$Y = \{y \in V(G) : \ d_G(u, y) \text{ is odd}\}$$

Clearly, $u \in X$. Since G is strongly connected, $Y \neq \emptyset$ if $v \geq 2$. We need to only prove that $\{X, Y\}$ is a bipartition of G.

To the end, we first show that $G[Y]$ is empty. The assertion is true if $|Y| = 1$. Suppose $|Y| \geq 2$ below. Arbitrarily choose two distinct vertices y and z in Y. Let P_1 and Q_1 be shortest (u, y)-path and (u, z)-path, P_2 and Q_2 be shortest (y, u)-path and (z, u)-path, respectively, in G. (Strongly connectedness of G ensures the existence of these four directed paths.) Then the length of P_1 is odd by definition of Y. Note that $P_1 \oplus P_2$ is a directed circuit on G of even length. Therefore the length of P_2 is odd. Similarly, the length of Q_2 is odd too. Thus, if, to the contrary, there exists a directed edge e joining from z to y, then $P_2 \oplus Q_1 \oplus \{e\}$ contains an odd directed circuit, which contradicts to our hypothesis. Similarly, there exists no edges joining from y to z. Therefore, no two vertices in Y are adjacent, that is, $G[Y]$ is empty.

By the same argument, we can prove that $G[X]$ is empty too. ∎

In the proof of Theorem 1.6, the strongly connectedness of G is no necessary for "only if", but is necessary for "if". For instance, the two tournaments of order three shown in Figure 1.6, the former is not strongly connected and contains no directed circuit, but it is not bipartite clearly.

Corollary 1.6.1 A strongly connected digraph is bipartite if and only if it contains no odd directed cycle.

Proof Suppose that G is a strongly connected bipartite digraph. Then G contains no odd directed circuit by Theorem 1.6. Thus, G contains no odd directed cycle since any odd directed circuit certainly contains an odd directed cycle.

Conversely, suppose that G contains no odd directed cycle. In order to prove that G is bipartite, we need to only show that G contains no odd directed circuit. Suppose to the contrary that C is an odd directed circuit in G. Then C is not a directed cycle and it can be expressed as the union of $k\,(\geq 2)$ edge-disjoint directed cycles $C_1, C_2, \cdots, C_k$, of which at least one is odd, contrary to the hypothesis. ∎

Corollary 1.6.2 (König [113], 1936) An undirected graph is bipartite if and only if it contains no odd cycle.

Corollary 1.6.3 A digraph is bipartite if and only if it contains no odd cycle.

The proofs of these two corollaries are left to the reader as exercises (the exercise 1.7.1).

We conclude this section with several examples.

Example 1.7.3 An undirected graph G has a balanced oriented graph if and only if G contains no vertex of odd degree.

Proof The condition is necessary clearly. By induction on the number of edges ε, we prove the condition is sufficient. If $\varepsilon = 0$ there is nothing to prove so suppose $\varepsilon > 0$ and the assertion holds for any undirected graph with the number of edges $\varepsilon \leq m$. Let G is an undirected graph without vertices of odd degree and $\varepsilon(G) = m + 1$. Let

$$S = \{x \in V(G) : \ d_G(x) = 0\},$$

and let $G_1 = G - S$. Then G_1 contains no vertex of odd degree and $\delta(G_1) \geq 2$. Thus G_1 contains a cycle by Example 1.7.1. Let C be a cycle in G_1. Then we obtain a directed cycle C' by assigning each edge in C an orientation whose direction agrees with one of C. Let $G_2 = G - E(C)$, then G_2 contains no vertex of odd degree and

$$\varepsilon(G_2) = \varepsilon(G) - \varepsilon(G_1) < \varepsilon(G) = m + 1.$$

By the induction hypothesis, there exists a balanced oriented graph D' of G_2. It follows that $D = D' \oplus C'$ is a balanced oriented graph of G. ∎

Example 1.7.4 Any strongly connected digraph with an odd circuit contains an odd directed circuit, and, hence contains an odd directed cycle.

Proof The assertion can be easily deduced from Corollary 1.6.3, the detail is left to the reader. We here give a direct proof.

Suppose that G is a strongly connected digraph and

$$C = x_1 e_1 x_2 e_2 \cdots x_i e_i x_{i+1} \cdots x_{2k+1} e_{2k+1} x_1$$

be an odd circuit in G, where $x_i \in V(G)$, $e_i \in E(G)$. Use P_i to denote a shortest (x_i, x_{i+1})-path for $i = 1, 2, \cdots, 2k$; and P_{2k+1} a shortest (x_{2k+1}, x_1)-path in G. Strongly connectedness of G ensures existence of these directed paths.

If there is some P_i with even length, then $\psi_G(e_i) = (x_{i+1}, x_i)$ by the shortness of P_i. Thus, $P + e_i$ is an odd directed cycle in G.

Suppose that the length of P_i is odd for every $i = 1, 2, \cdots, 2k+1$. Let $W = P_1 \oplus P_2 \cdots \oplus P_{2k+1}$. Then W is a closed directed walk of odd length, and it can be expressed as the union of several edge-disjoint directed circuits, of which at least one is odd. ∎

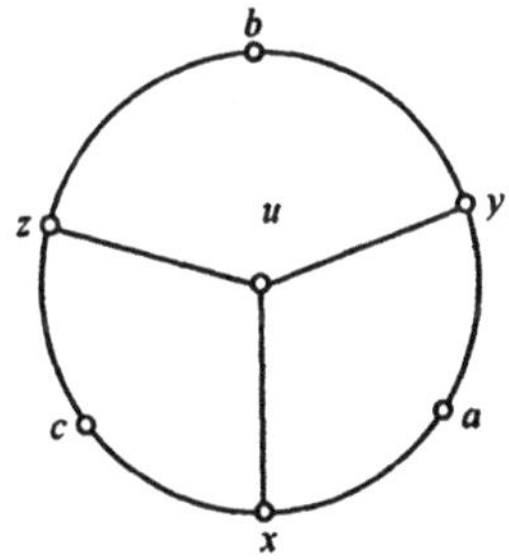

Figure 1.19: An illustration in the proof of Example 1.7.5

Example 1.7.5 Let G be a non-bipartite undirected graph. If G is simple and $\varepsilon > \frac{1}{4}\,(v-1)^2 + 1$, then G contains a triangle.

Proof Since G is non-bipartite, G contains odd cycle by Corollary 1.6.2. Let C be a shortest cycle of odd length with the set S of k vertices in G. Suppose to the contrary that k is at least five since G is simple. We first show that

$$|(S, \overline{S})| \leq 2(v-k). \tag{1.7}$$

Otherwise, there exists some $u \in \overline{S}$ such that $|N_G(u) \cap S| \geq 3$. Let $x, y, z \in N_G(u) \cap S$. Since G contains no triangle, there are $a, b, c \in S \setminus \{x, y, z\}$ and three cycles (see Figure 1.19) $C_1 = (u, x, \cdots, a, \cdots, y, u)$, $C_2 = (u, y, \cdots, b, \cdots, z, u)$ and $C_3 = (u, z, \cdots, c, \cdots, x, u)$, all of their lengths are less than k and at least one is odd, contrary to the choice of C. Therefore the inequality (1.7) holds.

By the choice of C we have $G[S] = C$ and, hence, $\varepsilon(G[S]) = k \geq 5$. It follows from (1.7) and Example 1.3.1 that

$$\begin{aligned}\varepsilon(G) &= \varepsilon(G[S]) + |(S, \overline{S})| + \varepsilon(G[\overline{S}]) \\ &\leq k + 2(v-k) + \tfrac{1}{4}\,(v-k)^2 \leq \tfrac{1}{4}\,(v-1)^2 + 1,\end{aligned}$$

which contradicts our hypothesis. Therefore, G contains a triangle. ∎

Exercises

1.7.1 Prove Corollary 1.6.2 and Corollary 1.6.3.

1.7.2 Prove that

(a) any graph with $\varepsilon \geq v$ contains a cycle;

(b) any connected 2-regular undirected graph is a cycle;

(c) any strongly connected 1-regular digraph is a directed cycle.

1.7.3 Suppose that G is a simple digraph.

(a) Prove that if $k = \max\{\delta^+, \delta^-\} > 0$, then G contains a directed cycle of length at least $k + 1$.

(b) Prove that if $\varepsilon > \frac{1}{2}\, v(v-1)$, then G contains a directed cycle.

(c) Construct a simple digraph with $\varepsilon = \frac{1}{2}\, v(v-1)$ such that it contains no directed cycle.

1.7.4 Suppose that T is a tournament. Prove that

(a) if $k = \max\{\delta^+, \delta^-\} > 0$, then T contains a directed cycle of length at least $2k + 1$;

(b) if T is strongly connected and $v \geq 4$, then there exists $S \subseteq V(T)$ such that $|S| \geq 2$ and $T - x$ is strongly connected for any $x \in S$;

(c) if T contains a directed k-cycle, then T contains a directed l-cycle for each $l = 3, 4, \cdots, k$.

1.7.5 Prove that if G is a simple undirected graph with $\delta \geq 3$, then G contains even cycle and the greatest common factor of all lengths of cycles in G is either 1 or 2.

1.7.6 Prove that if G is a connected simple undirected graph with $v > 2\delta$, then G contains a path of length at least 2δ.

1.7.7 Let G be a non-bipartite simple undirected graph and k be a given integer. Prove that if $k \geq 2$ and $\delta > \left\lfloor \dfrac{2v}{2k+1} \right\rfloor$, then G contains an odd cycle of length at most $(2k - 1)$.

1.7.8 Prove that a simple graph G of order $v \geq 4$ contains two different cycles with exactly one edge in common if it satisfies one of the following conditions:

(a) $\delta(G) \geq 3$; (b) $\varepsilon(G) = 2v - 3$.

1.7.9 A Δ-regular undirected graph of diameter k with the largest order is called a *maximum* (Δ, k)*-graph*. Used $n(\Delta, k)$ to denote the order of a maximum (Δ, k)-graph. Exercise 1.6.5 gives an upper bound of $n(\Delta, k)$ for $\Delta \geq 2$. Prove that

(a) $n(\Delta, 1) = \Delta + 1$ and a complete graph $K_{\Delta+1}$ is, to up isomorphism, the unique maximum $(\Delta, 1)$-graph;

(b) $n(2, k) = 2k + 1$ and a $(2k + 1)$-cycle is, to up isomorphism, the unique maximum $(2, k)$-graph;

(c) $n(3, 2) = 10$ and the Petersen graph is a maximum $(3, 2)$-graph.

(In addition, Hoffman and Singleton [97] showed $n(7, 2) = 50$, and Elspas [53] showed $n(3, 3) = 20$, $n(4, 2) = 15$ and $n(5, 2) = 24$. These are the only known exact values of $n(d, k)$ so far.)

1.7.10 A Δ-regular undirected graph with girth at least g having the least order is called a (Δ, g)*-cage*. Used $f(\Delta, g)$ to denote order in a (Δ, g)-cage. When $g \geq 3$, Example 1.7.2 gives a lower bound of $f(\Delta, g)$.

(a) Prove that the diameter $d(G) \leq g$ and the girth $g(G) = g$.

(b) Complete the proof of Example 1.7.2 for the case that g is even.

(c) Prove that $v(G) \leq \frac{\Delta}{\Delta-2}(\Delta - 1)^g$.

(d) Prove that $f(2, g) = g$ and a g-cycle is, to up isomorphism, the unique $(2, g)$-cage.

(e) Verify that, to up isomorphism, $K_{\Delta+1}$ is the unique $(\Delta, 3)$-cage and $K_{\Delta,\Delta}$ is the unique $(\Delta, 4)$-cage.

(f) The Petersen graph is a $(3, 5)$-cage.

(g) Can you depict $(3, 6)$-, $(3, 7)$-, $(4, 5)$-cages and more?

1.7.11 Suppose that G is a digraph without a directed cycle. Prove that

(a) $\delta^- = 0$;

(b) there is an ordering $x_1, x_2, \cdots, x_v$ of V such that every directed edge of G with head x_i has its tail in $\{x_1, x_2, \cdots, x_{i-1}\}$ for each $i = 1, 2, \cdots, v$.

1.7.12 The *converse* $\overleftarrow{G}$ of a digraph G is a digraph obtained from G by reversing the orientation of each edge.

(a) Prove that (i) $d^+_{\overleftarrow{G}}(x) = d^-_G(x)$; (ii) $d_{\overleftarrow{G}}(x, y) = d_G(y, x)$.

(b) By using part (ii) of (a), deduce from the exercise 1.7.11 (a) that if G is a digraph without a directed cycle, then $\delta^+ = 0$.

1.7.13 Let $G_1, G_2, \cdots, G_\omega$ be all strongly connected components of a digraph G. The *condensation* $\hat{G}$ of G is a simple digraph with ω vertices $u_1, u_2, \cdots, u_\omega$ and $(u_i, u_j) \in E(\hat{G})$ if and only if $E_G(V(G_i), V(G_j)) \neq \emptyset$. Prove that

(a) $\hat{G}$ contains no directed cycle;

(b) a simple digraph G contains no directed cycle if and only if $G \cong \hat{G}$.

1.7.14 A tournament T is called to be *transmissible* if, whenever (x, y) and (y, z) are edges of T, then (x, z) is also an edge of T. A sequence $(s_1, s_2, \cdots, s_n)$ of nonnegative integers is called a *score sequence* of a tournament if there exists a tournament T of order n whose vertices can be labelled as $x_1, x_2, \cdots, x_n$ such that $d^+_T(x_i) = s_i$ for $i = 1, 2, \cdots, n$. Prove that

(a) a tournament T is transmissible if and only if T contains no directed cycle;

(b) if a tournament T is transmissible, then $d^+_T(x) \neq d^+_T(y)$ and $d^-_T(x) \neq d^-_T(y)$ for any two vertices x and y of T;

(c) A non-decreasing sequence S of $n\,(\geq 1)$ nonnegative integers is a score sequence of a transmissible tournament of order n if and only if $S = (0, 1, 2, \cdots, n-1)$;

(d) there exists exactly one transmissible tournament of order n;

(e) there exists exactly one tournament of order n without a directed cycle;

(f) the condensation of a tournament is transmissible;

(g) every transmissible tournament contains exactly one Hamilton directed path;

(h) if T is not a transmissible tournament, then T contains at least three Hamilton directed paths;

(i) any tournament of order 2^{n-1} contains a transmissible tournament as its subgraph.

1.8 Eulerian Graphs

A (directed) trail that traverses every edge and every vertex of G is called an *Euler (directed) trail*. A closed Euler (directed) trail is called an *Euler (directed) circuit*. A (di)graph is *eulerian* if it contains an Euler (directed) circuit, and *noneulerian* otherwise.

Euler trails and Euler circuits are named after Leonhard Euler (1707–1783), who in 1736 characterized those graphs which contain them in the earliest known paper on graph theory [57]. At the time Euler was a professor of mathematics in St. Petersburg, and was led to the problem by the puzzle of the seven bridges on the Pregel (see Figure 1.20 (a)) in the ancient Prussian city of Königsberg (the birthplace and home of Immanuel Kant (1724-1805), a great German philosopher, and the seat of a great German university, which was take over by the USSR and renamed Kaliniggrad in 1946; since the collapse of the Soviet Union it has belonged to Russia). The good burghers of Königsberg wondered whether it was possible to plan a walk in such a way that passed through each bridge exactly once and that ended at the place where the walk began. This is the well-known Königsberg seven bridges problem.

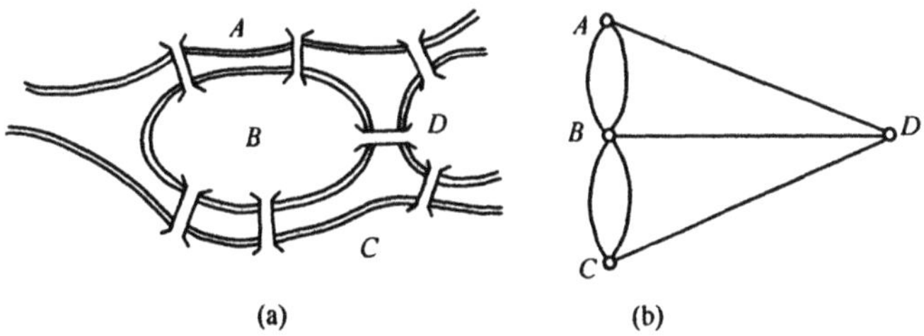

Figure 1.20: The Königsberg bridges and the corresponding graph

Euler abstracted the Königsberg seven bridges as the graph shown in Figure 1.20 (b), where vertices represent the four lands and edges represent the seven bridges, formalized the Königsberg seven bridges problem to the question whether such a graph contains an Euler circuit.

We now give a characterization of eulerian graphs.

Theorem 1.7 A digraph is eulerian if and only if it is connected and balanced.

Prof Suppose that G is an Euler digraph and let C be an Euler directed circuit of G. Then G is connected since C traverses every vertex of G by the definition. Arbitrarily choose $x \in V(C)$. It each time occurs as an internal vertex of C, two edges are incident with x, one out-going edge and another in-coming edge. Thus, $d^+_G(x) = d^-_G(x)$.

Conversely, suppose to the contrary that a connected and balanced digraph is not eulerian. Choose such a digraph with the number of edges as few as possible. Then G contains directed cycle since $\delta^+ = \delta^- \neq 0$ (the exercise 1.7.3). Let C be a directed circuit of maximum length in G. By assumption, C is not an Euler directed circuit of G, and so $G - E(C)$ contains a connected component G' with $\varepsilon(G') > 0$. Since C is itself balanced, thus the connected graph D' is also balanced. Since $\varepsilon(G') < \varepsilon(G)$, it follows from the choice of G that G' contains an Euler directed circuit C'. Since G is connected, $V(C) \oplus V(C') \neq \emptyset$. Thus, $C \oplus C'$ is a directed circuit of G with length larger than $\varepsilon(C)$, contradicting the choice of C. ∎

Corollary 1.7.1 A connected digraph G contains an Euler directed trail from x to y if and only if G satisfies the conditions:

$$\begin{array}{l} d^+_G(x) - d^-_G(x) = 1 = d^-_G(y) - d^+_G(y); \text{ and} \\ d^+_G(u) = d^-_G(u), \ \forall\, u \in V \setminus \{x, y\}. \end{array} \tag{1.8}$$

Proof Let G' be a digraph obtained from G by adding a new edge a from y to x.

Let T be an Euler directed trail from x to y of G. Then $T + a$ is an Euler directed circuit of G'. Thus, G' is balanced by Theorem 1.7. As a result, we have that

$$\begin{array}{l} d^+_G(u) = d^+_{G'}(u) = d^-_{G'}(u) = d^-_G(u), \ \forall\, u \in V \setminus \{x, y\}, \\ d^+_G(x) = d^+_{G'}(x) = d^-_{G'}(x) = d^-_G(x) + 1, \\ d^-_G(y) = d^-_{G'}(y) = d^+_{G'}(y) = d^+_G(y) + 1. \end{array}$$

Thus G satisfies the condition (1.8).

Conversely, suppose that G satisfies the condition (1.8). Then G' constructed above is balanced, and, thus, contains an Euler directed circuit by Theorem 1.7. Let C be an Euler directed circuit of G'. Then $C - a$ is an Euler directed trail from x to y in G. ∎

The necessary condition for an undirected graph to be eulerian was first found by Euler in 1736 when he studied the Königsberg seven bridges problem, while the sufficient condition was found in 1873 by Hierholzer [93]. We can state this as the following corollary.

Corollary 1.7.2 An undirected graph is eulerian if and only if it is connected and has no vertex of odd degree.

Proof Suppose that G is an eulerian undirected graph and C is an Euler circuit of G. Let D be an oriented graph of G by assigning every edge of G a direction in order of C. Then D is eulerian, connected and balanced by Theorem 1.7. Thus,

$$d_G(x) = d_D^+(x) + d_D^-(x) = 2d_D^+(x), \ \forall \ x \in V(G),$$

which means that G contains no odd vertex.

Conversely, suppose that G is a connected undirected graph without vertices of odd degree. By Example 1.7.3, there is a balanced oriented graph D of G. Thus, D is eulerian by Theorem 1.7. Let C be an Euler directed circuit of D. Then, we can obtain an Euler circuit of G by deleting the orientation of all edges of C. Thus, G is eulerian. ∎

Corollary 1.7.3 An undirected graph G contains an Euler trail if and only if G is connected and contains at most two vertices of odd degree.

Proof Suppose that T is an Euler trail with end-vertices x and y. If $x = y$, then T is an Euler circuit of G. Thus G is connected and contains no vertex of odd degree by Corollary 1.7.2. If $x \neq y$, then the closed trail T' obtained from T by adding a new edge e joining x and y is an Euler circuit of $G + e$. It follows from Corollary 1.7.2 that $G + e$ contains no vertex of odd degree. Thus, G contains exactly two vertices of odd degree, x and y.

Conversely, suppose that G is connected and contains at most two vertices of odd degree. If G contains no vertex of odd degree, then G contains an Euler circuit by Corollary 1.7.2, and, hence, contains an Euler trail. If G contains a vertex of odd degree, then G contains exactly two vertices of odd degree by Corollary 1.1.2. Let x and y be two vertices of odd degree of G. Then the graph G' obtained from G by adding a new edge e joining x and y contains no vertex of odd degree. It follows from Corollary 1.7.2 that G' contains an Euler circuit C, and, hence $C - e$ is an Euler trail of G. ∎

There are several other characterizations of an eulerian digraph, some of them, for example, are stated in the exercises at the end of this section and Corollary 4.2 in Section 4.2.

We conclude this section with two important classes of eulerian digraphs, the de Bruijn digraphs and the Kautz digraphs, which occur in the literature and textbooks of graph theory frequently.

Example 1.8.1 For any integers d and n with $n \geq 1$ and $d \geq 2$, we have, in Example 1.4.4, defined the n-dimensional d-ary *de Bruijn digraph* $B(d,n)$ as $(n-1)$th iterated line digraph $L^{n-1}(K_d^+)$, where K_d^+ denotes the digraph obtained from a complete digraph K_d of order d by appending one loop at each vertex. We have known that it is a d-regular connected digraph and so an eulerian digraph by Theorem 1.7.

We here give another equivalent definition of de Bruijn digraph $B(d,n)$. Its vertex-set

$$V = \{x_1x_2\cdots x_n : \ x_i \in \{0,1,\cdots,d-1\}, \ i = 1,2,\cdots,n\},$$

and edge-set E, where for $x, y \in V$, if $x = x_1x_2\cdots x_n$, then

$$(x,y) \in E \Longleftrightarrow y = x_2x_3\cdots x_n\alpha, \quad \alpha \in \{0,1,\cdots,d-1\}.$$

The de Bruin digraph $B(2,n)$ was first proposed by de Bruijn [35] and Good [75], independently, in 1946. Three smaller de Bruijn digraphs $B(2,n)$ are depicted in Figure 1.21 according to this kind of definition. ∎

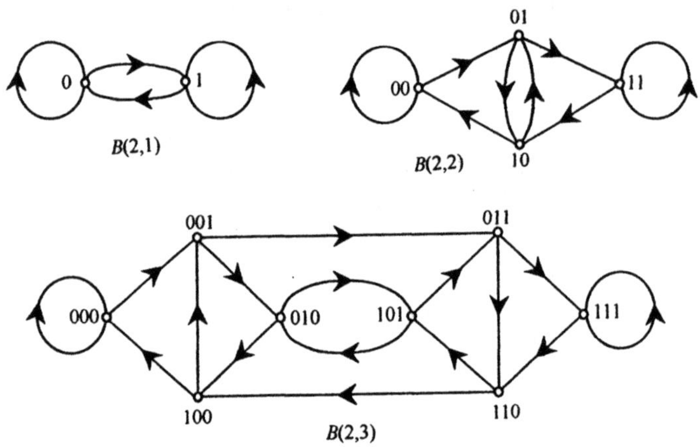

Figure 1.21: de Bruijn digraphs $B(2,1)$, $B(2,2)$ and $B(2,3)$

We now present another important property of de Bruijn digraph, due to Imase, Soneoka and Okada [100], the proof given here due to Xu, Tao and Xu [182].

Example 1.8.2 For any two distinct vertices x and y of $B(d,n)$, there are $d-1$ internally disjoint (x,y)-paths of length at most $n+1$.

Proof We proceed by induction on $n \geq 1$. Since $B(d,1) = K_d^+$, the theorem is true for $n = 1$ clearly. Suppose $n \geq 2$ and the theorem holds for any two vertices of $B(d, n-1)$. Assume that x and y are two distinct vertices of $B(d, n)$.

First, we consider the case that (x, y) is not an edge of $B(d, n)$. Since $B(d, n) = L(B(d, n-1))$, x and y correspond to two edges in $B(d, n-1)$. Let such two edges be $x = (u, u')$ and $y = (v, v')$. Then $u' \neq v$ since (x, y) is not an edge of $B(d, n)$. By induction hypothesis, there are $d-1$ internally disjoint (u', v)-paths of length at most n in $B(d, n-1)$, from which we can induce $d-1$ internally disjoint (x, y)-paths of length at most $n+1$ in $B(d, n)$.

We consider the case that (x, y) is an edge of $B(d, n)$. Then x and y can be written as

$$x = x_1 x_2 \cdots x_{n-1} x_n, \quad y = x_2 x_3 \cdots x_n y_n.$$

We construct $d-1$ (x, y)-walks $P_1, P_2, \cdots, P_{d-1}$ as follows.

$$\begin{aligned} P_1 &= x_1 x_2 \cdots x_{n-1} x_n y_n, \\ P_j &= x_1 x_2 \cdots x_n u_j x_2 x_3 \cdots x_n y_n, \ j = 2, 3, \cdots, d-1, \end{aligned}$$

where $u_2, \cdots, u_{d-1}$ are $d-2$ distinct elements in $\{0, 1, \cdots, d-1\} \setminus \{x_1, y_1\}$. It is obviously that P_1 is of length one and P_j is of length $n+1$ for each $j = 2, 3, \cdots, d-1$. In order to prove these (x, y)-walks are internally disjoint, it is sufficient to prove $P_2, P_3, \cdots, P_{d-1}$ are internally disjoint.

Suppose to the contrary that there are i and j $(2 \leq i \neq j \leq d-1)$ such that P_i and P_j have common vertices rather than x and y. Let u be the first internally common vertex of P_i and P_j from x to y. Assume that the subwalk $P_i(x, u)$ is of length a and $P_j(x, u)$ is of length b. Then $2 \leq a, b \leq n-1$. Let u' and u'' be in-neighbors of u on P_i and P_j, respectively. Then $u' \neq u''$. Since u can be reached in a steps from x along P_i and in b steps from x along P_j, then it can be written as

$$\begin{aligned} u &= x_{a+1} x_{a+2} \cdots x_n u_i x_2 \cdots x_a \\ &= x_{b+1} x_{b+2} \cdots x_n u_j x_2 \cdots x_b. \end{aligned}$$

From this expression, we have $x_a = x_b$ since $2 \leq a, b \leq n-1$, i.e.,

$$u' = x_a x_{a+1} \cdots x_n u_i x_2 \cdots x_{a-1} = x_b x_{b+1} \cdots x_n u_j x_2 \cdots x_{b-1} = u'',$$

a contradiction. Note that $P_2, \cdots, P_{d-1}$ may be not paths, but each of them must contain a path as its subgraph, and, thus, the conclusion follows. ∎

Example 1.8.3 For any given integers d and n with $n \geq 1$ and $d \geq 2$, we have, also in Example 1.4.4, defined the n-dimensional d-ary *Kautz* *digraph* $K(d,n)$ as $(n-1)$th iterated line digraph $L^{n-1}(K_{d+1})$, where K_{d+1} is a complete digraph of order $d+1$. We have known that it is a d-regular connected digraph and so an eulerian digraph by Theorem 1.7.

We here give another equivalent definition of Kautz digraph $K(d,n)$. Its vertex-set

$$V = \{x_1x_2\cdots x_n : x_i \in \{0,1,\cdots,d\},\ x_{i+1} \neq x_i,\ i=1,2,\cdots,n-1\},$$

and the edge-set E, where for $x,y \in V$, if $x = x_1x_2\cdots x_n$, then

$$(x,y) \in E \Longleftrightarrow y = x_2x_3\cdots x_n\alpha, \quad \alpha \in \{0,1,\cdots,d\} \setminus \{x_n\}.$$

This kind of definition was first proposed by Kautz [103] in 1969. Three smaller Kautz digraphs $K(d,n)$ are depicted in Figure 1.22.

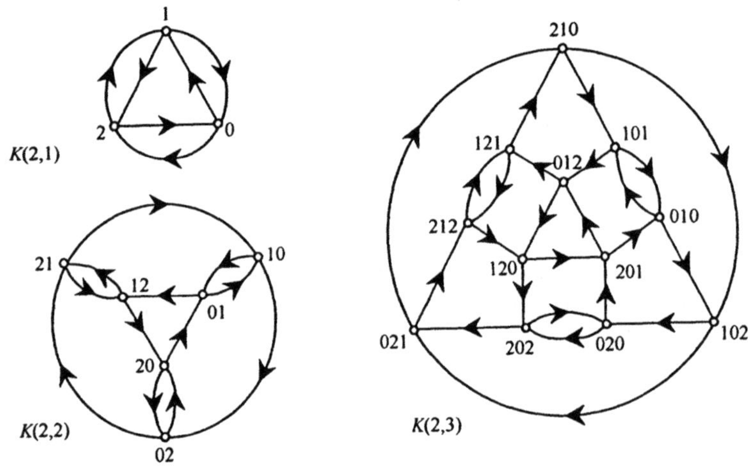

Figure 1.22: Kautz digraphs $K(2,1)$, $K(2,2)$ and $K(2,3)$

Both de Bruijn digraph $B(d,n)$ and Kautz digraph $K(d,n)$ are d-regular and have diameter n. Moreover, $B(d,n)$ has d^n vertices, while $K(d,n)$ has $d^n + d^{n-1}$ vertices.

The Kautz digraph also has property similar to that of de Bruijn digraph in Example 1.8.2, see the exercise 1.8.9 for the details. ∎

The definitions of de Bruijn digraph and Kautz digraph by line digraph are proposed by Fiol, Yebra, and Alergre [62]. This kind of definition is very useful and convenient for further studying other properties of de Bruijn digraphs and Kautz digraphs.

Exercises

1.8.1 Prove that a digraph G is eulerian if and only if G is connected and there are edge-disjoint directed cycles $C_1, C_2, \cdots, C_m$ such that $G = C_1 \oplus C_2 \oplus \cdots \oplus C_m$.

1.8.2 Suppose that G is a connected digraph and satisfies the condition

$$\sum_{x \in V} |d_G^+(x) - d_G^-(x)| = 2l, \quad l \geq 1.$$

Prove that there are l edge-disjoint directed trails $T_1, T_2, \cdots, T_l$ such that $G = T_1 \oplus T_2 \oplus \cdots \oplus T_m$.

1.8.3 Suppose that G is a connected digraph, x and y be two distinct vertices of G. Prove that if

$$d_G^+(x) - d_G^-(x) = l = d_G^-(y) - d_G^+(y); \text{ and}$$
$$d_G^+(u) = d_G^-(u), \ \forall\, u \in V \setminus \{x, y\},$$

then there are at least l edge-disjoint (x, y)-paths in G.

1.8.4 Suppose that G is an undirected graph. Prove that there is an oriented graph D of G such that $|d_D^+(x) - d_D^-(x)| \leq 1$ for any $x \in V$.

1.8.5 Prove that the cartesian product of two eulerian graphs is an eulerian graph, hence $2n$-cube Q_{2n} is an eulerian graph.

1.8.6 (a) Prove that if G is a connected digraph, $|d_G^+(x) - d_G^-(x)| \leq 1$ for any $x \in V$ and any edge of G is not contained in odd number of directed cycles, then G is eulerian.

(b) Give an example to show that the converse of (a) is not true.

1.8.7 Prove that a connected undirected G is eulerian if and only if every edge of G lies on an odd number of cycles. (The necessity is due to Toida [160] and the sufficiency to McKee [127])

1.8.8 Let G be an eulerian graph and x be a vertex of G. Prove that every trail of G with origin x (maybe closed trail) can be extended to an Euler circuit of G if and only if $G - x$ contains no cycle.

1.8.9 By making a little modification of the way in Example 1.8.2, prove that there are d internally disjoint (x, y)-paths for any two distinct vertices x and y of $K(d, n)$, one of length at most n, $d - 2$ of length at most $n + 1$, and one of length at most $n + 2$. (Du, Hsu and Lyuu [43])

1.9 Hamiltonian Graphs

A (directed) cycle that contains every vertex of a (di)graph G is called a *Hamilton (directed) cycle.* A (di)graph is *hamiltonian* if it contains a Hamilton (directed) cycle, and *non-hamiltonian* otherwise.. The problem determining whether a given graph is hamiltonian is called the *Hamilton problem.*

The name "hamiltonian" is derived from Sir William Rowan Hamilton (1805-1865), a well-known Irish mathematician. In 1857 (see [10]), Hamilton introduced a game known as "The Traveller's Dodecahedron" or "A Voyage Round the World". In this game, each corner of a regular solid dodecahedron was marked by a peg to represent an important city of the time (see Figure 1.23 (a)). The aim of the game was to find a path along the edges of the dodecahedron that passed through each city exactly once and that ended at the city where the route began. In order for the player to recall which cities in a route he had already visited, the string could be used to connect the appropriate pegs in the appropriate order. A cycle passing through each city once only was called a "voyage round the world".

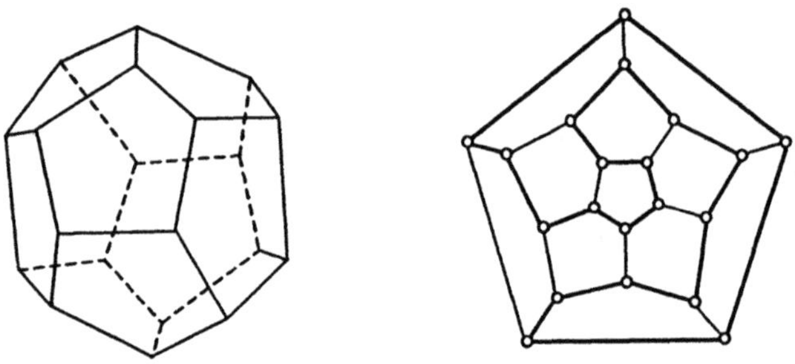

(a) (b)

Figure 1.23: The dodecahedron and the corresponding graph

The dodecahedron can be expressed as a graph shown in Figure 1.23 (b). The object of Hamilton's game may be described in graphical terms, namely, to determine whether the graph of the dodecahedron has a cycle containing each of its vertices, called a Hamilton cycle. It is from this that we get the term "Hamilton". A solution of the Hamilton's game is illustrated by the heavy edges in Figure 1.23 (b).

It is interesting to note that in 1855, two years before Hamilton introduced his game, the English mathematician Thomas P. Kirkman posed the following question in a paper [107] submitted to the Royal Society. Given the graph of a polyhedron, can one always find a cycle that passes through each vertex once and only once? Thus, Kirkman apparently introduced the general study of "hamiltonian graphs" although Hamilton's game generated interest in the problem.

Although the Hamilton problem is quite difficult to solve, it is easy to show the following result.

Example 1.9.1 The Petersen graph is non-hamiltonian.

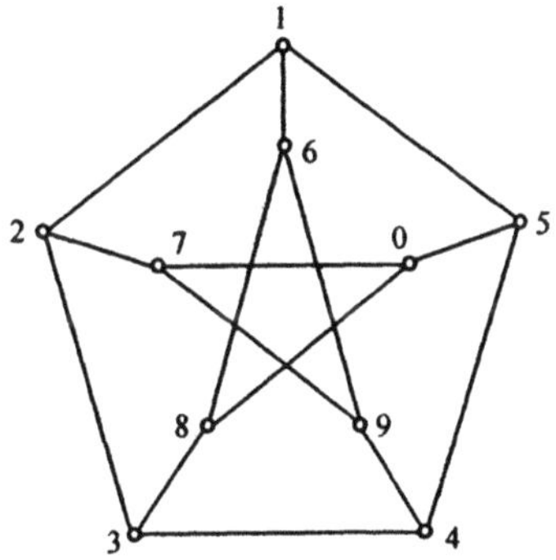

Figure 1.24: The Petersen graph is non-hamiltonian

Proof Suppose that G is the Petersen graph, and suppose to the contrary that G is hamiltonian. We label the vertices of G with the digits $1, 2, \cdots, 9, 0$ shown as Figure 1.24. Let $T = \{16, 27, 38, 49, 50\}$ be the subset of edges of G. Then $G - T$ is disconnected. Thus, any Hamilton cycle of G must contain even edges in T. It is not difficult to see that any cycle containing exactly two edges in T is not hamiltonian. It follows that every Hamilton cycle of G must contain four edges in T. Without loss of generality, suppose that C is a Hamilton cycle containing the edges $27, 38, 49, 50$. Then C must contain the edges $12, 15, 68, 69$, and does not contain the edges $23, 45$. Since the vertices 3 and 4 are contained in C, thus, the edge 34 must be contained in C. It follows that the set of edges $\{34, 49, 96, 68, 83\}$ can form a cycle that is contained in C, which is impossible. Therefore, Petersen graph is non-hamiltonian. ∎

It is not difficult to see that there exists a closed relationship between eulericity and hamiltonicity of a digraph that a digraph G is Eulerian if and only if $L(G)$ is hamiltonian (the exercise 1.9.8 (a)). However, in contrast with the case of eulerian graphs, no nontrivial necessary and sufficient condition for a graph to be hamiltonian is known; in fact, the problem of finding such a condition is one of the most main unsolved problems in graph theory. In this section, we introduce several necessary (but not sufficient) and sufficient (but not necessary) conditions.

An obvious and simple necessary condition is that any hamiltonian digraph must be strongly connected; any hamiltonian undirected graph must contains no cut-vertex (the exercise 1.9.1).

Theorem 1.8 If G is hamiltonian, then

$$\omega(G - S) \leq |S|, \quad \forall\, S \subset V. \tag{1.9}$$

Proof Let C be a Hamilton cycle of G. Then

$$\omega(C - S) \leq |S|, \quad \forall\, S \subset V(C) = V(G).$$

Since $C - S$ is a spanning subgraph of $G - S$,

$$\omega(G - S) \leq \omega(C - S) \leq |S|, \quad \forall\, S \subset V$$

as required. ∎

Example 1.9.2 The Petersen graph shows that the condition (1.9) is not sufficient. To the end, let G be Petersen graph. It is not difficult to verify that

$$\omega(G - S)\begin{cases} = 1 \leq |S|, & \text{for } |S| \leq 2; \\ \leq 2 < |S|, & \text{for } |S| = 3; \\ \leq 3 < |S|, & \text{for } |S| = 4; \\ \leq 5 \leq |S|, & \text{for } |S| \geq 5, \end{cases} \qquad \forall\, S \subset V(G).$$

The implies that the Petersen graph satisfies the condition (1.9), however, it has been prove to be non-hamiltonian in Example 1.9.1. ∎

We now introduce several sufficient conditions for a graph to be hamiltonian. Since parallel edges and loops do not affect whether a graph is hamiltonian, it suffices to limit our discussion to simple graphs. We start with a result due to Ore [141].

Theorem 1.9 Let G be a simple undirected graph of order $v \geq 3$. If

$$d_G(x) + d_G(y) \geq v, \text{ for any } x, y \in V(G), xy \notin E(G), \tag{1.10}$$

then G is hamiltonian.

Proof By contradiction. Suppose that the theorem is false, and let G be a maximal non-hamiltonian simple graph satisfying the condition (1.10). Since $v \geq 3$, G is not a complete graph. Let x and y be nonadjacent vertices in G. By the choice of G, $G + xy$ is hamiltonian. Moreover, since G is non-hamiltonian, each Hamilton cycle of $G + xy$ must contain the edge xy. Thus there is a Hamilton path connecting x and y in G. Let $P = (x_1, x_2, \cdots, x_v)$ be a Hamilton path, where $x = x_1$ and $y = x_v$. Set

$$S = \{x_i \in V(P) : \ xx_{i+1} \in E(G),\ 1 \le i \le v-2\},$$
$$T = \{x_j \in V(P) : \ yx_j \in E(G),\ 2 \le j \le v-1\}.$$

Since G is simple and $y \notin S \cup T$, we have

$$|S| = d_G(x),\ |T| = d_G(y) \quad \text{and} \quad |S \cup T| \le v - 1.$$

We now show that $|S \cap T| = 0$. In fact, if there is some $x_i \in S \cap T$, then $C = (x, x_{i+1}, x_{i+2}, \cdots, x_{v-1}, y, x_i, \cdots, x_2, x)$ is a Hamilton cycle in G (see Figure 1.25), contrary to our assumption. It follows that

$$d_G(x) + d_G(y) = |S| + |T| = |S \cup T| \le v - 1.$$

But this contradicts the condition (1.10), and so G is hamiltonian. ∎

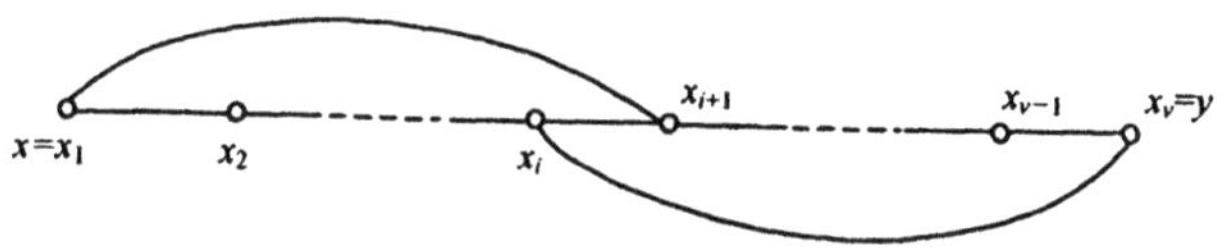

Figure 1.25: An illustration the proof of Theorem 1.9

We now present another proof of Theorem 1.9, which is useful to understand the proof of Theorem 1.10 later.

Another proof of Theorem 1.9 It is easy to see that any graph G satisfying the condition (1.10) is connected (the exercise 1.5.6 (c)) and contains a cycle. Let $C = (x_1, x_2, \cdots, x_k, x_1)$ be a longest cycle in G. Suppose $k < v$ and let $R = V(G) \setminus V(C)$. Since G is connected, there is $y \in R$ adjacent to some vertex in C, without loss of generality, say x_k. Because of the choice of C, we have $x_1 y \notin E(G)$. Let

$$S = \{x_i \in V(C) : \ x_1 x_{i+1} \in E(G),\ 1 \le i \le k-1\},$$
$$T = \{x_j \in V(C) : \ yx_j \in E(G),\ 2 \le j \le k\}.$$

Then it is easy to verify that

$$|S \cup T| \le k \quad \text{and} \quad |S \cap T| = 0. \tag{1.11}$$

By (1.11), the number of neighbors of $\{x_1, y\}$ in C

$$d_C(x_1) + d_C(y) = |S| + |T| = |S \cup T| \le k. \tag{1.12}$$

On the other hand, by the choice of C, at most one of x_1z and yz is an edge of G for any $z \in R \setminus \{y\}$. Therefore, the number of neighbors of $\{x_1, y\}$ in R

$$d_R(x_1) + d_R(y) \leq v - k - 1. \tag{1.13}$$

Summing up the inequalities (1.12) and (1.13) yields

$$d_G(x_1) + d_G(y) \leq k + v - k - 1 = v - 1.$$

But this contradicts the condition (1.10). Thus, $k = v$, that is, C is a Hamilton cycle in G. ∎

Corollary 1.9 (Dirac [38]) Every simple graph with $v \geq 3$ and $\delta \geq \frac{1}{2}\,v$ is hamiltonian. ∎

Last, we present a necessary condition for a digraph to be non-hamiltoian, due to Bondy and Thomassen [19], from many interesting sufficient conditions for a digraph to be hamiltonian can be deduced immediately.

For a digraph G with a vertex x, we will write $d_G(x) = d_G^+(x) + d_G^-(x)$.

Theorem 1.10 Let $C = (x_1, x_2, \cdots, x_k, x_1)$ be a longest directed cycle in a strongly connected digraph G. If $k < v$, then there are $x \in V(G) \setminus V(C)$ and two integers $a \in [1, k]$ and $b \in [1, k-1]$ such that

(i) $(x_a, x) \in E(G)$;

(ii) x_{a+i} and x are nonadjacent for any $i \in [1, b]$;

(iii) $d_G(x) + d_G(x_{a+b}) \leq 2v - 1 - b$, where the subscripts are modulo k.

Proof Let $S = V(C)$. Since G is strongly connected and $|S| < v$, there are $x_i, x_j \in S$ and an (x_i, x_j)-path P in G that does not contain any vertex in S except x_i and x_j. We will say P is an S-path if $x_i \neq x_j$ and an S-cycle otherwise.

Case 1 G contains no S-path.

Let $P = (x_a, y_1, \cdots, y_t, x_a)$ be an S-cycle with $x_a \in S$. Obviously, we have

$$|E_G[x_{a+1}, S]| \leq 2(k-1). \tag{1.14}$$

Let $x = y_1$. Since G contains no S-path, x and x_i are nonadjacent in G for any $i \in [1, k]$ and $i \neq a$. We have

$$|E_G[x, S]| \leq 2. \tag{1.15}$$

Since G contains no S-path, at most one of (x_{a+1}, y) and (y, x), and at most one of (x, y) and (y, x_{a+1}) is an edge of G for any $y \in V \setminus (S \cup \{x\})$.

Thus, we have

$$|E_G[y, \{x_{a+1}, x\}]| \leq 2, \ \forall\, y \in V \setminus (S \cup \{x\}). \tag{1.16}$$

Summing up the inequalities (1.14), (1.15) and (1.16) yields

$$d_G(x) + d_G(x_{a+1}) \leq 2(k-1) + 2 + 2(v-k-1) = 2v - 2.$$

Take $b = 1$ as required.

Case 2 G contains an S-path.

Choose an S-path $P = (x_a, y_1, \cdots, y_t, x_{a+r})$ such that r is as small as possible. Then $r \geq 2$ by the assumption of C. Let $x = y_1$.

By the choice of r, x and x_{a+i} are nonadjacent in G for any $i \in [1, r-1]$. By the assumption of C, at most one of (x_i, x) and (x, x_{i+1}) is an edge of G for any $i \in [1, k-1]$. Thus, we have

$$|E_G[x, S]| \leq k - r + 2. \tag{1.17}$$

And for any $y \in V \setminus (S \cup \{x\})$ and any $i \in [1, k-1]$, we have

$$|E_G[y, \{x, x_{a+1}\}]| \leq 2. \tag{1.18}$$

Let b be the maximum $i \in [1, r-1]$ for which $G[S]$ contains an (x_{a+r}, x_a)-path P' and $V(P') = \{x_{a+r}, x_{a+r+1}, \cdots, x_{a-1}, x_a, x_{a+1}, \cdots, x_{a+i-1}\}$.

By the choice of b, $x_{a+b} \notin V(P')$ and $|V(P')| = k - r + b$, we have

$$\begin{aligned} |E_G[x_{a+b}, V(P')]| &\leq k - r + b + 1; \\ |E_G[x_{a+b}, S \setminus V(P')]| &\leq 2(r - b - 1). \end{aligned} \tag{1.19}$$

Summing up the inequalities (1.17), (1.18) and (1.19) yields

$$\begin{aligned} d_G(x) + d_G(x_{a+b}) &\leq (k - r + 2) + (k - r + b + 1) \\ &\quad + 2(r - b - 1) + 2(v - k - 1) \\ &= 2v - b - 1, \end{aligned}$$

as required and the theorem follows. ∎

Corollary 1.10.1 (Meyniel [132]) Let G be a strongly connected simple digraph. If $d_G(x) + d_G(y) \geq 2v - 1$ for any two nonadjacent vertices x and y of G, then G is hamiltonian.

Corollary 1.10.2 (Ghouila-Houri [73]) Let G be a strongly connected simple digraph. If $d_G(x) \geq v$ for any vertex x of G, then G is hamiltonian.

Corollary 1.10.3 (Nash-Williams [139]) Let G be a simple digraph. If $\delta \geq \frac{1}{2}\, v > 1$, then G is hamiltonian.

Corollary 1.10.4 (Camion [23]) Every strongly connected tournament is hamiltonian.

Corollary 1.10.5 (Woodall [178]) Let G be a simple digraph. If either $(x, y) \in E(G)$ or $d_G^+(x) + d_G^-(y) \geq v$ for any two vertices x and y of G, then G is hamiltonian.

Corollary 1.10.6 (Rédel [150]) Every tournament contains a Hamilton path.

The proofs of these corollaries are left to the reader as an exercise (the exercise 1.9.2).

Exercises

1.9.1 (a) Prove that every hamiltonian graph contains no cut-vertex.

(b) Prove that every hamiltonian digraph is strongly connected.

(c) Prove that every hamiltonian bipartite graph is an equally bipartite graph.

(d) Give examples to show that the converse propositions of (a), (b) and (c) are not true.

1.9.2 (a) Prove the corollaries 1.10.1 $\sim$ 1.10.6 making use of Theorem 1.10.

(b) Give examples to show that the converse propositions of the corollaries 1.10.1 $\sim$ 1.10.6 are not true.

(c) Give examples to show that the degree-conditions in the corollaries 1.10.1 $\sim$ 1.10.5 can be improved.

1.9.3 (a) Prove that every strongly connected simple digraph with $v \geq 3$ and $\varepsilon > (v-1)(v-2)+2$ is hamiltonian.

(b) Construct a strongly connected and non-hamiltonian digraph with $v \geq 3$ and $\varepsilon = (v-1)(v-2)+2$.

1.9.4 (a) Prove that every connected simple undirected graph with $v \geq 3$ and $\varepsilon > \frac{1}{2}\,(v-1)(v-2)+1$ is hamiltonian.

(b) Construct a connected and non-hamiltonian undirected graph with $v \geq 3$ and $\varepsilon = \frac{1}{2}\,(v-1)(v-2)+1$.

1.9.5 The *closure* $c(G)$ of an undirected graph G is the graph obtained from G by recursively joining pairs of nonadjacent vertices whose degree sum is at least v until no such pair remains. Prove that $c(G)$ is well defined and a simple graph G is hamiltonian if and only if $c(G)$ is hamiltonian.

(Bondy and Chvatal [17])

1.9.6 Let G be a simple undirected graph with $v \geq 3$ and without a cut-vertex. Prove, using the exercise 1.9.5, that if $\max\{d_G(x), d_G(y)\} \geq \frac{1}{2} v$ for any two vertices x and y with distance two in G, then G is hamiltonian. (Genghua Fan [59])

1.9.7 Let G be an equally bipartite simple undirected graph of order $2n$ ($n \geq 2$). Prove that G is hamiltonian if it satisfies one of the following conditions:

(a) $d_G(x) + d_G(y) > n$ for any two nonadjacent vertices x and y of G;

(b) $\varepsilon > n^2 - n + 1$;

(c) $\delta > \frac{1}{2} n$.

1.9.8 (a) Prove that a digraph G is Eulerian if and only if its line digraph $L(G)$ is hamiltonian.

(b) Prove that the de Bruijn digraph $B(d, n)$ and the Kautz digraph $K(d, n)$ are hamiltonian.

(c) Construct a non-eulerian undirected graph G whose line graph $L(G)$ is hamiltonian.

1.9.9 Let n be an integer, G be a simple undirected graph, and $F \subseteq E(G)$ with $|F| = n$. Prove that if $d_G(x) + d_G(y) \geq \frac{1}{2}(v + n)$ for any two nonadjacent vertices x and y of G, and $G[F]$ is the union of several disjoint paths, then G contains a Hamilton cycle C with $F \subseteq E(C)$.

1.9.10 Prove that the following five problems are equivalent: To determine whether or not (Nash-Williams [139])

(a) an undirected graph contains a Hamilton cycle;

(b) an undirected graph contains a Hamilton path;

(c) a digraph contains a Hamilton directed cycle;

(d) a digraph contains a Hamilton directed path;

(e) a bipartite graph contains a Hamilton cycle.

1.10 Matrix Presentation of Graphs

In this section, we introduce two kinds of *matrix presentations* of a graph, that is, the adjacency matrix and incidence matrix of the graph.

A graph G with the vertex-set $V(G) = \{x_1, x_2, \cdots, v_v\}$ can be described by means of matrices. The *adjacency matrix* of G is a $v \times v$ matrix

$$\mathbf{A}(G) = (a_{ij}), \quad \text{where } a_{ij} = \mu(x_i, x_j) = |E_G(x_i, x_j)|.$$

For example, for the digraph D and the undirected graph G shown in Figure 1.26, their adjacency matrices $\mathbf{A}(D)$ and $\mathbf{A}(G)$ are as follows.

$$\mathbf{A}(D) = \begin{pmatrix} 0 & 2 & 0 & 0 \\ 0 & 0 & 0 & 0 \\ 1 & 1 & 0 & 1 \\ 1 & 0 & 0 & 1 \end{pmatrix}, \quad \mathbf{A}(G) = \begin{pmatrix} 0 & 2 & 1 & 1 \\ 2 & 0 & 1 & 0 \\ 1 & 1 & 0 & 1 \\ 1 & 0 & 1 & 1 \end{pmatrix}.$$

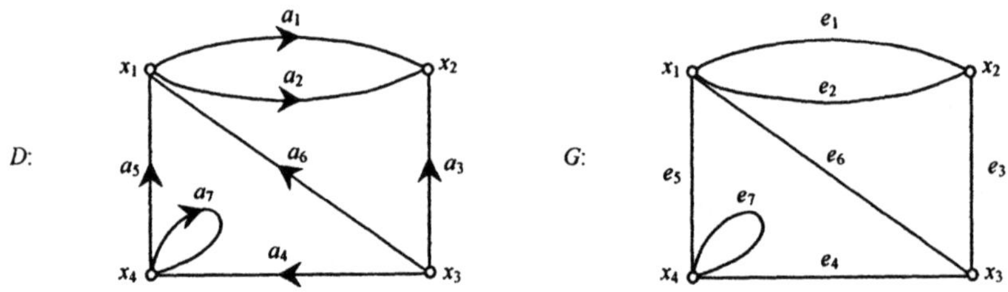

Figure 1.26: A digraph D and an undirected graph G

The *incidence matrix* of a loopless graph G is a $v \times \varepsilon$ matrix

$$\mathbf{M}(G) = (m_x(e)), \quad x \in V(G) \text{ and } e \in E(G),$$

where, if G is directed, then

$$m_x(e) = \begin{cases} 1, & \text{if } x \text{ is the tail of } e; \\ -1, & \text{if } x \text{ is the head of } e; \\ 0, & \text{otherwise,} \end{cases}$$

and if G is undirected, then

$$m_x(e) = \begin{cases} 1, & \text{if } e \text{ is incident with } x; \\ 0, & \text{otherwise.} \end{cases}$$

For example, for the digraph D and the undirected graph G shown in Figure 1.26, the incidence matrix $\mathbf{M}(D-a_7)$ and $\mathbf{M}(G)$ are as follows.

$$\mathbf{M}(D-a_7)=\begin{pmatrix} 1 & 1 & 0 & 0 & -1 & -1 \\ -1 & -1 & -1 & 0 & 0 & 0 \\ 0 & 0 & 1 & 1 & 0 & 1 \\ 0 & 0 & 0 & -1 & 1 & 0 \end{pmatrix},$$

$$\mathbf{M}(G-e_7)=\begin{pmatrix} 1 & 1 & 0 & 0 & 1 & 1 \\ 1 & 1 & 1 & 0 & 0 & 0 \\ 0 & 0 & 1 & 1 & 0 & 1 \\ 0 & 0 & 0 & 1 & 1 & 0 \end{pmatrix}.$$

The adjacency matrix or the incidence matrix of a graph is another representation of the graph, and it is this form that a graph can be commonly stored in computers. The matrix representation of a graph is often convenient if one intends to use a computer to obtain some information or solve a problem concerning the graph. This kind of representation of a graph is conducive to study properties of the graph by means of algebraic methods.

Let

$$\sigma=\begin{pmatrix} 1 & 2 & \cdots & n \\ i_1 & i_2 & \cdots & i_n \end{pmatrix}$$

be a permutation of the set $\{1,2,\cdots,n\}$. Then we obtain an $n\times n$ *permutation matrix* $\mathbf{P}=(p_{ij})$ defined by

$$p_{ij}=\begin{cases} 1, & \text{if } j=\sigma(i); \\ 0, & \text{otherwise.} \end{cases}$$

It is not difficult to see that the adjacency matrices of two isomorphic graphs are permutation similar. In other words, assume that $\mathbf{A}$ and $\mathbf{B}$ are the adjacency matrices of two isomorphic graphs G and H, respectively, then there exists a $v\times v$ permutation matrix $\mathbf{P}$ such that $\mathbf{A}=\mathbf{P}^{-1}\mathbf{B}\mathbf{P}$.

Similarly, the incidence matrices of two isomorphic graphs are permutation equivalent. In other words, assume that $\mathbf{M}$ and $\mathbf{N}$ are the incidence matrices of two isomorphic graphs G and H, respectively, then there exist a $v\times v$ permutation matrix $\mathbf{P}$ and an $\varepsilon\times\varepsilon$ permutation matrix $\mathbf{Q}$ such that $\mathbf{M}=\mathbf{PNQ}$.

It is these properties that makes us convenient to study structures of graphs by using their matrix presentations. We now present a very useful result on the adjacency matrix of a graph as follows.

Theorem 1.11 Let $\mathbf{A}$ be the adjacency matrix of a digraph G with the vertex set $\{x_1, x_2, \cdots, x_v\}$. Then the entry in position (i, j) of $\mathbf{A}^k$ is the number of different (x_i, x_j)-walks of length k in G.

Proof The proof is by induction on k. The result is obvious for $k = 1$ since there exist a_{ij} (x_i, x_j)-walks of length one if and only if there exist a_{ij} edges from x_i to x_j in G. Let $\mathbf{A}^{k-1} = \left(a_{ij}^{(k-1)}\right)$ and assume that $a_{ij}^{(k-1)}$ is the number of different (x_i, x_j)-walks of length $k-1$ in G; furthermore, let $\mathbf{A}^k = \left(a_{ij}^{(k)}\right)$. Since $\mathbf{A}^k = \mathbf{A}^{k-1} \cdot \mathbf{A}$, we have

$$a_{ij}^{(k)} = \sum_{l=1}^{v} a_{il}^{(k-1)} \cdot a_{lj}. \tag{1.20}$$

Every (x_i, x_j)-walk of length k in G consists of an (x_i, x_l)-walk of length $k-1$, where x_l $(1 \leq l \leq v)$ is adjacent to x_j, followed by an edge from x_l to x_j. Thus by the induction hypothesis and the equation (1.20), we have the desired result. ∎

It should be noted that walks could not be replaced by paths in Theorem 1.11 in general.

It is easy to see that there is the unique (x, y)-walk of length n for any pair (x, y) of vertices in $B(d, n)$. We obtain from Theorem 1.11 immediately that if $\mathbf{A}$ is the adjacency matrix of $B(d, n)$, then $\mathbf{A}^n = \mathbf{J}$, where $\mathbf{J}$ is an n-square matrix all of whose entries are 1. Similarly, if $\mathbf{A}$ is the adjacency matrix of $K(d, n)$, then $\mathbf{A}^n + \mathbf{A}^{n-1} = \mathbf{J}$.

We will, in Section 1.11 this book, introduce an important application of the adjacency matrix of a graph, specially Theorem 1.11, in matrix theory. We here give two other examples, which are important results in graph theory, to show that adjacency and incidence matrices are very useful for studying graphs. The first example is a result on (Δ, k)-Moore digraphs, due to Plesnik and Znom [146], and rediscovered by Bridges and Toueg [20].

Example 1.10.1 There is no (Δ, k)-Moore digraph for $\Delta \geq 2$ and $k \geq 2$.

Proof Assume that G is a (Δ, k)-Moore digraph whose order n reaches the Moore bound defined in (1.6), and let $\mathbf{A}$ be the adjacency matrix of G. By the exercise 1.6.2, G is a Δ-regular and simple digraph. Furthermore, by Theorem 1.11, we have

$$\mathbf{I} + \mathbf{A} + \mathbf{A}^2 + \cdots + \mathbf{A}^k = \mathbf{J}, \tag{1.21}$$

where **I** is an identity square matrix. The expression (1.21) implies that **J** is a polynomial in **A**, and so the matrices **A** and **J** have a common set of eigenvectors.

It is not difficult to show that Δ is an eigenvalue of **A** (see the exercise 1.10.6). Let r be any eigenvalue other than Δ, and let **X** be an eigenvector corresponding to r. Noting that the zero, as an eigenvalue of **J**, has the multiplicity $n-1$, we have

$$\mathbf{AX} = r\mathbf{X}, \qquad \mathbf{JX} = 0.$$

By (1.21), we obtain the relation

$$1 + r + r^2 + \cdots + r^k = 0. \tag{1.22}$$

The expression (2.22) shows that r has the multiplicity $(k+1)$ as the unite root, i.e., $r^{k+1} = 1$. Let $r_1, r_2, \cdots, r_{n-1}$ be $n-1$ eigenvalues of **A** other than Δ. By Theorem 1.11, all the main diagonal entries of $\mathbf{A}^i (1 \le i \le k)$ are 0, that is,

$$\mathrm{Tr}\mathbf{A}^i = 0, \quad i = 1, 2, \cdots, k.$$

Thus the sum of the eigenvalues of $\mathbf{A}^i$

$$\Delta^i + \sum_{j=1}^{n-1} r_j^i = 0, \quad i = 1, 2 \cdots, k. \tag{1.23}$$

Since $r_j \bar{r}_j = |r_j|^2 = 1 = r_j^{k+1}$, it follows that $r_j^{-1} = \bar{r}_j = r_j^k$, where $\bar{r}_j$ is the conjugate complex number of r_j. Setting $i = 1$ and $i = k$ in (1.23), respectively, we have

$$-\Delta = \sum_{j=1}^{n-1} r_j, \qquad -\Delta^k = \sum_{j=1}^{n-1} r_j^k.$$

Taking the conjugates of the above expressions and noting (1.23), we have that

$$-\Delta = \sum_{j=1}^{n-1} r_j^{-1} = \sum_{j=1}^{n-1} r_j^k = -\Delta^k,$$

which holds if and only if either $k = 1$ or $\Delta = 1$. This contradicts to our assumption and, thus, the conclusion follows. ∎

By Example 1.10.1 and the expression (1.6), a digraph with the maximum degree Δ and diameter 2 has order at most $\Delta^2 + \Delta$. We have known that the

Kautz digraph $K(\Delta,2)$ had order $\Delta^2+\Delta$. Therefore, $K(\Delta,2)$ is a maximum $(\Delta,2)$-digraph, which is the unique known maximum $(\Delta,2)$-digraph up to now.

Example 1.10.2 Let $\mathbf{M}$ be the incidence matrix of a digraph G without loops, and let $\mathbf{M}_i$ be the matrix obtained from $\mathbf{M}$ by deleting the ith row. Then the algebraic cofactor of any entry in $\mathbf{M}\mathbf{M}^T$ is equal to the determinant $\det(\mathbf{M}_i\mathbf{M}_i^T)$, where $\mathbf{M}^T$ denotes the transpose of $\mathbf{M}$.

Proof Suppose that the vertex-set of G is $\{x_1, x_2, \cdots, x_v\}$, and let $\mathbf{N} = \mathbf{M}\mathbf{M}^T$. Suppose that the entry in position (i,j) of $\mathbf{N}$ is n_{ij}. Then $\mathbf{N}$ is symmetric and

$$n_{ij} = \begin{cases} d_G(x_i) = d_G^+(x_i) + d_G^-(x_i), & \text{for } j = i; \\ -\mu(x_i, x_j) - \mu(x_j, x_i), & \text{for } j \neq i. \end{cases}$$

Thus the sum of any row and the sum of any column of $\mathbf{N}$ all are 0. It is a routine algebraic exercise to show that the algebraic cofactors of all entries in $\mathbf{M}\mathbf{M}^T$ have the same value (the exercise 1.10.5). Let N_{ij} be the algebraic cofactors of n_{ij} in $\mathbf{M}\mathbf{M}^T$, and, without loss of generality, suppose

$$N_{ij} = N_{11}, \quad \forall\, 1 \leq i,j \leq v.$$

Let α_1 be the first row vector of $\mathbf{M}$. Then

$$\mathbf{N} = \mathbf{M}\mathbf{M}^T = \begin{pmatrix} \alpha_1 \\ \mathbf{M}_1 \end{pmatrix} \cdot \begin{pmatrix} \alpha_1^T & \mathbf{M}_1^T \end{pmatrix} = \begin{pmatrix} \alpha_1\alpha_1^T & \alpha_1\mathbf{M}_1^T \\ \mathbf{M}_1\alpha_1^T & \mathbf{M}_1\mathbf{M}_1^T \end{pmatrix}.$$

Thus, we have

$$N_{ij} = N_{11} = \det(\mathbf{M}_i\mathbf{M}_i^T), \quad \forall\, 1 \leq i,j \leq v$$

as desired. ∎

We conclude this section with some remarks. Let $\mathbf{A}$ be the adjacency matrix of an undirected graph G with the vertex-set $\{x_1, x_2, \cdots, x_v\}$, $\mathbf{M}$ the incidence matrix of any oriented graph D of G, and let $\mathbf{B}$ be the $v \times v$ diagonal matrix with the main diagonal entries $b_{ii} = d_G(x_i)$. It is easily shown that (the exercise 1.10.5) $\mathbf{M}\mathbf{M}^T = \mathbf{B} - \mathbf{A}$, which is called *Laplace matrix* in the literature and textbook on graph theory.

We have known from Example 1.10.2 that all the algebraic cofactors of entries in $\mathbf{M}\mathbf{M}^T$ have the same value, the determinant $\det(\mathbf{M}_i\mathbf{M}_i^T)$. This value is a very important invariant of isomorphic graphs. We will, in Section 2.3, know what this invariant is.

Exercises

1.10.1 Write out the adjacency and incidence matrices of the following graphs

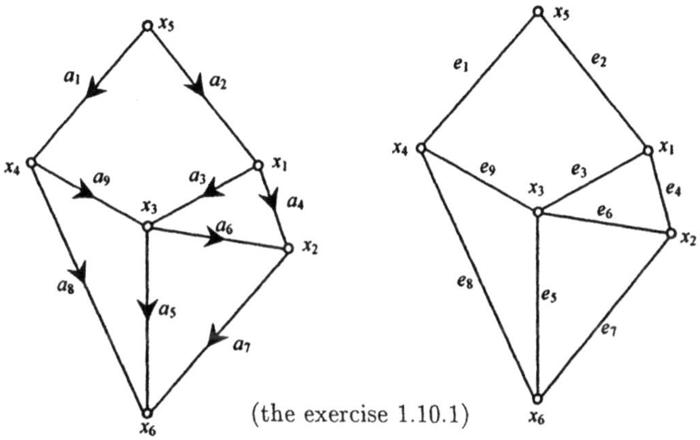

(the exercise 1.10.1)

1.10.2 Let **A** be the adjacency matrix of a graph. Question that

(a) what do the row sum and column sum of **A** represent, respectively?

(b) what does the sum of all elements in **A** represent?

1.10.3 Let **M** be the incidence matrix of a digraph D or an undirected graph G with the vertex-set $\{x_1, x_2, \cdots, x_v\}$. Prove that

(a) the sum of all positive (resp. negative) entries on ith row of $\mathbf{M}(D)$ is $d_D^+(x_i)$ (resp. $d_D^-(x_i)$); the sum of all entries on ith row of $\mathbf{M}(G)$ is $d_G(x_i)$;

(b) the jth column sum of $\mathbf{M}(D)$ is 0, while the jth column sum of $\mathbf{M}(G)$ is 2;

(c) rank $(\mathbf{M}) \leq v - \omega$;

(d) **M** is permutation equivalent to $\begin{pmatrix} \mathbf{M}_{11} & \mathbf{O} \\ \mathbf{O} & \mathbf{M}_{22} \end{pmatrix} \Longleftrightarrow D$ (or G) is disconnected.

1.10.4 Let **A** be the adjacency matrix of a digraph D or an undirected graph G. Prove that

(a) **A** is permutation similar to

(i) $\begin{pmatrix} \mathbf{O} & \mathbf{A}_{12} \\ \mathbf{A}_{21} & \mathbf{O} \end{pmatrix} \Longleftrightarrow D$ (or G) is bipartite,

(ii) $\begin{pmatrix} \mathbf{A}_{11} & \mathbf{O} \\ \mathbf{O} & \mathbf{A}_{22} \end{pmatrix} \Longleftrightarrow D$ (or G) is disconnected,

(iii) $\begin{pmatrix} \mathbf{A}_{11} & \mathbf{A}_{12} \\ \mathbf{O} & \mathbf{A}_{22} \end{pmatrix} \Longleftrightarrow D$ is not strongly connected,

(iv) an upper triangular matrix $\Longleftrightarrow D$ contains no directed cycle of length at least 2;

(b) D is strongly connected $\Longleftrightarrow \mathbf{I} + \mathbf{A} + \mathbf{A}^2 + \cdots + \mathbf{A}^{v-1} > 0$;

(c) if D is a strongly connected tournament with $v \geq 5$, then $\mathbf{A}^{d+3} > 0$, where $d = d(D)$ is diameter of D.

1.10.5 Let $\mathbf{A}$ be the adjacency matrix of an undirected graph G with the vertex-set $\{x_1, x_2, \cdots, x_v\}$, $\mathbf{M}$ the incidence matrix of any oriented graph D of G, and let $\mathbf{B}$ be the $v \times v$ diagonal matrix with main diagonal elements $b_{ii} = d_G(x_i)$.

(a) Prove $\mathbf{M}\mathbf{M}^T = \mathbf{B} - \mathbf{A}$.

(b) Prove that algebraic cofactors of all entries in $\mathbf{M}\mathbf{M}^T$ are identical.

(c) Verify (a) for the graph shown in Exercise 1.10.1 and count the algebraic cofactor of the entry $(1, 1)$ in $\mathbf{M}\mathbf{M}^T$ (the value is 66).

(**d**) Prove that $\mathbf{B} - \mathbf{A}$ is semi-positive definite, and that G is connected if and only if rank $(\mathbf{B} - \mathbf{A}) = v - 1$. (D.Raghavarao, 1977)

1.10.6 Let $\mathbf{A}$ be the adjacency matrix of a graph G (undirected or directed). The eigenvalues of $\mathbf{A}$ is referred to as the *eigenvalues* of G; the characteristic polynomial det $(\lambda \mathbf{I} - \mathbf{A})$ is referred to as the *characteristic polynomial* of G. Suppose that characteristic polynomial of G is

$$P_G(\lambda) = \det(\lambda \mathbf{I} - \mathbf{A}) = \lambda^v + c_1 \lambda^{v-1} + \cdots + c_{v-1}\lambda + c_v.$$

(a) Count the characteristic polynomials of the following two graphs.

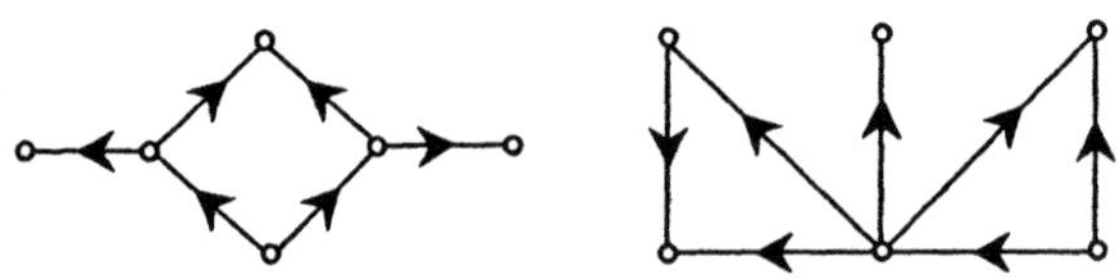

(the exercise 1.10.6)

(b) Prove

$$c_k = \sum_{H \in \mathcal{H}_k} (-1)^{\omega(H)}, \qquad k = 1, 2, \cdots, v,$$

where $\mathcal{H}_k$ is the set of (1-) 2-regular subgraphs of (di)graphs of G.

(M.Milic (1964), H. Sachs (1964), L.Spialter (1964))

(c) Prove that $c_1 = 0$; $-c_2 = \varepsilon$; and $-c_3$ is equal to twice the number of triangles in G.

(d) Prove that if $\lambda_1, \lambda_2, \cdots, \lambda_v$ are all eigenvalues of G, then

(i) $\lambda_1 + \lambda_2 + \cdots + \lambda_v = -c_1$;

(ii) the number of different directed walks of length k in G is $(\lambda_1^k + \lambda_2^k + \cdots + \lambda_v^k)$.

(e) Let λ be the maximum eigenvalue of G. Prove that

(i) $\delta^+ \leq \lambda \leq \Delta^+$, and $\delta^- \leq \lambda \leq \Delta^-$ (or $\delta \leq \lambda \leq \Delta$), and the equalities hold if and only if G is regular;

(ii) if G is strongly connected and regular, then λ has the multiplicity 1.

(f) Prove that a strongly connected digraph of diameter d has at leat $d + 1$ distinct eigenvalues.

1.10.7 (a) Let $\mathbf{A}$ be the adjacency matrix of a digraph G. Prove that there is a polynomial $p(x)$ such that $\mathbf{J} = p(\mathbf{A})$ if and only if G is strongly connected and regular.

(b) Let C_n be a directed cycle of order n, $\mathbf{A}$ be the adjacency matrix of C_n. Find a polynomial $p(x)$ such that $\mathbf{J} = p(\mathbf{A})$.

1.10.8 (a) Let G be an undirected graph of diameter 2. Prove that if $\Delta \neq 2, 3, 7$ or 57, then $v \leq \Delta^2$.

(This result is due to Hoffman and Singleton [97]. In fact, Erdös, Fajtlowicz and Hoffman [55] have shown $v \leq \Delta^2 - 1$.)

(b) Construct two undirected graphs with diameter 2 and the maximum degree $\Delta = 2$ and 3, respectively, such that $v = \Delta^2 + 1$.

(Hoffman and Singleton [97] have constructed such a graph for $\Delta = 7$. However, whether there exists an undirected graph of order $v = \Delta^2 + 1$ and maximum degree $\Delta = 57$ is unknown.)

Applications

1.11 Exponents of Primitive Matrices

In this section, as an application of graph theory to matrix theory, we consider the exponents of primitive matrices. The student who does not major in mathematics may skip it.

A real matrix $\mathbf{A}$ is called to be *nonnegative* (resp. *positive*) if all of its entries are nonnegative (resp. positive), denoted by $\mathbf{A} \geq 0$ (resp. $\mathbf{A} > 0$). A nonnegative square matrix $\mathbf{A}$ is called to be *primitive* if there is a positive integer k such that $\mathbf{A}^k > 0$.

Consider the two nonnegative matrices $\mathbf{A}$ and $\mathbf{B}$, where

$$\mathbf{A} = \begin{pmatrix} 1 & 1 & 0 & 0 \\ 0 & 0 & 1 & 1 \\ 0 & 0 & 0 & 1 \\ 1 & 1 & 0 & 0 \end{pmatrix}, \qquad \mathbf{B} = \begin{pmatrix} 0 & 1 & 0 \\ 0 & 0 & 1 \\ 1 & 0 & 0 \end{pmatrix}.$$

We have

$$\mathbf{A}^2 = \begin{pmatrix} 1 & 1 & 1 & 1 \\ 1 & 1 & 0 & 1 \\ 1 & 1 & 0 & 0 \\ 1 & 1 & 1 & 1 \end{pmatrix}, \qquad \mathbf{A}^3 = \begin{pmatrix} 2 & 2 & 1 & 2 \\ 2 & 2 & 1 & 1 \\ 1 & 1 & 1 & 1 \\ 2 & 2 & 1 & 2 \end{pmatrix}.$$

$$\mathbf{B}^2 = \begin{pmatrix} 0 & 0 & 1 \\ 1 & 0 & 0 \\ 0 & 1 & 0 \end{pmatrix}, \quad \mathbf{B}^3 = \begin{pmatrix} 1 & 0 & 0 \\ 0 & 1 & 0 \\ 0 & 0 & 1 \end{pmatrix}, \quad \mathbf{B}^4 = \begin{pmatrix} 0 & 1 & 0 \\ 0 & 0 & 1 \\ 1 & 0 & 0 \end{pmatrix} = \mathbf{B}.$$

This shows that the matrix $\mathbf{A}$ is primitive, while the matrix $\mathbf{B}$ is not.

Let $\mathbf{A}$ be a primitive square matrix. The smallest positive integer k with $\mathbf{A}^k > 0$ is called the *primitive exponent* of $\mathbf{A}$, denoted by $e(\mathbf{A})$. For example, $e(\mathbf{A}) = 3$, where $\mathbf{A}$ is defined above.

In 1950, Wielandt [177] established a tight upper bounded of $e(\mathbf{A})$, that is, $e(\mathbf{A}) \leq (n-1)^2 + 1$ for any primitive n-matrices $\mathbf{A}$. Surprisingly, this classical result of matrix theory can be referred to a direct consequence of a result of graph theory.

It is clear that the primitivity and the primitive exponent of a nonnegative matrix has no relation to whose concrete value of every positive entry. Thus, it suffices to limit our discussion to $(0,1)$-matrices.

Graph theory is a power tool to study combinatorial properties of $(0,1)$-matrices, see Liu and Lai [118]. In fact, as stated in Section 1.10, the adjacency matrix of every digraph that contains at most one loop at every vertex and no parallel edge is a $(0,1)$-matrix. Conversely, for any n-square $(0,1)$-matrix $\mathbf{A}$, there exists a digraph G of order n such that its adjacency matrix $\mathbf{A}(G) = \mathbf{A}$. Indeed, for an n-square $(0,1)$-matrix $\mathbf{A}= (a_{ij})$, we can construct a digraph without parallel edges $G = (V, E)$ as follows.

$$V(G) = \{x_1, x_2, \cdots, x_n\}, \quad (x_i, x_j) \in E(G) \Longleftrightarrow d_{ij} = 1.$$

Clearly, $\mathbf{A}$ is the adjacency matrix $\mathbf{A}(G)$ of G. So constructing digraph G is called the *associated digraph* with $\mathbf{A}$, denoted by $G(\mathbf{A})$.

The associated digraph $G(\mathbf{A})$ with a $(0,1)$-matrix $\mathbf{A}$ is called to be *primitive* if $\mathbf{A}$ is primitive, In other words, if a digraph G is primitive, then its adjacency matrix $\mathbf{A}$ is primitive. $e(\mathbf{A})$ is called the *primitive exponent* of G, denoted by $e(G)$. Thus, the problem of studying the primitivity of a $(0,1)$-matrix $\mathbf{A}$ can be reduced to one of studying the primitivity of the associated digraph $G(\mathbf{A})$ with $\mathbf{A}$.

For a digraph G, investigating the primitivity of G or determining $e(G)$ is quite difficult in general. However, if G is primitive, we then can obtain a tight upper of $e(G)$.

Before presenting our main results in this section, we recall the Frobenius set and the Frobenius number in number theory, which will be used in the proof of our main theorem.

Let $n_1, n_2, \cdots, n_l$ be positive integers, Z be the set of nonnegative integers. The set

$$F(n_1, n_2, \cdots, n_l) = \{z_1 n_1 + z_2 n_2 + \cdots + z_l n_l : \ z_i \in Z\}$$

is called the *Frobenius set.* It had been proved that if g.c.d.$(n_1, \cdots, n_l) = 1$, then there is a positive n_0 such that $n \in F(n_1, n_2, \cdots, n_l)$ for every integer n $(\geq n_0)$. The smallest number n_0 is called the *Frobenius number*, denoted by $\phi(n_1, n_2, \cdots, n_l)$. It has been determined that

$$\phi(n_1, n_2) = (n_1 - 1)(n_2 - 1).$$

However determining $\phi(n_1, n_2, \cdots, n_l)$ is quite difficult for general $l \geq 3$.

Theorem 1.12 (Rosenblatt [154]) A digraph G is primitive if and only if G is strongly connected and g.c.d.$(l_1, l_2, \cdots, l_c) = 1$, where $\{l_1, l_2, \cdots, l_c\}$ is the set of lengths of all cycles in G.

Proof Without loss of generality, suppose that G is a digraph without parallel edges. Since G is primitive if and only if the adjacency matrix $\mathbf{A}$ of G is primitive, there is a positive integer k such that $\mathbf{A}^k > 0$. By Theorem 1.11, there are an (x_i, x_j)-walk of length k and an (x_j, x_i)-walk of length k for any two vertices x_i and x_j in G. This means that G is strongly connected.

Let x_i and x_j be two vertices in G. Suppose that a is an edge of G with $\psi_G(a) = (x_i, x_l)$ and W_{lj} is an (x_l, x_j)-walk of length k in G. Then $a \oplus W_{lj}$ is an (x_i, x_j)-walk of length $k+1$ in G. In particular, there is an (x_i, x_i)-walk of length k and an (x_i, x_i)-walk of length $k+1$ in G.

Note that any (x_i, x_i)-walk W_{ii} is the union of several edge-disjoint cycles. It follows that the length of W_{ii} must be a multiple of g.c.d.$(l_1, l_2, \cdots, l_c)$. This means that k and $k+1$ both are multiples of g.c.d.$(l_1, l_2, \cdots, l_c)$, and so $1 = (k+1) - k$ is also a multiple of g.c.d.$(l_1, l_2, \cdots, l_c)$, namely, g.c.d.$(l_1, l_2, \cdots, l_c) = 1$.

Conversely, suppose that G is a strongly connected digraph with g.c.d.$(l_1, l_2, \cdots, l_c) = 1$. In order to show that G is primitive, by Theorem 1.11, we need to only prove that there are a positive integer k and an (x, y)-walk of length k for any two vertices x and y in G.

Let x and y be two vertices in G. Since G is strongly connected, by Example 1.5.5, there is an (x, y)-walk W_{xy} containing all vertices in G. Denote the length of W_{xy} by k_{xy}.

Since W_{xy} and any directed cycle have vertices in common, any (x, y)-walk obtained from W_{xy} by adding any directed cycle several times is still an (x, y)-walk. This means that $k_{xy} + r$ is the length of some (x, y)-walk in G for any $r \in F(l_1, l_2, \cdots, l_c)$. Since g.c.d.$(l_1, l_2, \cdots, l_c) = 1$, thus $\phi(l_1, l_2, \cdots, l_c)$ is well-defined. Choose

$$k = \max\{k_{xy} : x, y \in V(G)\} + \phi(l_1, l_2, \cdots, l_c).$$

Then $k \geq k_{xy} + \phi(l_1, l_2, \cdots, l_c)$ for any $x, y \in V(G)$. Let $k = k_{xy} + r$, where $r \geq \phi(l_1, l_2, \cdots, l_c)$. Then $r \in F(l_1, l_2, \cdots, l_c)$ by the definition of $\phi(l_1, l_2, \cdots, l_c)$. Thus, there is an (x, y)-walk of length k in G as desired. ∎

Corollary 1.12.1 Any primitive digraph of order $n \geq 2$ must contain a directed cycle of length less than n.

Proof Let G be a primitive digraph of order n. By Theorem 1.12, G is strongly connected. Thus there are an (x, y)-path P and a (y, x)-path Q, and so $P \cup Q$ contains a directed cycle. Suppose to the contrary that all lengths of directed cycles are n, then g.c.d.$(l_1, l_2, \cdots, l_c) = n \geq 2$. This contradicts Theorem 1.12 and the result follows. ∎

Corollary 1.12.2 Suppose that T_n is a tournament of order $n \geq 3$. If T_n is primitive, then T_n is strongly connected and $n \geq 4$. Conversely, if T_n is strongly connected and $n \geq 4$, then T_n is primitive. Moreover, $e(T_4) = 9$, and $e(T_n) \leq n + 2$ for $n \geq 5$.

Proof Suppose that T_n is a tournament. If T_n is primitive, then by Theorem 1.12, T_n is strongly connected. If $n = 3$, then since T_3 is a directed cycle, and by Corollary 1.12.1, T_3 is not primitive. This means $n \geq 4$.

Conversely, suppose that T_n is strongly connected and $n \geq 4$. For $n = 4$, all tournaments, up to isomorphism, are shown in Figure 1.6, from which we easily see that only one is strongly connected and primitive, with primitive exponent is equal to 9. For $n \geq 5$, by Corollary 1.5 and Theorem 1.11, the adjacency matrix $\mathbf{A}$ of T_n satisfies $\mathbf{A}^{d+3} > 0$, where d is diameter of T_n. Thus T_n is primitive and

$$e(T_n) \leq d + 3 \leq (n - 1) + 3 = n + 2$$

as desired. ∎

We will known from the exercise 1.11.3 that there is a tournament T_n ($n \geq 5$) such that $e(T_n) = n + 2$. In fact, Moon and Pallman [134] showed that for any integer $k \in [3, n + 2]$ there is a tournament T_n such that $e(T_n) = k$ if $n \geq 7$ (see the exercise 1.11.4).

Theorem 1.13 (Dulmage and Mendelsohn [46]) Let G be a primitive digraph of order n. If G contains a directed cycle of length s, then

$$e(G) \leq n + s(n - 2).$$

Proof Suppose that $\mathbf{A}$ is the adjacency matrix of a primitive digraph G. In order to show the theorem, by Theorem 1.11, we need to only prove that there is an (x_i, x_j)-walk of length exactly $n + s(n - 2)$ for any pair (x_i, x_j) of two vertices x_i and x_j in G.

Let C is a directed cycle of length s in G. Since G is primitive if and only if $\mathbf{A}$ is primitive, it follows that $\mathbf{A}^s$ is primitive. By Theorem 1.12, G and $G^s = G(\mathbf{A}^s)$ both are strongly connected. Note that $V(G^s) = V(G)$

and $(x, y) \in E(G^s)$ if and only if G contains an (x, y)-walk of length s by Theorem 1.11. Thus G^s contains a loop at every vertex x_k in C. This shows that G^s contains an (x_k, x_j)-walk of length exactly $(n-1)$, that is, G contains an (x_k, x_j)-walk of length exactly $s(n-1)$, denote such a walk by W_{kj}.

If x_i is in C, then choose a vertex x_k of C that can be reached in $(n-s)$ steps from x_i along the order of C, and denoted such an (x_i, x_k)-walk in C by C_{ik}. Thus $C_{ik} \oplus W_{kj}$ is an (x_i, x_j)-walk of length exactly $(n-s)+s(n-1) = n + s(n-2)$ in G as desired.

If x_i is not in C, then choose a shortest directed path from x_i to C. Let P_{il} be such a path from x_i to some vertex x_l of C and let the length of P_{il} be l. Then $1 \le l \le n-s$. Suppose that x_k is a vertex of C that can be reached in $(n-s-l)$ steps from x_l along the order of C, and denoted such an (x_l, x_k)-walk in C by C_{lk}. Thus $P_{il} \oplus C_{lk} \oplus W_{kj}$ is an (x_i, x_j)-walk of length exactly $l + (n-s-l) + s(n-1) = n + s(n-2)$ in G as desired. ∎

Corollary 1.13 (Wielandt's theorem) If $\mathbf{A}$ is a primitive n-square matrix, then

$$e(\mathbf{A}) \le (n-1)^2 + 1.$$

Proof It is known that $\mathbf{A}$ is primitive if and only if the associated digraph $G(\mathbf{A})$ with $\mathbf{A}$ is primitive. By Corollary 1.12.1, $G(\mathbf{A})$ contains a directed cycle of length less than n. Let C be a directed cycle of length s $(\le n-1)$ in G. It follows from Theorem 1.13 that

$$e(\mathbf{A}) = e(G) \le n + (n-1)(n-2) = (n-1)^2 + 1$$

as desired. ∎

Dulmage and Mendelsohn gave an example to show that the upper bound $(n-1)^2+1$ can be reached.

Example 1.11.1 Let $n \ge 2$ and matrix

$$\mathbf{A}_2 = \begin{pmatrix} 0 & 1 \\ 1 & 1 \end{pmatrix}, \quad \mathbf{A}_n = \begin{pmatrix} 0 & 1 & 0 & \cdots & 0 & 0 \\ 0 & 0 & 1 & \cdots & 0 & 0 \\ \cdot & \cdot & \cdot & \cdots & \cdot & \cdot \\ 0 & 0 & 0 & \cdots & 0 & 1 \\ 0 & 1 & 1 & \cdots & 0 & 0 \end{pmatrix} \quad \text{for } n \ge 3.$$

Then $e(\mathbf{A}_n) = (n-1)^2 + 1$.

Proof Let G be the associated digraph $G(\mathbf{A}_n)$ with $\mathbf{A}_n$, as shown in Figure 1.27. Clearly, G is strongly connected. Since G contains only two different directed cycles whose lengths are n and $(n-1)$, respectively, and g.c.d.$(n, n-1) = 1$, thus, by Theorem 1.12, G is primitive. It follows from Corollary 1.13 that $e(G) \leq (n-1)^2 + 1$.

We now prove $e(G) \geq (n-1)^2 + 1$. Consider the vertex x_1 in G. Let W be a closed (x_1, x_1)-walk of nontrivial length, which consists of several directed cycles. Note that G contains only one directed cycle of length n and only one directed cycle of length $(n-1)$ that does not pass through x_1. Therefore, W contains at least one directed cycle of length n. Thus length of W is equal to $n + z_1 n + z_2(n-1)$, where z_1 and z_2 are nonnegative integers. Since $z_1 n + z_2(n-1) \in F(n, n-1)$ and $\phi(n, n-1) - 1 \notin F(n, n-1)$, thus, G contains no closed (x_1, x_1)-walk of length $(n + \phi(n, n-1) - 1)$. Therefore, we have

$$e(G) \geq n + \phi(n, n-1) = n + (n-1)(n-2) = (n-1)^2 + 1$$

as desired. ∎

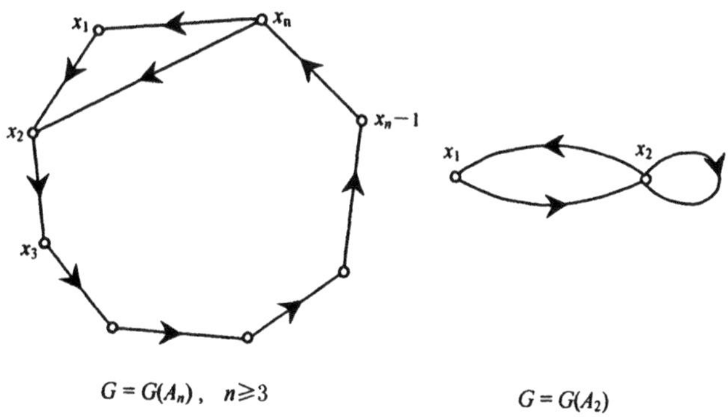

Figure 1.27: The associated digraph G with the matrix $\mathbf{A}_n$

The following example shows that there is a primitive n-square matrix $\mathbf{B}_n$ such that $e(\mathbf{B}_n) = (n-1)^2$ for any $n \geq 2$.

Example 1.11.2 Let $n \geq 2$ and matrix

$$\mathbf{B}_2 = \begin{pmatrix} 1 & 1 \\ 1 & 1 \end{pmatrix}, \qquad \mathbf{B}_3 = \begin{pmatrix} 0 & 1 & 0 \\ 1 & 0 & 1 \\ 1 & 1 & 0 \end{pmatrix},$$

$$\mathbf{B}_n = \begin{pmatrix} 0 & 1 & 0 & 0 & \cdots & 0 & 0 & 0 \\ 0 & 0 & 1 & 0 & \cdots & 0 & 0 & 0 \\ 0 & 0 & 0 & 1 & \cdots & 0 & 0 & 0 \\ \cdot & \cdot & \cdot & \cdot & \cdots & \cdot & \cdot & \cdot \\ 0 & 0 & 0 & 0 & \cdots & 0 & 1 & 0 \\ 1 & 0 & 0 & 0 & \cdots & 0 & 0 & 1 \\ 1 & 1 & 0 & 0 & \cdots & 0 & 0 & 0 \end{pmatrix} \quad \text{for } n \geq 4.$$

Then $e(\mathbf{B}_n) = (n-1)^2$.

Proof Let H be the associated digraph $G(\mathbf{B_n})$ with $\mathbf{B}_n$, as shown in Figure 1.28. Clearly, H is strongly connected. Since H contains only three different directed cycles, one of their lengths is n, two are $(n-1)$, and g.c.d.$(n, n-1) = 1$, thus, by Theorem 1.12, H is primitive.

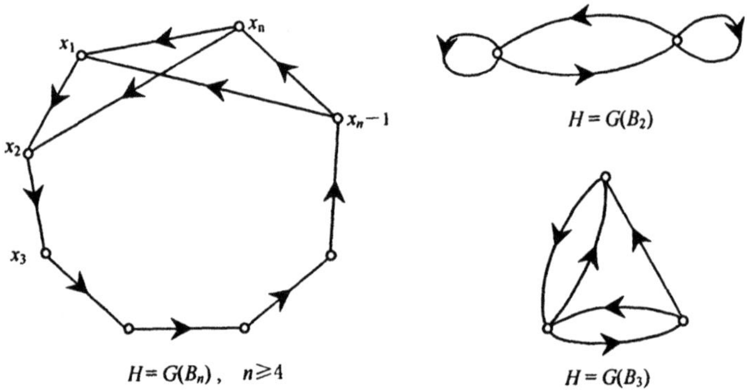

Figure 1.28: The associated digraph H with the matrix $\mathbf{B}_n$

Arbitrarily choose two vertices x and y in H. Let P_{xy} be a shortest (x, y)-path in H, and denote its length by p_{xy}. Then $p_{xy} \leq n-1$. Since every vertex in H is included in two directed cycles whose lengths are n and $(n-1)$, respectively, a directed walk obtained from P_{xy} by adding several directed cycles of length n and/or $(n-1)$ is still an (x, y)-walk. This shows that there is an (x, y)-walk of length $p_{xy} + r$ in H for any $r \in F(n, n-1)$. Let

$$q = \phi(n, n-1) + n - 1 - p_{xy}.$$

Then $q \in F(n, n-1)$ since $q \geq \phi(n, n-1)$. Thus, there exists an (x, y)-walk of length $q + p_{xy}$ in H. Because of arbitrariness of x, y and $\phi(n, n-1) =$

$(n-1)(n-2)$, we have

$$e(H) \leq q + p_{xy} = (n-1)(n-2) + (n-1) = (n-1)^2.$$

On the other hand, H contains the unique (x_1, x_n)-path of length $(n-1)$. Therefore, length of every (x_1, x_n)-walk in H can be expressed as $n - 1 + r$, where $r \in F(n, n-1)$. Because of $\phi(n, n-1) - 1 \notin F(n, n-1)$, H contains no directed walk of length $(n - 1 + \phi(n, n-1) - 1)$. Thus, we have

$$e(H) \geq \phi(n, n-1) + n - 1 = (n-1)^2$$

as desired. ∎

For $n = 4$, for example, H is a strongly connected tournament T_4. Thus, $e(T_4) = 9$ by Example 1.11.2, which is identical with the result in Corollary 1.12.2.

Graph theory provides an important and powerful tool for studying nonnegative matricies by a combinatorial method. The reader who is interested in combinatorial matrix theory can be referred to Liu and Lai's book Matrices in Combinatorics and Graph Theory [118].

Exercises

1.11.1 Let matrices

$$\mathbf{A} = \begin{pmatrix} 0 & 1 & 0 & 0 \\ 1 & 0 & 1 & 0 \\ 0 & 0 & 0 & 1 \\ 1 & 0 & 0 & 0 \end{pmatrix}, \qquad \mathbf{B} = \begin{pmatrix} 0 & 1 & 0 & 0 \\ 0 & 0 & 1 & 0 \\ 1 & 0 & 0 & 1 \\ 1 & 1 & 0 & 0 \end{pmatrix}.$$

(a) Drawing associated digraphs with $\mathbf{A}$ and $\mathbf{B}$, respectively.

(b) Prove that $\mathbf{A}$ is not primitive, but $\mathbf{B}$ is primitive and $e(\mathbf{B}) = 9$.

1.11.2 Prove that

(a) every strongly connected digraph with a loop must be primitive;

(b) $e(G) \leq d(G)$, and hence $e(G) \leq n - 1$, for any strongly connected digraph G with a loop at every vertex, where $d(G)$ is diameter of G of order n.

1.11.3 Prove $e(\mathbf{A}) = n+2$, where $\mathbf{A}$ is an $n(\geq 5)$-square matrix

$$\mathbf{A} = \begin{pmatrix} 0 & 0 & 1 & 1 & 1 & 1 & \cdots & 1 & 1 & 1 \\ 1 & 0 & 0 & 1 & 1 & 1 & \cdots & 1 & 1 & 1 \\ 0 & 1 & 0 & 0 & 1 & 1 & \cdots & 1 & 1 & 1 \\ 0 & 0 & 1 & 0 & 0 & 1 & \cdots & 1 & 1 & 1 \\ \cdot & \cdot & \cdot & \cdot & \cdot & \cdot & \cdots & \cdot & \cdot & \cdot \\ 0 & 0 & 0 & 0 & 0 & 0 & \cdots & 1 & 0 & 0 \\ 0 & 0 & 0 & 0 & 0 & 0 & \cdots & 0 & 1 & 0 \end{pmatrix}.$$

1.11.4 Let T_n be a strongly connected tournament of order n (≥ 4). Prove that

(a) $e(T_n) \geq 3$;

(b) $e(T_n) \neq 3$ for $n \leq 6$;

(c) there is T_5 such that $e(T_5) = k$ for any k with $4 \leq k \leq 7$;

(d) there is T_6 such that $e(T_6) = k$ for any k with $4 \leq k \leq 8$;

(e) if there is T_n such that $e(T_n) = k$, then there are T_{n+1} and T'_{n+1} such that $e(T_{n+1}) = k$ and $e(T'_{n+1}) = k+1$ for any n and k with $n \geq 5$ and $3 \leq k \leq n+2$;

(f) there is T_n such that $e(T_n) = k$ for any n and k with $n \geq 7$ and $3 \leq k \leq n+2$.

1.11.5 An n-square matrix $\mathbf{A}$ is called to be *reducible* if there is a permutation matrix $\mathbf{P}$ such that

$$\mathbf{PAP}^T = \begin{pmatrix} \mathbf{A}_{11} & \mathbf{O} \\ \mathbf{A}_{21} & \mathbf{A}_{22} \end{pmatrix},$$

where $\mathbf{A}_{11}$ is an l-square matrix, $1 \leq l \leq n-1$; and to be *irreducible* otherwise. Prove that

(a) a nonnegative $n(>1)$-square matrix $\mathbf{A}$ is irreducible if and only if $G(\mathbf{A})$ is strongly connected;

(b) if $\mathbf{A}$ is irreducible nonnegative n-square matrix with at least k (≥ 1) non-zero diagonal elements, then $\mathbf{A}$ is primitive and $e(\mathbf{A}) \leq 2n-k-1$.

Chapter 2

Trees and Graphic Spaces

In this chapter, we will discuss a class of the simplest connected graphs without cycles, that is, trees, called by almost all authors. Trees will play an important role in studying structural properties and applications of graphs.

We will first present some basic properties of trees and spanning trees, then discuss relationship between spanning trees and cycles as well as between cotrees and bonds of a graph. The relations between the set of cycles and the set of bonds of a graph will be further explored with the aid of linear algebraic theory by establishing the graphic spaces. We are particularly interested in the edge-space of a digraph G as well as its two subspaces, the cycle-space and the bond-space, consisting of all cycle-vectors and bond-vectors in the edge-space, respectively. With aid of the incidence matrix of G we will show that these two subspaces are orthogonally complementary in the edge-space of G. By a spanning tree in G, we can easily find the basic matrices of the cycle-space and the bond-space. In view of importance of spanning trees and as an application of the edge-space theory, we will deduce several formulae for counting the number of spanning in a connected graph without loops.

As applications of spanning trees, we will present two efficient algorithms to solve the minimum connector problem and the shortest path problem, that is, to find a minimum spanning tree and an optimal spanning tree rooted at a given vertex of a weighted graph, respectively. At the end of this chapter, we will present an application of the edge-space theory in the electrical network theory.

2.1 Trees and Spanning Trees

A graph is called a *forest* if it contains no cycle. A connected forest is called a *tree*. A forest is also called an *acyclic graph* in the literature and textbooks on graph theory.

Clearly, forests and trees both are bipartite simple graphs. The concepts of forests and trees are not related to orientations of edges. Therefore, in discussing structural properties of forests or trees, we will restrict ourselves to undirected graphs. The graph shown in Figure 2.1 (a) is a forest consisting of two trees, and one in (b) is a tree.

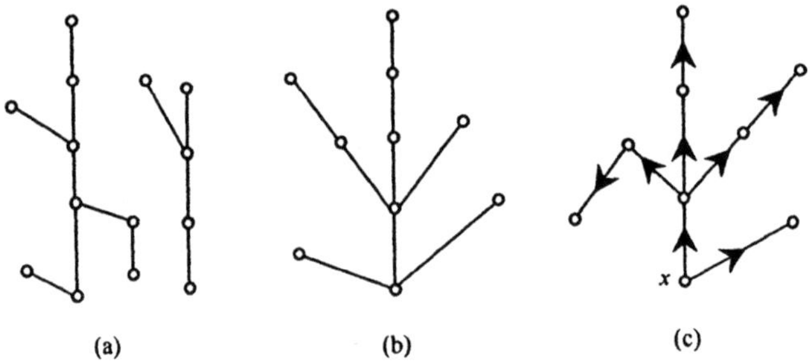

Figure 2.1: (a) A forest; (b) a tree; (c) an out-tree rooted at x

The following theorem describes some fundamental characterizations of a tree.

Theorem 2.1 A graph G is a tree if and only if G satisfies one of the following conditions:

(a) G has no loop and there is a unique path between any two vertices;
(b) G is connected and $\omega(G-e)=2$ for any edge e of G;
(c) G is connected and $\varepsilon=v-1$.

Proof Let G be a tree. Then G is a connected simple graph by definition of a tree.

Suppose that there are two distinct paths P_1 and P_2 joining some two vertices in G. Since $P_1 \neq P_2$, there is an edge $e=xy$ in G such that it is in P_1 but not in P_2. Clearly, the subgraph $(P_1 \cup P_2)-e$ is connected, and contains a xy-path P. But then $P+e$ is a cycle of G, which contradicts the hypothesis that G is a tree, and so the condition (a) is necessary.

Let $e=xy$ be any edge of G. Clearly $\omega(G-e) \leq 2$ since G is connected. On the other hand, xey is a unique xy-path in G by (a). This shows that x and

y are in different connected components of $G - e$, which means $\omega(G - e) \geq 2$, and the condition (b) is necessary.

We prove $\varepsilon = v - 1$ by induction on the number edges $\varepsilon \geq 0$. There is noting to do if $\varepsilon = 0$. Suppose that the condition (c) is necessary for any tree with ε fewer than m, and let G be a tree with $\varepsilon = m \geq 1$. Arbitrarily choose an edge e of G. Then $\omega(G - e) = 2$ by (b). Let G_1 and G_2 be two components of $G - e$. Then G_1 and G_2 both are trees, and so $\varepsilon(G_i) < m$ for each $i = 1, 2$. By the induction hypothesis, we have $\varepsilon(G_i) = v(G_i) - 1$ for each $i = 1, 2$. It follows that

$$\varepsilon(G) = \varepsilon(G_1) + \varepsilon(G_2) + 1 = v(G_1) + v(G_2) - 1 = v(G) - 1.$$

By the principle of induction the condition (c) is necessary.

We now show that each of the three conditions is sufficient for a graph G to be a tree.

Let G be a loopless graph and there is a unique path between any two vertices of G. Suppose to the contrary that G contains a cycle C. Then $v(C) \geq 3$. Arbitrarily choose two distinct vertices x and y of C. Then x and y partition C into two xy-paths of G, a contradiction. G therefore is a tree.

Suppose that G satisfies the condition (b) and contains a cycle C. Let e be an edge of C with two end-vertices x and y. Since G is connected and $\omega(G - e) = 2$, the edge e is not a loop. This shows that x and y are in different connected components of $G - e$. But, $C - e$ is an xy-path in $G - e$, a contradiction. Therefore, G is a tree.

Suppose that G satisfies the condition (c). We prove that G contains no cycle by induction on $v \geq 1$. If $v = 1$, then $\varepsilon = 0$. Thus, G is trivial and hence contains no cycle. Suppose that any connected graph of order n (≥ 1) and size $\varepsilon = n - 1$ contains no cycle. Let G be a connected graph of order $n + 1$ and size $\varepsilon = n$. Then $\delta(G) \geq 1$. If $\delta(G) \geq 2$, then, by Corollary 1.1, we can easily deduce a contradiction as follows

$$2n = 2\varepsilon = \sum_{x \in V} d_G(x) \geq 2(n + 1).$$

Therefore, there is a vertex x of G with $d_G(x) = 1$. Thus, $G - x$ is a connected graph of order n and size $n - 1$. By the induction hypothesis, $G - x$ contains no cycle, and so does G. ∎

Corollary 2.1 A graph G is a forest if and only if $\varepsilon = v - \omega$. ∎

The result stated in Corollary 2.1 is simple, but very useful. We now give two examples as its applications.

Example 2.1.1 A forest without isolated vertices contains at least 2ω vertices of degree one.

Proof Let G be a forest without isolated vertices. Then $\delta(G) \geq 1$. Let R be the set of vertices of degree one and let $r = |R|$. By Corollary 2.1 and Corollary 1.1.1, we have

$$\begin{aligned} 2(v-\omega) &= 2\varepsilon = \sum_{x \in V} d_G(x) = \sum_{x \in V \setminus R} d_G(x) + r \\ &\geq 2(v-r) + r = 2v - r, \end{aligned}$$

which means that $r \geq 2\omega$, as desired. ∎

Example 2.1.2 (Bondy [15]) Let $\mathcal{A} = \{A_1, A_2, \cdots, A_n\}$ be a family of n distinct subsets of $X = \{1, 2, \cdots, n\}$. Then there is an element $x \in X$ such that the sets $A_1 \setminus \{x\}, A_2 \setminus \{x\}, \cdots, A_n \setminus \{x\}$ are all distinct.

Proof We first note that if $A, B \in X, A \neq B$ and $A \setminus \{i\} = B \setminus \{i\}$ then either $A = B \cup \{i\}$ or $B = A \cup \{i\}$. Thus, in either case, $A \Delta B = \{i\}$, where $A \Delta B = (A \setminus B) \cup (B \setminus A)$ is the symmetric difference of the sets A and B.

Suppose to the contrary that the assertion is false. Then $n \geq 2$ and our aim is to arrive at a contradiction. Thus, for every $i \in X$ there are $k = k(i)$ and $l = l(i)$, $1 \leq k < l \leq n$, such that $A_k \setminus \{i\} = A_l \setminus \{i\}$, i.e., $A_k \Delta A_l = \{i\}$. Construct a simple undirected graph G as follows. $V(G) = X$, and two vertices $k(i)$ and $l(i)$ linked by an undirected edge if and only if $A_k \Delta A_l = \{i\}$ for every $i = 1, 2, \cdots, n$. By our hypothesis, we have $\varepsilon(G) \geq n$. Thus, G contains a cycle by Corollary 2.1. Suppose that $(i_1, i_2, \cdots, i_s, i_1)$ is a cycle in G. We may, without loss of generality, suppose that $i_j = j$ and $k(j) = i_j$. Then $i_{j+1} = l(j)$ for $j = 1, 2, \cdots, s$. But then

$$\begin{aligned} \{s\} &= A_1 \Delta A_s = (A_1 \Delta A_2) \Delta (A_2 \Delta A_3) \Delta \cdots \Delta (A_{s-1} \Delta A_s) \\ &= \bigcup_{j=1}^{s-1} (A_j \Delta A_{j+1}) = \{1, 2, \cdots, s-1\}, \end{aligned}$$

a blatant contradiction. This completes the proof. ∎

Let G be a digraph that is a tree. If there is a vertex x of G such that the unique xy-path is an (x, y)-path for any y other than x in G, then G is called an *out-tree* rooted at x, where x is called a *root* of G. An *in-tree* can be defined similarly. Out-trees and in-trees both are called *rooted trees*. The digraph shown in Figure 2.1 (c) is an out-tree rooted at x.

Let F be a spanning subgraph of a graph G with $\omega(F) = \omega(G)$. F is called a *spanning forest* of G if F is a forest; F is called a *spanning tree* of G if F is a tree.

The concepts of spanning forests and spanning trees are not related to orientations of edges. Therefore, in discussing its properties , we will restrict ourselves to undirected graphs. Figure 2.2 indicates a spanning forest and a spanning tree (induced by the heavy edges).

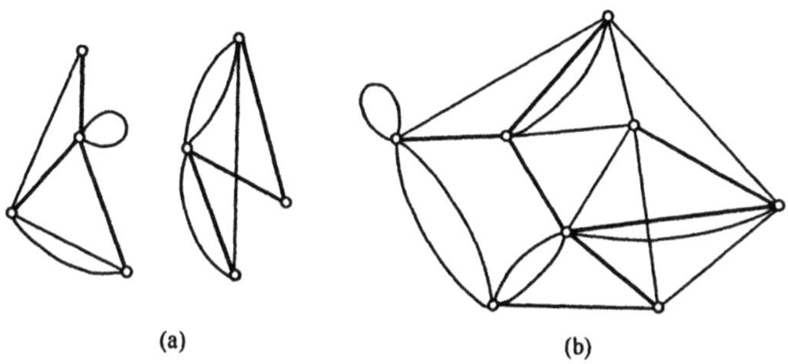

Figure 2.2: (a) A spanning forest; (b) a spanning tree

Theorem 2.2 A graph contains a spanning tree if and only if it is connected.

Proof It is clear that if a graph contains a spanning tree then it is connected. Conversely, let G be a connected graph and T be a connected spanning subgraph with edges as few as possible. Then it is clear from the choice of T that $\omega(T) = 1$ and $\omega(T - e) = 2$ for any edge e of T. Therefore, T is a tree by Theorem 2.1. ∎

Corollary 2.2.1 Every graph contains a spanning forest and every connected graph contains a spanning tree. ∎

Corollary 2.2.2 $\varepsilon \geq v - \omega$ for every graph.

Proof Let G be a graph. Then G contains a spanning forest F, that is, every connected component of G contains a spanning tree, by Corollary 2.2.1. It follows from Corollary 2.1 that

$$\varepsilon(F) = v(F) - \omega(F) = v(G) - \omega(G) = v - \omega$$

Thus $\varepsilon = \varepsilon(G) \geq \varepsilon(F) = v - \omega$. ∎

Theorem 2.3 If F is a spanning forest of a graph G and $E(G) \setminus E(F)$ nonempty, then $F + e$ contains a unique cycle for every edge e of G not in F.

Proof By definition of a spanning forest, F contains no cycle and $F + e$ contains a cycle. Let x and y be two end-vertices of e and P the unique xy-path in F by Theorem 2.1. Then $P + e$ is a unique cycle in F. ∎

Corollary 2.3 Any loopless graph contains at least $\varepsilon - v + \omega$ distinct cycles. ∎

Let G be a graph. A nonempty subset B of $E(G)$ is called a *cut* of G if there is a nonempty proper subset S of $V(G)$ such that $B = [S, \overline{S}]$, which denotes the set of edges of G with one end-vertex in S and the other in $\overline{S}$, where $\overline{S} = V(G) \setminus S$. A cut of G is called to be a*minimal* if none of its nonempty proper subsets is a cut of G. A minimal cut is also called a *bond*. Figure 2.3 indicates a cut and a bond in an undirected graph.

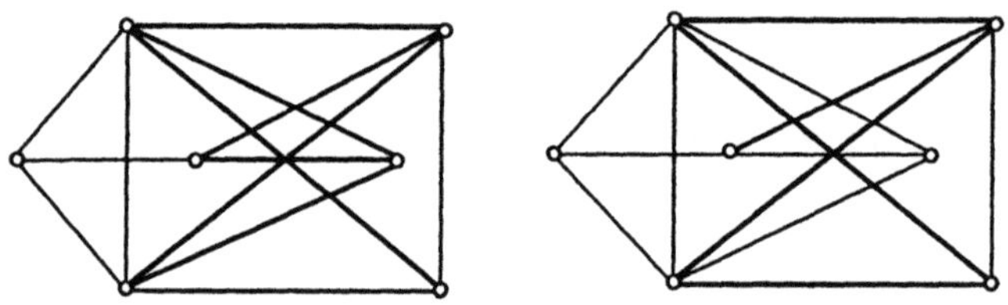

(a) (b)

Figure 2.3: (a) A cut; (b) a bond

It is clear that $\omega(G - B) > \omega(G)$ if $B \subseteq E(G)$ is a cut of G. The converse, however, is not always true. It is also not difficult to see that $\omega(G - B) = \omega(G) + 1$ if B is a bond of G, and a cut B is a bond if and only if $\omega(G - B) = \omega(G) + 1$.

If F is a spanning subgraph of G, then the subgraph $G - E(F)$ is called the *cograph* of F in G, denoted by $\overline{F}(G)$. It is clear that the cograph $\overline{F}(K_v)$ of F in K_v is the complement F^c of F (the complement of a graph is defined in the exercise 1.2.5). If F is a spanning forest (resp. tree), then $\overline{F}(G)$ is called the *coforest* (resp. *cotree*) of G. For the convenience' sake, we sometimes write $\overline{F}$ for $\overline{F}(G)$ when G is a unique graph under discussion. Moreover we usually also write $\overline{F}$ instead of $E(\overline{F})$.

Because these concepts have no relation to orientations of edges, in discussing their properties, we will restrict ourselves to undirected graphs.

Theorem 2.4 Let F be a spanning forest of a nonempty graph G, and let e be any edge of F. Then $\overline{F}$ contains no bond of G, while $\overline{F}+e$ contains a unique bond of G.

Proof Let B be any bond of G. Then $\omega(G-B)-1=\omega(G)=\omega(F)$. This implies that $E(F)\cap B\neq\emptyset$, and so $B\not\subseteq\overline{F}$.

In order to prove the second assertion, we denote by S the vertex-set of some component of $F-e$. Then $B=[S,\overline{S}]$ is a cut of G, thus, $\overline{F}+e$ contains a cut of G.

Suppose that $\overline{F}+e$ contains two distinct bonds B and B' of G. Then $e\in B\cap B'$ since $\overline{F}$ contains no bond by the first assertion. Thus, $(B\cup B')-e$ contains a bond, and so $\overline{F}$ contains a bond of G, contrary to the first assertion. It follows that $\overline{F}+e$ contains a unique bond of G. ∎

Corollary 2.4 Any loopless graph contains at least $\upsilon-\omega$ distinct bonds. ∎

Theorem 2.3 explores a relationship between cycles and spanning forests or trees, and, Theorem 2.4 explores a relationship between bonds and coforests or cotrees. The relationship between spanning forests (resp. spanning trees) and coforests (resp. cotrees) is complementary in a graph. The relationship between cycles and bonds is analogous to that between forests and coforests, and will be further explored in the next section, in which we will see that Theorem 2.3 and Theorem 2.4 play an important role.

In addition, we will, in the following sections of this book, see that spanning trees are widely used in various algorithms for solving some applications by methods of graph theory.

Exercises

2.1.1 Let G be a digraph and x be a vertex of G. Prove that

(a) G is an out-tree rooted at x if and only if G contains no directed cycle, $d^-_G(x)=0$ and $d^-_G(y)=1$ for any y other than x in G;

(b) G is an in-tree rooted at x if and only if G contains no directed cycle, $d^+_G(x)=0$ and $d^+_G(y)=1$ for any y other than x in G.

2.1.2 Let G be a nontrivial tree. Prove that

(a) two end-vertices of any longest path in G are of degree one;

(b) all longest paths have at least one vertex in common;

(c) G contains at least $(v-k)$ paths of length at least k if $d(G) \geq 2k-3$ and $k \geq 2$.

2.1.3 Let G be a nontrivial tree, v_i denote the number of i-degree vertices in G. Prove that

(a) $v_1 \geq \Delta(G)$ and G is a path if $v_1 = 2$;

(b) $v_1 \geq v_i + 2$ for each $i = 3, 4, \cdots, \Delta$ if $\Delta \geq 3$;

(c) $v_1 = 2 + \sum\limits_{x \in U} (d_G(x) - 2)$, where $U = \{x \in V(G) : \ d_G(x) \geq 3\}$.

2.1.4 Let T be a tree. Prove that if $\{X, Y\}$ is a bipartition of G with $|X| \geq |Y| = k+1$, then there are at least $(k+1)$ vertices of degree one in X.

2.1.5 Let G_1 and G_2 be two distinct trees with the same vertex-set V. Prove that $d(G_1) = d(G_2)$ if $G_1 - x \cong G_2 - x$ for any $x \in V$. (P.J.Kelly, 1957)

2.1.6 Let G be a forest with exactly $2k\,(k \geq 1)$ vertices of odd degree. Prove that there are k edge-disjoint paths $P_1, P_2, \cdots, P_k$ in G such that $E(G) = E(P_1) \cup E(P_2) \cup \cdots \cup E(P_k)$.

2.1.7 Let G be a nontrivial tree of order k. Prove that any simple undirected graph H with $\delta(H) \geq k-1$ contains a subgraph isomorphic to G.

2.1.8 Let G be a tree and $G_i = (V_i, E_i)\,(i = 1, 2, \cdots, k)$ be sub-trees of G, $B = V_1 \cap V_2 \cap \cdots \cap V_k$. Prove that

(a) $B \neq \emptyset$ if $V_i \cap V_j \neq \emptyset$ for $1 \leq i \neq j \leq k$;

(b) $G[B]$ is a tree if $B \neq \emptyset$.

2.1.9 Prove that any graph contains at least $\varepsilon - v + \omega$ distinct cycles and $v - \omega$ distinct bonds.

2.1.10 Prove that if G is a strongly connected digraph and x is any vertex of G then there are a spanning out-tree and a spanning in-tree rooted at x in G.

2.1.11 Prove that if G is a connected graph and S is a nonempty proper subset of $V(G)$ then the cut $[S, \overline{S}]$ is minimal if and only if both $G[S]$ and $G[\overline{S}]$ are connected.

2.1.12 Prove that every cut can be expressed as a union of several edge-disjoint bonds.

2.1.13 Prove that the symmetric difference $B\Delta B'$ of two distinct cuts B and B' is still a cut, and so contains a bond.

2.1.14 Prove that any complete graph of order at least four contains at least two edge-disjoint spanning trees.

2.1.15 The *tree graph* of a connected graph G, denoted by $T(G)$, is a simple undirected graph whose vertex-set consists of all spanning trees of G, and two spanning trees T_i and T_j are linked by an undirected edge if and only if T_i and T_j have exactly $v - 2$ edges in common. Prove that $T(G)$ is connected for any graph G.

2.1.16 Prove that if a graph G contains k edge-disjoint spanning trees then, for any partition $\{V_1, V_2, \cdots, V_n\}$ of $V(G)$, the number of edges whose end-vertices are in different parts of the partition is at least $k(n-1)$.

(W.T.Tutte (1961) and C.St.J.A.Nash-Williams (1961) proved that this necessary condition for G to contain k edge-disjoint spanning trees is also sufficient.)

2.1.17 A spanning subgraph T of a connected graph G is called to be *distance-preserving* from a vertex x in G if $d_T(x, y) = d_G(x, y)$ for every vertex y. Prove that for every vertex x of a connected graph G, there exists a spanning tree T that is distance-preserving from x. (O.Ore, 1962)

2.1.18 Let G be a connected undirected simple graph and x be a vertex of G. Prove that there is an oriented graph D that contains a spanning out-tree T rooted at x satisfying the following conditions:

(a) $T + a$ contains a directed cycle for any directed edge a of the cotree $\overline{T}$;

(b) there is a directed edge a of $\overline{T}$ such that $C \subseteq T + a$ for any directed cycle C of D.

2.2 Vector Spaces of Graphs

Let G be a graph with the vertex-set $V = \{u_1, u_2, \cdots, u_v\}$ and the edge-set $E = \{a_1, a_2, \cdots, a_\varepsilon\}$, and $\mathcal{R}$ be the set of all real numbers. The *vertex-space*, $\mathcal{V}(G)$, of G is the vector space of all functions from $V(G)$ into $\mathcal{R}$. Similarly, the *edge-space*, $\mathcal{E}(G)$, of G is the vector space of all functions from $E(G)$ into $\mathcal{R}$. It is clear that

$$\dim \mathcal{V}(G) = v, \quad \dim \mathcal{E}(G) = \varepsilon.$$

Let $\mathbf{f}$ be an element in $\mathcal{V}(G)$, and let $\mathbf{f}(u_i) = x_i$. Then $\mathbf{f}$ is usually written a formal sum of the vertices

$$\mathbf{f} = \sum_{i=1}^{v} x_i \mathbf{u}_i.$$

If we think of $\mathbf{u}_i$ as the vector in $\mathcal{V}(G)$ which satisfies

$$\mathbf{u}_i(u_j) = \begin{cases} 1, & \text{for } j = i; \\ 0, & \text{for } j \neq i, \end{cases}$$

then $\{\mathbf{u}_1, \mathbf{u}_2, \cdots, \mathbf{u}_v\}$ is a basis of $\mathcal{V}(G)$ and the formal sum above can be expressed in terms of the basis. We call $\{\mathbf{u}_1, \mathbf{u}_2, \cdots, \mathbf{u}_v\}$ the standard basis of the vertex-space $\mathcal{V}(G)$. $x_1, x_2, \cdots, x_v$ is the coordinates of $\mathbf{f}$ under the standard basis. We write $\mathbf{f} = (x_1, x_2, \cdots, x_v)$.

Similarly, let $\mathbf{g}$ be an element in $\mathcal{E}(G)$, and let $\mathbf{g}(a_i) = y_i$. Then $\mathbf{g}$ is written a formal sum of the edges

$$\mathbf{g} = \sum_{i=1}^{\varepsilon} y_i \mathbf{a}_i.$$

If we think of $\mathbf{a}_i$ as the vector in $\mathcal{E}(G)$ which satisfies

$$\mathbf{a}_i(a_j) = \begin{cases} 1, & \text{for } j = i; \\ 0, & \text{for } j \neq i, \end{cases}$$

then $\{\mathbf{a}_1, \mathbf{a}_2, \cdots, \mathbf{a}_\varepsilon\}$ is a basis of $\mathcal{E}(G)$ and called the standard basis of the edge-space $\mathcal{E}(G)$. $y_1, y_2, \cdots, y_\varepsilon$ is the coordinates of $\mathbf{g}$ under the standard basis. We write $\mathbf{g} = (y_1, y_2, \cdots, y_\varepsilon)$.

We can endow these spaces with the inner product (i.e., scalar multiplication) in which the standard bases are orthogonal.

In this section we are only concerned with the edge-space $\mathcal{E}(G)$ of a loopless digraph G. We will define two subspaces which turn out to be orthogonal complements of each other.

To start with some symbols. Let G be a loopless graph and let $\mathbf{w}$ be a vector in $\mathcal{E}(G)$. The ordered binary $(G,\mathbf{w})$ is called a *weighted graph*, where $\mathbf{w}$ is called a *weighted function* and $\mathbf{w}\ (a)$ is called a *weight* of the edge a of G. It is often convenient to write $\mathbf{w}(x,y)$ for $\mathbf{w}((x,y))$ if $(x,y) \in E(G)$.

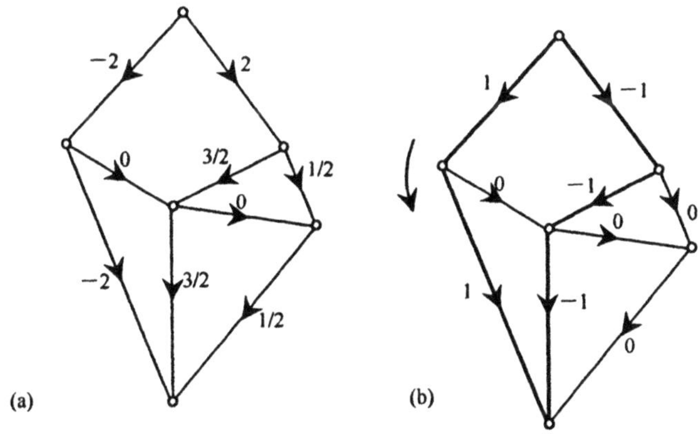

Figure 2.4: (a) A cycle-vector; (b) a cycle-vector associated with a cycle

Two weighted digraphs are shown in Figure 2.4. Weighted graphs frequently occur in various applications of graph theory, in which weights may be distance, costs and so on.

Let B be a subset of edges of G in $(G,\mathbf{w})$. We write

$$\mathbf{w}\ (B) = \sum_{a \in B} \mathbf{w}\ (a).$$

If S is a nonempty proper subset of $V(G)$, then we write

$$\mathbf{w}^+(S) = \mathbf{w}(E_G^+(S)), \quad \mathbf{w}^-(S) = \mathbf{w}(E_G^-(S)).$$

A vector $\mathbf{f}$ in $\mathcal{E}(G)$ is called a *cycle-vector* if it satisfies the condition:

$$\mathbf{f}^+(u) = \mathbf{f}^-(u), \quad \forall\ u \in V(G). \tag{2.1}$$

Figure 2.4 (a) shows a cycle-vector in the digraph. If we think of G as an electrical network, then a cycle-vector represents a circulation of currents in the network. Thus a cycle-vector is also called a *circulation* in G. It is easy to verify that all cycle-vectors in G form a subspace of $\mathcal{E}(G)$ (the exercise 2.2.2), denoted by $\mathcal{C}(G)$.

There are certain cycle-vectors of special interest. These are associated with cycles in G. Let C be a cycle in G with a given cyclic orientation and let C^+ denote the set of the edges of C whose direction agree with this orientation. We associate with C a vector $\mathbf{f}_c \in \mathcal{E}(G)$ defined by

$$\mathbf{f}_c(a) = \begin{cases} 1, & \text{if } a \in C^+; \\ -1, & \text{if } a \in C \setminus C^+; \\ 0, & \text{if } a \notin C. \end{cases}$$

Clearly, $\mathbf{f}_c$ satisfies (2.1) and hence is a cycle-vector, which is called a *cycle-vector associated with* C. Figure 2.4 (b) depicts a cycle-vector associated with a cycle consisting of the heavy edges.

We will, a little later, see that each vector in $\mathcal{C}(G)$ can be expressed as a linear combination of cycle-vectors associated with some cycles. For this reason we call $\mathcal{C}(G)$ the *cycle-space* of G.

We now turn our attention to another class of vectors in $\mathcal{E}(G)$. For a given a vector $\mathbf{p}$ in $\mathcal{V}(G)$, a vector $\delta_{\mathbf{p}}$ in $\mathcal{E}(G)$ can be defined by the following rule:

$$\delta_{\mathbf{p}}(a) = \mathbf{p}(x) - \mathbf{p}(y), \ \forall\, a \in E(G) \text{ with } \psi_G(a) = (x, y). \tag{2.2}$$

The vector $\delta_{\mathbf{p}}$ is called a *bond-vector*. If G is thought of an electrical network with potential $\mathbf{p}(x)$ at every vertex x, then, by (2.2), the bond-vector $\delta_{\mathbf{p}}$ represents the potential difference along the wires of the network. For this reason a bond-vector $\delta_{\mathbf{p}}$ is also called a *potential difference* in G. Figure 2.5 (a) shows a digraph with a given vector $\mathbf{p} \in \mathcal{V}(G)$ and the corresponding bond-vector $\delta_{\mathbf{p}}$.

As with cycle-vectors, all bond-vectors in G form a subspace of $\mathcal{E}(G)$ (the exercise 2.2.3), denoted by $\mathcal{B}(G)$.

Analogous to the cycle-vector $\mathbf{f}_c$ associated with a cycle C, there is a vector $\mathbf{g}_B$ associated with a bond B. In fact, since for any bond B of G there exists a nonempty proper subset S of $V(G)$ such $B = [S, \overline{S}]$, we define $\mathbf{g}_B$ by

$$\mathbf{g}_B(a) = \begin{cases} 1, & \text{if } a \in (S, \overline{S}); \\ -1, & \text{if } a \in (\overline{S}, S); \\ 0, & \text{if } a \notin B. \end{cases}$$

If we set

$$\mathbf{p}(u) = \begin{cases} 1, & \text{if } u \in S; \\ 0, & \text{if } u \notin S, \end{cases}$$

then it can be easily verified that $\mathbf{g}_B = \delta_\mathbf{p}$. Thus $\mathbf{g}_B$ is a bond-vector, which is called a *bond-vector associated with* B. Figure 2.5 (b) depicts a bond-vector with a bond consisting of the heavy edges.

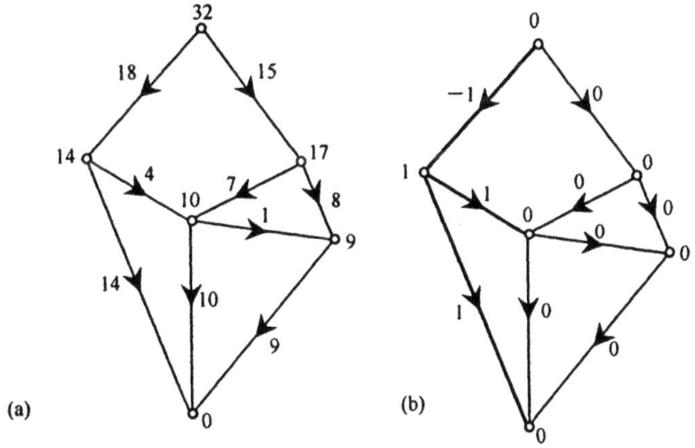

Figure 2.5: (a) A given $\mathbf{p} \in \mathcal{V}(G)$ and the corresponding bond-vector; (b) a bond-vector associated with a bond

We will, a little later, see that each vector in $\mathcal{B}$ can be expressed as a linear combination of the bond-vectors associated with some bonds. For this reason we call $\mathcal{B}(G)$ the *bond-space* of G.

Theorem 2.5 Let $\mathbf{M}$ be the incidence matrix of a digraph G. Then $\mathcal{B}$ is the row space $\mathcal{M}$ of $\mathbf{M}$ and $\mathcal{C}$ is its orthogonal complement in $\mathcal{M}$.

Proof Let $\mathbf{g} = \mathbf{M}_x = (m_x(a_1), m_x(a_2), \cdots, m_x(a_\varepsilon))$ be the row vector of $\mathbf{M}$ corresponding to a vertex x of G. Clearly, $\mathbf{g} \in \mathcal{E}(G)$. Let

$$\mathbf{p}(u) = \begin{cases} 1, & \text{if } u = x; \\ 0, & \text{if } u \neq x. \end{cases}$$

Then $\mathbf{p} \in \mathcal{V}(G)$ and $\mathbf{g} = \delta_\mathbf{p}$, and hence $\mathbf{g} \in \mathcal{B}(G)$, that is, $\mathcal{M} \subseteq \mathcal{B}(G)$.

Conversely, let $\mathbf{g} \in \mathcal{B}(G)$. Then there is $\mathbf{p} \in \mathcal{V}(G)$ satisfying (2.2) such that $\delta_\mathbf{p} = \mathbf{g}$. Thus, for any $a \in E(G)$ with $\psi_G(a) = (x, y)$, we have

$$\mathbf{g}(a) = \delta_\mathbf{p}(a) = \mathbf{p}(x) - \mathbf{p}(y) = \sum_{x \in V} \mathbf{p}(x) \cdot \mathbf{M}_x(a).$$

Thus $\mathbf{g}$ is a linear combination of the rows of $\mathbf{M}$, which implies that $\mathcal{B} \subseteq \mathcal{M}$.

Let $\mathbf{f} \in \mathcal{B}(G)$. Then $\mathbf{f}$ satisfies (2.1) if and only if

$$\sum_{a \in E} \mathbf{f}(a) \cdot \mathbf{M}_x(a) = 0, \ \forall\, x \in V(G).$$

This implies that $\mathbf{f}$ is a cycle-vector if and only if it is orthogonal to each row of $\mathbf{M}$. Hence, $\mathcal{C}$ is the orthogonal complement of $\mathcal{B}$. ∎

Corollary 2.5 $\mathcal{E}(G) = \mathcal{C}(G) + \mathcal{B}(G)$ for any loopless digraph G. ∎

Let $\mathbf{f}$ be a nonzero vector in $\mathcal{E}(G)$. The *support* of $\mathbf{f}$ in G, denoted by $G_{\mathbf{f}}$, is the subgraph of G induced by the set of edges at which the value of $\mathbf{f}$ is nonzero.

For instance, for $\mathbf{f}$ given in Figure 2.4 (b), its support $G_{\mathbf{f}}$ is the cycle induced by the heavy edges; for $\mathbf{g}$ given in Figure 2.5 (b), its support $G_{\mathbf{g}}$ is the bond induced by the heavy edges. Generally, we have the following results.

Lemma 2.6 Let $\mathbf{f}$ and $\mathbf{g}$ be two nonzero vectors in $\mathcal{E}(G)$. Then
(a) the support $G_{\mathbf{f}}$ contains a cycle if $\mathbf{f} \in \mathcal{C}(G)$;
(b) the support $G_{\mathbf{g}}$ contains a bond if $\mathbf{g} \in \mathcal{B}(G)$.

Proof (a) Suppose that $\mathbf{f}$ is a nonzero vector in $\mathcal{E}(G)$. Then, $G_{\mathbf{f}}$ is nonempty, and $\delta(G_{\mathbf{f}}) \geq 2$ by (2.1). Thus $G_{\mathbf{f}}$ contains a cycle by Example 1.7.1.

(b) Suppose that $\mathbf{g}$ is a nonzero vector in $\mathcal{E}(G)$. Then $G_{\mathbf{g}}$ is nonempty and there is a nonzero and nonidentical vector $\mathbf{p} \in \mathcal{V}(G)$ such that $\mathbf{g} = \delta_{\mathbf{p}}$. Arbitrarily choose a vertex u of $G_{\mathbf{g}}$. For this fixed u, let

$$S = \{w \in V(G) : \ \mathbf{p}(w) = \mathbf{p}(u)\}.$$

Then S is a nonempty proper subset of $V(G)$ since $\mathbf{p}$ is not an identical vector in $\mathcal{V}(G)$. Thus the cut $[S, \overline{S}] \neq \emptyset$, and $\mathbf{g}(a) \neq 0$ for any $a \in [S, \overline{S}]$. These facts imply that $[S, \overline{S}] \subseteq E(G_{\mathbf{g}})$, and so $G_{\mathbf{g}}$ contains a bond by the exercise 2.1.12. ∎

To state our main result in this section, we need the following notation. If $\mathbf{B}$ is a matrix whose rows are vectors in $\mathcal{E}(G)$, and if $R \subseteq E(G)$, we denote by $\mathbf{B}|R$ the submatrix of $\mathbf{B}$ consisting of those columns of R. For instance, consider the digraph G shown in Figure 2.6 with the edge-set $\{a_1, a_2, \cdots, a_9\}$.

Let the matrix

$$\mathbf{B} = \begin{array}{c} \begin{matrix} a_1 & a_2 & a_3 & a_4 & a_5 & a_6 & a_7 & a_8 & a_9 \end{matrix} \\ \begin{pmatrix} 0 & 0 & -1 & -1 & 1 & 0 & 0 & 0 & 0 \\ 0 & 0 & 1 & 1 & 0 & 1 & 0 & 0 & 0 \\ -1 & 1 & 1 & 1 & 0 & 0 & 1 & 0 & 0 \\ 1 & -1 & 0 & 0 & 0 & 0 & 0 & 1 & 0 \\ 0 & 1 & 1 & 0 & 0 & 0 & 0 & 0 & 1 \end{pmatrix} \end{array}, \tag{2.3}$$

which consists of the following five rows in $\mathcal{E}(G)$

$$\begin{array}{rl} \mathbf{g}_5 = & (\ 0, \ \ 0, \ -1, \ -1, \ 1, \ 0, \ 0, \ 0, \ 0) \\ \mathbf{g}_6 = & (\ 0, \ \ 0, \ \ 1, \ \ 1, \ 0, \ 1, \ 0, \ 0, \ 0) \\ \mathbf{g}_7 = & (-1, \ \ 1, \ \ 1, \ \ 1, \ 0, \ 0, \ 1, \ 0, \ 0) \\ \mathbf{g}_8 = & (\ 1, \ -1, \ \ 0, \ \ 0, \ 0, \ 0, \ 0, \ 1, \ 0) \\ \mathbf{g}_9 = & (\ 0, \ \ 1, \ \ 1, \ \ 0, \ 0, \ 0, \ 0, \ 0, \ 1) \end{array} \tag{2.4}$$

If $R = \{a_5, a_6, a_7, a_8, a_9\}$, then $\mathbf{B}|R = \mathbf{I}_5$, an identity matrix of order 5.

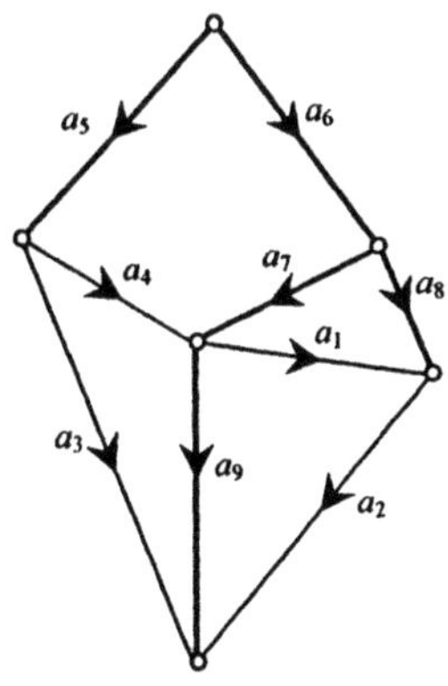

Figure 2.6: A digraph G with labelled edges

A matrix $\mathbf{B}$ is called a *basis matrix* of the bond-space $\mathcal{B}$ if the rows of $\mathbf{B}$ form a basis for $\mathcal{B}$; a basis matrix of the cycle-space $\mathcal{C}$ is similarly defined.

Theorem 2.6 Let $\mathbf{B}$ and $\mathbf{C}$ be basis matrices of the bond-space $\mathcal{B}$ and the cycle-space $\mathcal{C}$ of a digraph G, respectively. Then for any $R \subseteq E(G)$,

(a) the columns of $\mathbf{B}|R$ are linearly independent if and only if $G[R]$ contains no cycle, and

(b) the columns of $\mathbf{C}\,|R$ are linearly independent if and only if $G[R]$ contains no bond.

Proof (a) Denote by $\mathbf{B}_a$ the column-vector of $\mathbf{B}$ corresponding to the edge a. Suppose to the contrary that the columns of $\mathbf{B}|R$ are linearly dependent. Then there is a nonzero vector in $\mathcal{E}(G)$

$$\mathbf{f}(a) = \begin{cases} \neq 0, & \text{if } a \in R; \\ = 0, & \text{if } a \notin R \end{cases} \quad \text{such that} \quad \sum_{a \in R} \mathbf{f}(a)\mathbf{B}_a = 0.$$

Note that

$$0 = \sum_{a \in R} \mathbf{f}(a)\mathbf{B}_a = \sum_{a \in E(G)} \mathbf{f}(a)\mathbf{B}_a = \mathbf{B}\mathbf{f}^T$$

means that $\mathbf{f}$ is orthogonal to each row of $\mathbf{B}$, and hence orthogonal to each vector in $\mathcal{B}$. This implies $\mathbf{f} \notin \mathcal{B}$, and so $\mathbf{f} \in \mathcal{C}$ by Corollary 2.5. By Lemma 2.6, $G_{\mathbf{f}}$ contains a cycle, and so does $G[R]$ since $G_{\mathbf{f}} \subseteq G[R]$. However, this contradicts our hypothesis. It follows that the column-vectors of $\mathbf{B}|R$ are linearly independent.

Conversely, suppose to the contrary that $G[R]$ contains a cycle C. Let $\mathbf{f}_c$ be a cycle-vector associated with C. Then $\mathbf{0} \neq \mathbf{f}_c \in \mathcal{C}(G)$. Thus $\mathbf{f}_c$ is orthogonal to each row of $\mathbf{B}$. On the one hand, we have

$$\sum_{a \in E} \mathbf{f}_c(a)\mathbf{B}_a = \mathbf{B}\mathbf{f}_c^T = 0.$$

On the other hand, we have

$$\sum_{a \in E(G)} \mathbf{f}_c(a)\mathbf{B}_a = \sum_{a \in R} \mathbf{f}_c(a)\mathbf{B}_a.$$

Thus, the nonzero vector $\mathbf{f}_c$ satisfies

$$\sum_{a \in R} \mathbf{f}_c(a)\mathbf{B}_a = 0.$$

This shows that the column-vectors of $\mathbf{B}|R$ are linearly dependent, which contradicts our hypothesis. It follows that $G[R]$ contains no cycle.

A similar argument using Lemma 2.6 yields a proof of (b), the detail is left to the reader as an exercise (the exercise 2.2.1). ∎

Corollary 2.6 The dimensions of $\mathcal{B}$ and $\mathcal{C}$ are given by

$$\dim \mathcal{B} = \upsilon - \omega \quad \text{and} \quad \dim \mathcal{C} = \varepsilon - \upsilon + \omega.$$

Proof Let **B** be a basis matrix of $\mathcal{B}$. By Theorem 2.6 (a),

$$\text{rank}\,(\mathbf{B}) = \max\{\,|R| :\ R \subseteq E \text{ and } G[R] \text{ contains no cycle}\}.$$

The above maximum can be attained only when $G[R]$ is a spanning forest of G, and is equal to $v-\omega$ by Corollary 2.1. Since dim $\mathcal{B}$ = rank (**B**), this proves the first assertion. The second follows immediately since $\mathcal{C}$ is the orthogonal complement of $\mathcal{B}$. ∎

Let F be a spanning forest of a loopless digraph G. Label the edges of G by $a_1, a_2, \cdots, a_\varepsilon$ such that

$$E(\overline{F}) = \{a_1, a_2, \cdots, a_{\varepsilon-v+\omega}\}, \quad E(F) = \{a_{\varepsilon-v+\omega+1}, \cdots, a_\varepsilon\}.$$

For instance, see Figure 2.6, a spanning tree F is induced by the set of edges $\{a_5, a_6, a_7, a_8, a_9\}$.

By Theorem 2.3, for any $a_i \in E(\overline{F})$, $F + a_i$ contains a unique cycle of G, denoted by C_i, which is called a *fundamental cycle* associated with the forest F. Let $\mathbf{f}_i$ denote the cycle-vector associated with the cycle C_i, defined so that $\mathbf{f}_i(a_i) = 1$. The $(\varepsilon - v + \omega) \times \varepsilon$ matrix $\mathbf{C}_F$ whose rows are

$$\mathbf{f}_1, \mathbf{f}_2, \cdots, \mathbf{f}_{\varepsilon-v+\omega} \tag{2.5}$$

can be certainly represented in the form

$$\mathbf{C}_F = (\mathbf{I}_{\varepsilon-v+\omega}\ \mathbf{C}_2),$$

where $\mathbf{I}_{\varepsilon-v+\omega} = \mathbf{C}_F|\overline{F}$ is an identity matrix of order $\varepsilon-v+\omega$, and $\mathbf{C}_2 = \mathbf{C}_F|F$ is an $(\varepsilon - v + \omega) \times (v - w)$ matrix. Since rank $(\mathbf{C}_F) = \varepsilon - v + \omega$, the vectors in (2.5) is a basis of the cycle-space $\mathcal{C}(G)$. We refer to the matrix $\mathbf{C}_F$ as the *basis matrix of $\mathcal{C}(G)$ corresponding to the spanning forest F.*

For instance, consider the digraph G shown in Figure 2.6, in which $v = 6, \varepsilon = 9$ and a spanning tree F is determined by the heavy edges a_5, a_6, a_7, a_8, a_9, and the cotree $\overline{F}$ is determined by the edges a_1, a_2, a_3, a_4. Four fundamental cycles associated with F are $C_i = F + a_i$ $(i = 1, 2, 3, 4)$. Four cycle-vectors associated with these fundamental cycle C_i's are

$$\begin{array}{l} \mathbf{f}_1 = (1,\ 0,\ 0,\ 0,\ 0,\ 0,\ 1,\ -1,\ 0) \\ \mathbf{f}_2 = (0,\ 1,\ 0,\ 0,\ 0,\ 0,\ -1,\ 1,\ -1) \\ \mathbf{f}_3 = (0,\ 0,\ 1,\ 0,\ 1,\ -1,\ -1,\ 0,\ -1) \\ \mathbf{f}_4 = (0,\ 0,\ 0,\ 1,\ 1,\ -1,\ -1,\ 0,\ 0) \end{array} \tag{2.6}$$

A basis matrix $\mathbf{C}_F$ of $\mathcal{C}(G)$ consisting of these four cycle-vectors is

$$\mathbf{C}_F = \begin{array}{c} \begin{array}{ccccccccc} a_1 & a_2 & a_3 & a_4 & a_5 & a_6 & a_7 & a_8 & a_9 \end{array} \\ \begin{pmatrix} 1 & 0 & 0 & 0 & 0 & 0 & 1 & -1 & 0 \\ 0 & 1 & 0 & 0 & 0 & 0 & -1 & 1 & -1 \\ 0 & 0 & 1 & 0 & 1 & -1 & -1 & 0 & -1 \\ 0 & 0 & 0 & 1 & 1 & -1 & -1 & 0 & 0 \end{pmatrix} \end{array} = (\mathbf{I}_4, \mathbf{C}_2), \tag{2.7}$$

where

$$\mathbf{I}_4 = \mathbf{C}_F|\overline{F}, \quad \mathbf{C}_2 = \mathbf{C}_F|T = \begin{pmatrix} 0 & 0 & 1 & -1 & 0 \\ 0 & 0 & -1 & 1 & -1 \\ 1 & -1 & -1 & 0 & -1 \\ 1 & -1 & -1 & 0 & 0 \end{pmatrix}$$

The cycle-vector $\mathbf{f}$ shown in Figure 2.4 can be expressed as follows.

$$\begin{aligned} \mathbf{f} = & \ (\mathbf{f}(a_1), \mathbf{f}(a_2), \mathbf{f}(a_3), \cdots, \mathbf{f}(a_8), \mathbf{f}(a_9)) \\ = & \left(0,\ \frac{1}{2},\ -2,\ 0,\ -2,\ 2,\ \frac{3}{2},\ \frac{1}{2},\ \frac{3}{2}\right) \\ = & \ 0 \cdot \mathbf{f}_1 + \frac{1}{2} \cdot \mathbf{f}_2 + (-2) \cdot \mathbf{f}_3 + 0 \cdot \mathbf{f}_4 \\ = & \sum_{i=1}^{4} \mathbf{f}(a_i)\, \mathbf{f}_i. \end{aligned}$$

This shows that $\mathbf{f}$ is uniquely determined by $\mathbf{f}|\overline{F}$. This fact is true for any cycle-vector in $\mathcal{C}$ (the exercise 2.2.6).

Similarly, by Theorem 2.4, for any $a_j \in E(F)$, $\overline{F} + a_j$ contains a unique bond of G, denoted by B_j, which is called a *fundamental bond* associated with the forest F. Let $\mathbf{g}_j$ denote the bond-vector associated with the fundamental bond B_j, defined so that $\mathbf{g}_j(a_j) = 1$. The $(\upsilon - \omega) \times \varepsilon$ matrix $\mathbf{B}_F$ whose rows are

$$\mathbf{g}_{\varepsilon-\upsilon+\omega+1}, \mathbf{g}_{\varepsilon-\upsilon+\omega+2}, \cdots, \mathbf{g}_{\varepsilon} \tag{2.8}$$

certainly has the following form:

$$\mathbf{B}_F = (\mathbf{B}_1\ \mathbf{I}_{\upsilon-\omega}),$$

where $\mathbf{B}_1 = \mathbf{B}_F|\overline{F}$ is $(\upsilon - \omega) \times (\varepsilon - \upsilon + \omega)$ matrix, and $\mathbf{I}_{\upsilon-\omega} = \mathbf{B}_F|F$ is an identity matrix of order $\upsilon - \omega$. Since rank $(\mathbf{B}_F) = \upsilon - \omega$, the vectors in (2.8) is a basis of the bond-space $\mathcal{B}(G)$. We refer to the matrix $\mathbf{B}_F$ as the *basis matrix of* $\mathcal{B}(G)$ *corresponding to the spanning forest* F.

For instance, consider the digraph G shown in Figure 2.6. Five fundamental bonds associated with F are $B_j = F + a_j$ $(i = 5, 6, 7, 8, 9)$. Five bond-vectors associated with these fundamental bonds B_j's are shown in (2.4). A basis matrix $\mathbf{B}_F = (\mathbf{B}_1\ \mathbf{I}_5)$ of $\mathcal{B}(G)$ consisting of these five bond-vectors are shown in (2.3), where

$$\mathbf{B}_1 = \mathbf{B}_F|\overline{T} = \begin{pmatrix} 0 & 0 & -1 & -1 \\ 0 & 0 & 1 & 1 \\ -1 & 1 & 1 & 1 \\ 1 & -1 & 0 & 0 \\ 0 & 1 & 1 & 0 \end{pmatrix} \quad \mathbf{I}_5 = \mathbf{B}_F|F.$$

The bond-vector $\mathbf{g}$ shown in Figure 2.5 can be expressed as follows.

$$\begin{aligned} \mathbf{g} = &\ (1,\ 9,\ 14,\ 4,\ 18,\ 15,\ 7,\ 8,\ 10) \\ = &\ 18 \cdot \mathbf{g}_5 + 15 \cdot \mathbf{g}_6 + 7 \cdot \mathbf{g}_7 + 8 \cdot \mathbf{g}_8 + 10\mathbf{g}_9 \\ = &\ \sum_{j=5}^{9} \mathbf{g}(a_j)\, \mathbf{g}_j. \end{aligned}$$

This shows that $\mathbf{g}$ is uniquely determined by $\mathbf{g}|F$. This fact is true for any bond-vector in $\mathcal{B}$ (the exercise 2.2.6).

The relation between cycles and bonds that has become apparent from the foregoing discussion finds its proper setting in the theory of matroids. The interested reader is referred to Tutte [164]. We conclude this section with two examples.

Example 2.2.1 Let $\mathbf{B}$ be a basis matrix of the bond-space $\mathcal{B}(G)$, F be spanning forest of G and $\mathbf{B}_F$ be the basis matrix of $\mathcal{B}(G)$ corresponding to F. Then $\mathbf{B}$ is uniquely determined by $\mathbf{B}|F$, and $\mathbf{B} = (\mathbf{B}|F)\mathbf{B}_F$.

Proof Since both $\mathbf{B}$ and $\mathbf{B}_F$ are basis matrices of the bond-space $\mathcal{B}(G)$, there is a unique invertible matrix $\mathbf{P}$ such that $\mathbf{B} = \mathbf{P}\mathbf{B}_F$. Thus, $\mathbf{B}$ is uniquely determined by $\mathbf{P}$. Note that $\mathbf{B}|F = (\mathbf{P}\mathbf{B}_F)|F = \mathbf{P}(\mathbf{B}_F|F) = \mathbf{P}$. Therefore, $\mathbf{B} = \mathbf{P}\mathbf{B}_F = (\mathbf{B}|F)\mathbf{B}_F$. ∎

Example 2.2.2 Let $\mathbf{K}$ be a matrix obtained from the incidence matrix $\mathbf{M}$ of a connected loopless digraph G by deleting any one of its rows. Then $\mathbf{K}$ is a basis matrix of $\mathcal{B}(G)$.

Proof Since G is connected, by Corollary 2.6, dim $\mathcal{B} = v - 1$. By Theorem 2.5, $\mathcal{B}$ is the space $\mathcal{M}$ consisting of all row-vectors of $\mathbf{M}$. Since $\mathbf{K}$ is a submatrix of $\mathbf{M}$, rank $(\mathbf{K}) \leq v - 1$ by definition of $\mathbf{K}$. Therefore, we need to only prove that rank $(\mathbf{K}) \geq v - 1$.

Let $\beta_1, \beta_2, \cdots, \beta_v$ be row-vectors of $\mathbf{M}$. Thus, by definition of $\mathbf{M}$,

$$\beta_1 + \beta_2 + \cdots + \beta_v = \mathbf{0}. \tag{2.9}$$

We, without loss of generality, suppose that $\beta_1, \beta_2, \cdots, \beta_{v-1}$ are row-vectors of $\mathbf{K}$. If $\beta_1, \beta_2, \cdots, \beta_{v-1}$ are linearly dependent, then there are scalars $\lambda_1, \lambda_2, \cdots, \lambda_{v-1}$, not all zero, without loss of generality, say $\lambda_1 = -1$, such that

$$\beta_1 = \lambda_2\beta_2 + \lambda_3\beta_3 + \cdots + \lambda_{v-1}\beta_{v-1}. \tag{2.10}$$

It follows from (2.9) and (2.10) that

$$\beta_v = -(1 + \lambda_2)\beta_2 - (1 + \lambda_3)\beta_3 - \cdots - (1 + \lambda_{v-1})\beta_{v-1}.$$

Thus, each of $\beta_1, \beta_2, \cdots, \beta_v$ can be expressed as a linear combinations of $\beta_2, \beta_3, \cdots, \beta_{v-1}$. This implies that rank $\mathbf{M} \leq v - 2$, which contradicts the fact rank $(\mathbf{M})$ = dim $\mathcal{M}$ = dim $\mathcal{B} = v - 1$. Therefore, $\beta_1, \beta_2, \cdots, \beta_{v-1}$ are linearly independent, and so rank $(\mathbf{K}) \geq v - 1$, as required. ∎

Exercises

2.2.1 Give a proof of Theorem 2.6 (b).

2.2.2 Let C be a cycle in a loopless digraph G. Prove that

(a) $\mathbf{f}_c \in \mathcal{E}(G)$ is a cycle-vector in G;

(b) all cycle-vectors in G form a subspace of $\mathcal{E}(G)$.

2.2.3 Let B be a bond in a digraph G. Prove that

(a) $\mathbf{g}_B \in \mathcal{E}(G)$ is a bond-vector in G;

(b) all bond-vectors in G form a subspace of $\mathcal{E}(G)$.

2.2.4 In the following digraph G, the heavy edges indicate a spanning tree T of G.

(a) Extend the weights on T to a bond-vector $\mathbf{g}$ in G, write out the basis matrix $\mathbf{B}_T$ of $\mathcal{B}(G)$ corresponding to T and the expression of $\mathbf{g}$ by using this basis.

(b) Extend the weights on $\overline{T}$ to a cycle-vector $\mathbf{f}$ in G, write out the basis matrix $\mathbf{C}_T$ of $\mathcal{C}(G)$ corresponding to T and the expression of $\mathbf{f}$ by using this basis.

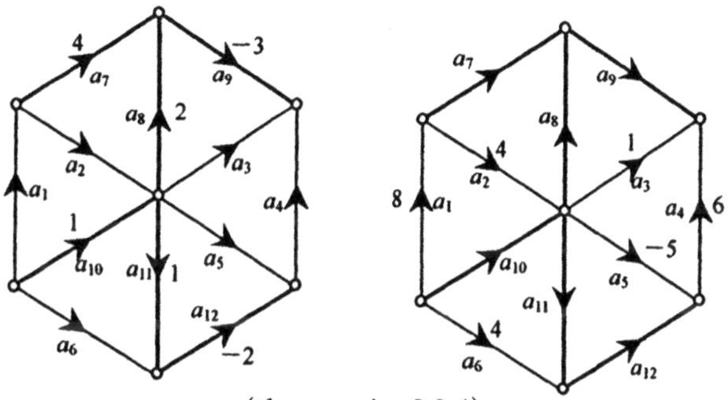

(the exercise 2.2.4)

2.2.5 Let F be a spanning forest in a digraph G, $\mathbf{B}_F = (\mathbf{B}_1\ \mathbf{I}_{v-\omega})$ and $\mathbf{C}_F = (\mathbf{I}_{\varepsilon-v+\omega}\ \mathbf{C}_2)$ be the basis matrices of $\mathcal{B}(G)$ and, respectively, $\mathcal{C}(G)$ corresponding to F. Prove that

(a) $\mathbf{C}_F(\mathbf{B}_F)^T = \mathbf{O}$;

(b) $\mathbf{C}_F = \left(\mathbf{I}_{\varepsilon-v+\omega}\ -\mathbf{B}_1^T\right)$;

(c) $\mathbf{B}_F = \left(-\mathbf{C}_2{}^T\ \mathbf{I}_{v-\omega}\right)$.

2.2.6 Let $\mathbf{f}$ be a cycle-vector and $\mathbf{g}$ be a bond-vector in $\mathcal{E}(G)$, and let F be a spanning forest of G, $\mathbf{B}_F$ and $\mathbf{C}_F$ be the basis matrices of $\mathcal{B}(G)$ and $\mathcal{C}(G)$ corresponding to F, respectively. Prove that

(a) $\mathbf{f}$ is uniquely determined by $\mathbf{f}|\overline{F}$ and $\mathbf{f} = (\mathbf{f}|\overline{F})\mathbf{C}_F$;

(b) $\mathbf{g}$ is uniquely determined by $\mathbf{g}|F$ and $\mathbf{g} = (\mathbf{g}|F)\mathbf{B}_F$.

2.2.7 Let $\mathbf{C}$ be a basis matrix of the cycle space $\mathcal{C}(G)$, F be spanning forest of G and $\mathbf{C}_F$ be the basis matrix of $\mathcal{C}(G)$ corresponding to F. Prove that $\mathbf{C}$ is uniquely determined by $\mathbf{C}\,|\overline{F}$, and $\mathbf{C} = (\mathbf{C}\,|\overline{F})\mathbf{C}_F$.

2.2.8 Let G be a loopless digraph. Prove that

(a) any cycle of G can be represented into symmetric differences of several fundamental cycles;

(b) any bond of G can be represented into symmetric differences of several fundamental bonds.

2.3 Enumeration of Spanning Trees

As we have seen in the preceding section, spanning forests or trees of a graph play an important role in discussing the spaces of graphs. In this section we are concerned the number of distinct spanning trees of a connected graph, and we will present several formulae for counting the number, which are derived from the theory on the spaces of graphs.

Let G be a connected loopless graph. We use the symbol $\varsigma(G)$ to denote the number of spanning trees in G. Let $\mathbf{B}$ be a basis matrix of the bond-space $\mathcal{B}(G)$. From Theorem 2.6, if R is a subset of $E(G)$ with $|R| = v - 1$ then the square submatrix $\mathbf{B}|R$ is invertible if and only if the subgraph $G[R]$ induced by R is a spanning tree of G. Thus, it is clear that $\varsigma(G)$ is equal to the number of invertible submatrices of $\mathbf{B}$ of order $v - 1$. In order to obtain formulae for counting this number, we need the following notion.

A matrix is said to be *unimodular* if all of its full square submatrices have determinants $0, +1$ or -1.

Example 2.3.1 Let G be a loopless digraph, F be a spanning forest of G. Then the basis matrices of $\mathcal{B}(G)$ and $\mathcal{C}(G)$ corresponding to F both are unimodular.

Proof Let $\mathbf{B}_F$ be the basis matrix of $\mathcal{B}(G)$ corresponding to F and let $\mathbf{P}$ be a full square submatrix of $\mathbf{B}_F$. Then $\mathbf{P}$ has order $v - \omega$ by Corollary 2.6. Let R be the set of edges of G corresponding to the columns of $\mathbf{P}$, and let $F' = G[R]$. Thus $\mathbf{P} = \mathbf{B}_F|F'$. If F' contains a cycle, then det $\mathbf{P} = 0$ by Theorem 2.6 (a). Suppose that F' contains no cycle below. Let $\mathbf{B}'$ be the basis matrix of $\mathcal{B}$ corresponding to F'. It follows from Example 2.2.1 that

$$(\mathbf{B}_F|F')\,\mathbf{B}' = \mathbf{B}_F.$$

Restricting both sides to F, we obtain

$$(\mathbf{B}_F|F')\,(\mathbf{B}'|F) = \mathbf{B}_F|F.$$

Taking determinants and noting that $\mathbf{B}_F|F$ is an identity matrix, we have

$$\det(\mathbf{B}_F|F') \cdot \det(\mathbf{B}'|F) = 1. \tag{2.11}$$

Two determinants in (2.11), being determinants of integral matrices, are themselves integers. It follows that

$$\det \mathbf{P} = \det(\mathbf{B}_F|F') = \pm 1.$$

A similar argument gives a proof of another assertion. ∎

Example 2.3.2 Let **K** be a matrix obtained from the incidence matrix **M** of a loopless digraph G by deleting any one of its rows. Then **K** is unimodular.

Proof Let **P** be a full square submatrix of **K**. To show that **K** is unimodular, it suffices to prove that the determinant of **P** has value $0, +1$ or -1. In fact, we could show a stronger result than what we need, that is, all n-square submatrices of **P** have determinants $0, +1$ or -1 by induction on $n \geq 1$. If $n = 1$, the assertion holds clearly. Suppose that all n-square submatrices of **P** have determinants $0, +1$ or -1. We want to prove that all $(n+1)$-square submatrices of **P** have determinants $0, +1$ or -1.

Let

$$\mathbf{Q} = (q_{ij}) \quad 1 \leq i, j \leq n+1$$

be an $(n+1)$-square submatrix of **P**. By definition of the matrix **M**, there are at most two nonzero entries in every column of **Q**. If every column of **Q** has exactly two nonzero entries, the sum of all row vectors is equal to zero, which means that $\det \mathbf{Q} = 0$.

Suppose below that at least one column of **Q** has at most one nonzero entry. We suppose that q_{ij} is such an entry. Then $q_{r,j} = 0$ for all r's other than i. Let $\mathbf{Q}_{ij}$ be the algebraic cofactor of q_{ij} in **Q**. Then $\mathbf{Q}_{ij}$ is an n-square submatrices of **P**. By the induction hypothesis, $\det \mathbf{Q}_{ij} = 0, 1$ or -1. It follows from properties of determinants that

$$\det \mathbf{Q} = (-1)^{i+j} q_{ij} \cdot \det (\mathbf{Q}_{ij}) = 0, +1 \text{ or } -1$$

as desired. ∎

Theorem 2.7 Let G be a connected loopless digraph, **B** and **C** be unimodular basis matrices of $\mathcal{B}(G)$ and $\mathcal{C}(G)$, respectively. Then

$$\varsigma(G) = \det (\mathbf{B}\mathbf{B}^T) = \det (\mathbf{C}\mathbf{C}^T).$$

Proof Using Binet-Cauchy's formula for the determinant of the product of two rectangular matrices, we obtain

$$\det (\mathbf{B}\mathbf{B}^T) = \sum_{\substack{R \subseteq E(G) \\ |R| = v-1}} (\det (\mathbf{B}|R))^2. \tag{2.12}$$

Thus, on the one hand, by Theorem 2.6 (a), the number of nonzero terms in (2.12) is equal to $\varsigma(G)$. On the other hand, since G is connected and $\mathbf{B}$ is unimodular, each such a term has value 1. This shows that $\varsigma(G) = \det(\mathbf{B}\mathbf{B}^T)$.

In a similar argument, we can prove another formula by replacing $\mathbf{B}$ by $\mathbf{C}$. ∎

Several enumeration formulae on the number of spanning trees in a given graph can be easily deduced from Theorem 2.7.

Corollary 2.7.1 Let G be a connected loopless digraph, $\mathbf{B}$ and $\mathbf{C}$ be unimodular basis matrices of $\mathcal{B}(G)$ and $\mathcal{C}(G)$, respectively. Then

$$\varsigma(G) = \pm\det\begin{pmatrix}\mathbf{B}\\ \mathbf{C}\end{pmatrix}.$$

Proof By Theorem 2.7, we have

$$\varsigma(G)^2 = \det(\mathbf{B}\mathbf{B}^T)\cdot\det(\mathbf{C}\mathbf{C}^T) = \det\begin{pmatrix}\mathbf{B}\mathbf{B}^T & \mathbf{O}\\ \mathbf{O} & \mathbf{C}\mathbf{C}^T\end{pmatrix}.$$

Note that $\mathbf{C}\mathbf{B}^T = \mathbf{B}\mathbf{C}^T = \mathbf{O}$ since $\mathcal{B}$ and $\mathcal{C}$ are orthogonal by Corollary 2.5. Thus,

$$\begin{aligned}\varsigma(G)^2 = &\det\begin{pmatrix}\mathbf{B}\mathbf{B}^T & \mathbf{B}\mathbf{C}^T\\ \mathbf{C}\mathbf{B}^T & \mathbf{C}\mathbf{C}^T\end{pmatrix} = \det\left(\begin{pmatrix}\mathbf{B}\\ \mathbf{C}\end{pmatrix}\cdot\begin{pmatrix}\mathbf{B}^T & \mathbf{C}^T\end{pmatrix}\right)\\ = &\det\begin{pmatrix}\mathbf{B}\\ \mathbf{C}\end{pmatrix}\cdot\det\begin{pmatrix}\mathbf{B}^T & \mathbf{C}^T\end{pmatrix} = \left(\det\begin{pmatrix}\mathbf{B}\\ \mathbf{C}\end{pmatrix}\right)^2\end{aligned}$$

as desired. ∎

Corollary 2.7.2 Let G be a connected loopless digraph, T be a spanning tree of G, $\mathbf{B}_T$ and $\mathbf{C}_T$ be the basis matrices of $\mathcal{B}(G)$ and, respectively, $\mathcal{C}(G)$ corresponding to T, and let $\mathbf{K}$ be a matrix obtained from the incidence matrix $\mathbf{M}$ of G by deleting any one of its rows. Then

$$\varsigma(G) = \det\left(\mathbf{B}_T\mathbf{B}_T^T\right) = \det\left(\mathbf{C}_T\mathbf{C}_T^T\right) = \det\left(\mathbf{K}\mathbf{K}^T\right).$$

Proof By Example 2.3.1, both $\mathbf{B}_T$ and $\mathbf{C}_T$ are unimodular; by Example 2.2.2 and Example 2.3.2, $\mathbf{K}$ is a basis matrix of $\mathcal{B}$ and is unimodular. Thus, by Theorem 2.7, the those three formulae follow immediately. ∎

Example 2.3.3 Consider the digraph G shown in Figure 2.6. Its incidence matrix is

$$\mathbf{M} = \begin{pmatrix} 0 & 0 & 0 & 0 & 0 & -1 & 1 & 1 & 0 \\ -1 & 1 & 0 & 0 & 0 & 0 & 0 & -1 & 0 \\ 1 & 0 & 0 & -1 & 0 & 0 & -1 & 0 & 1 \\ 0 & 0 & 1 & 1 & -1 & 0 & 0 & 0 & 0 \\ 0 & 0 & 0 & 0 & 1 & 1 & 0 & 0 & 0 \\ 0 & -1 & -1 & 0 & 0 & 0 & 0 & 0 & -1 \end{pmatrix}$$

Let $\mathbf{K}$ be a matrix obtained from $\mathbf{M}$ by deleting the last row. Then

$$\det \mathbf{K}\mathbf{K}^T = \det \begin{pmatrix} 3 & -1 & -1 & 0 & -1 \\ -1 & 3 & -1 & 0 & 0 \\ -1 & -1 & 4 & -1 & 0 \\ 0 & 0 & -1 & 3 & -1 \\ -1 & 0 & 0 & -1 & 2 \end{pmatrix} = 66$$

Let T be a spanning tree of G with the edge-set $\{a_5, a_6, a_7, a_8, a_9\}$. The basis matrix $\mathbf{B}_T$ of $\mathcal{B}(G)$ and the basis matrix $\mathbf{C}_T$ of $\mathcal{C}(G)$ corresponding to T are shown in (2.3) and (2.7), respectively. Then

$$\mathbf{B}_T\mathbf{B}_T^T = \begin{pmatrix} 3 & -2 & -2 & 0 & -1 \\ -2 & 3 & 2 & 0 & 1 \\ -2 & 2 & 5 & -2 & 2 \\ 0 & 0 & -2 & 3 & -1 \\ -1 & 1 & 2 & -1 & 3 \end{pmatrix},$$

and

$$\mathbf{C}_T\mathbf{C}_T^T = \begin{pmatrix} 3 & -2 & -1 & -1 \\ -2 & 4 & 2 & 1 \\ -1 & 2 & 5 & 3 \\ -1 & 1 & 3 & 4 \end{pmatrix}.$$

Their determinants are equal to 66. ∎

Corollary 2.7.3 Let T_n denote a tournament of order $n\,(\geq 2)$. Then $\varsigma(T_n) = n^{n-2}$.

Proof By Corollary 2.7.2 and a simply count, we have

$$\begin{aligned} \varsigma(T_n) &= \det \mathbf{K}\mathbf{K}^T \\ &= \det \begin{pmatrix} n-1 & -1 & \cdots & -1 & -1 \\ -1 & n-1 & \cdots & -1 & -1 \\ \cdots & \cdots & \cdots & \cdots & \cdots \\ -1 & -1 & \cdots & n-1 & -1 \\ -1 & -1 & \cdots & -1 & n-1 \end{pmatrix} = n^{n-2} \end{aligned}$$

as required. ∎

Figure 2.7 shows a tournament T_4 and its 16 distinct spanning trees.

T_4

Figure 2.7: A tournament T_4 and all of its distinct spanning trees

Note that n^{n-2} is the number of distinct spanning trees of T_n, but not the number of non-isomorphic spanning trees of T_n. In fact, there are only 6 non-isomorphic spanning trees among the 16 distinct spanning trees (see Figure 2.8).

By considering an oriented graph of an undirected graph G, the following two enumeration formulae on the number of spanning trees in G can be immediately deduced from Corollary 2.7.2 and Corollary 2.7.3, left to the reader as an exercise for details.

Corollary 2.7.4 (Matrix-tree theorem, Kirchhoff [106]) Let G be a connected, labelled and loopless undirected graph. Then

$$\varsigma(G) = \det\left(\mathbf{K}\mathbf{K}^T\right),$$

where $\mathbf{K}$ is a matrix obtained from the incidence matrix $\mathbf{M}$ of an oriented graph of G by deleting any one of its rows. ∎

Corollary 2.7.5 (Cayley's formula, Cayley [26]) Let K_n be a labelled complete undirected graph. Then

$$\varsigma(K_n) = n^{n-2}, \quad \text{for } n \geq 2.$$ ∎

Figure 2.8: 6 non-isomorphic spanning trees of order 4

Exercises

2.3.1 Let $\mathbf{M}$ be the incidence matrix of a connected loopless digraph G, $\mathbf{K}$ be a matrix obtained from $\mathbf{M}$ by deleting any one of its rows, $\mathbf{C}$ be unimodular basis matrix of the cycle-space $\mathcal{C}(G)$. Prove that

(a) $\varsigma(G) = \pm\det\begin{bmatrix}\mathbf{K}\\ \mathbf{C}\end{bmatrix}$.

(b) the algebraic cofactor of any entry in $\mathbf{M}\mathbf{M}^T$ is equal to $\varsigma(G)$.

2.3.2 Let G be a connected, labelled and loopless undirected graph with the vertex-set $\{x_1, x_2, \cdots, x_v\}$ and $\mathbf{A}$ be the adjacency matrix of G. Let $\mathbf{B}$ be the $v \times v$ diagonal matrix with the main diagonal elements $b_{ii} = d_G(x_i)$ for each $i = 1, 2, \cdots, v$. Prove that the algebraic cofactor of any entry in $\mathbf{B} - \mathbf{A}$ is equal to $\varsigma(G)$.

2.3.3 A matrix is *totally unimodular* if all square submatrices have determinants $0, +1$ or -1. Prove that

(a) any basis matrix of $\mathcal{B}$ and $\mathcal{C}$ corresponding to a spanning forest is totally unimodular;

(b) the incidence matrix of a loopless digraph is totally unimodular;

(c) the incidence matrix of a simple undirected graph G is totally unimodular if and only if G is bipartite. (J.Egerváry, 1931)

2.3.4 Prove that the following labelled undirected graphs

(a) $\varsigma(C_n) = n$, where C_n is a cycle of order n;

(b) $\varsigma(K_n - e) = (n-2)n^{n-3}$, where $K_n - e$ denotes a subgraph of a complete graph K_n by deleting an edge e;

(c) $\varsigma(K_{m,n}) = m^{n-1}n^{m-1}$, where $K_{m,n}$ is a complete bipartite graph.

2.3.5 Let G be a nontrivial connected loopless graph, and e an edge of G. Prove that $\varsigma(G) = \varsigma(G \cdot e) + \varsigma(G - e)$.

2.3.6 Let G be a digraph without directed cycles, x a vertex of G, and let ς_x denote the number of spanning out-trees rooted at x of G.

(a) Prove that $\varsigma_x(G) = \prod\limits_{y \in V \setminus \{x\}} d_G^-(y)$.

(b) Verify the conclusion in (a) for the tournament T_4 in Figure 2.7.

2.3.7 Let G be a connected and loopless digraph with the vertex-set $\{x_1, x_2, \cdots, x_v\}$, $\mathbf{A}$ the adjacency matrix of G and $\mathbf{M}$ the incidence matrix of G. Let $\mathbf{S}$ be the $v \times v$ diagonal matrix with main diagonal elements $s_{ii} = d_G^-(x_i)$, $\mathbf{N} = \mathbf{S} - \mathbf{A}$, N_{ij} the algebraic cofactor of the entry n_{ij} in $\mathbf{N}$; $\mathbf{M}_i$ a matrix obtained from $\mathbf{M}$ by deleting the row x_i; $\overline{\mathbf{M}}_i$ a matrix obtained from $\mathbf{M}_i$ by replacing 0 by 1; let ς_i denote the number of spanning out-trees rooted at x_i of G. Prove that

(a) if $\varepsilon = v - 1$, then G is an out-tree rooted at x_i if and only if $N_{ii} = 1$ for every $i = 1, 2, \cdots, v$;

(b) $\varsigma_i(G) = N_{ii}$; (W.T.Tutte, 1964)

(c) $\varsigma_i(G) = \det(\overline{\mathbf{M}}_i \mathbf{M}_i^T)$; (W.T.Tutte, 1948)

(d) the total number of spanning out-trees of a complete digraph of order n is equal to n^{n-1}.

Applications

2.4 The Minimum Connector Problem

A roadway network connecting a number of towns is to be set up. Given the cost c_{ij} of constructing a direct link between two towns x_i and x_j, design such a network to minimize the total cost of construction. This is known as the *minimum connector problem.*

Many real-world situations can be referred to this problem, for example, the heating system, the power system in a city, the communication system in some area and so on.

Construct a weighted simple undirected graph $(G; \mathbf{w})$ as follows. Each town x_i is regarded as a vertex of G, and $x_i x_j \in E(G) \iff \mathbf{w}(x_i x_j) = c_{ij} < +\infty$. Thus, the minimum connector problem is just that of finding a connected spanning subgraph of G with the minimum weight. Since the weights represent costs, they are certainly positive, and we may therefore assume that such a minimum-weight spanning subgraph is a spanning tree of G. A minimum-weight spanning tree of a weighted graph will be called a *minimum tree.*

It is clear that G contains a spanning tree if and only if G is connected. By Corollary 2.7.5, if G is connected, then $\varsigma(G) \leq v^{v-2}$. Thus the minimum tree certainly exists in a connected graph G. Theoretically, we may find a minimum tree of a weighted graph $(G, \mathbf{w})$ by enumerating all spanning trees of G and, then, comparing their weights. Practically, it is impossible when the number of vertices is large. Thus, an efficient algorithm for finding a minimum tree is necessary.

There are a number of known efficient algorithms to find a minimum tree in a weighted connected graph. The best-known ones are Prim's algorithm [147] which will be described here, and Kruskal's algorithm [114] which will be outlined in the exercise 2.4.3.

We first consider a special case in which $\mathbf{w}(e) = 1$ for each edge e of the graph G. In this case, a minimum tree is then a spanning tree. Thus, we merely need to construct a spanning tree of G. Using the exercise 1.5.4 (a), a simple algorithm for finding such a tree is as follows.

1. Arbitrarily choose $x_0 \in V(G)$. Set $V_0 = \{x_0\}$ and $T_0 = x_0, k = 0$.

2. If V_{k-1} and T_{k-1} have been chosen, then choose $e_k \in [V_{k-1}, \overline{V}_{k-1}]$. Hence there is $u \in V_{k-1}, x_k \in \overline{V}_{k-1}$ such that $e_k = ux_k$. Set $V_k = V_{k-1} \cup \{x_k\}$, and $T_k = T_{k-1} + e_k$.

3. Stop when step 2 can not be implemented further.

This algorithm works because the subgraph constructed by every step is a subtree of G. Since G is connected, the algorithm must stop when $V_{v-1} = V(G)$. The subgraph T_{v-1} is a spanning tree of G by Theorem 2.1 because T_{v-1} is connected and has $v - 1$ edges.

Prim extended it to general connected weighted graphs, thereby solving the minimum connector problem. Prim intended, the k-th step in his algorithm, to find a function $\mathbf{l} \in \mathcal{V}(G)$, a vertex-subsets V_k and a subtrees T_k with the vertex-set V_k.

Prim's Algorithm

1. Arbitrarily choose $x_0 \in V(G)$. Set $\mathbf{l}(x_0) = 0$, $\mathbf{l}(x) = \infty$ $(x \neq x_0)$, $V_0 = \{x_0\}$ and $T_0 = x_0, k = 0$.

2. For any $x \in N_G(x_{k-1}) \cap \overline{V}_{k-1}$, if $\mathbf{w}(x_{k-1}x) < \mathbf{l}(x)$, then replace $\mathbf{l}(x)$ by $\mathbf{w}(x_{k-1}x)$. Choose $x_k \in \overline{V}_{k-1}$ such that $\mathbf{l}(x_k) = \min\{\mathbf{l}(x) : x \in \overline{V}_{k-1}\}$. Set $e_k = ux_k$, $u \in V_{k-1}$ such that $\mathbf{w}(e_k) = \mathbf{l}(x_k)$. Let $V_k = V_{k-1} \cup \{x_k\}$, and $T_k = T_{k-1} + e_k$.

3. Stop when step 2 can not be implemented further.

Example 2.4.1 As an example, we consider the weighted graph G shown in Figure 2.9 (a). The process of construction of a minimum tree by Prim's algorithm is illustrated in Figure 2.9, where (b) shows the first step of the algorithm; (c) - (h) show 6 iterations of the second step of the algorithm, each iteration obtains a new vertex x_k and a new edge e_k (denoted by an heavy edge). A minimum tree T_{v-1} is shown in (h), its weight $\mathbf{w}(T_{v-1}) = 21 = \sum_{x \in V} \mathbf{l}(x)$. ∎

Prim's algorithm clearly produces a spanning tree since it is connected and has $v - 1$ edges. The following theorem ensures that such a tree will always be minimum.

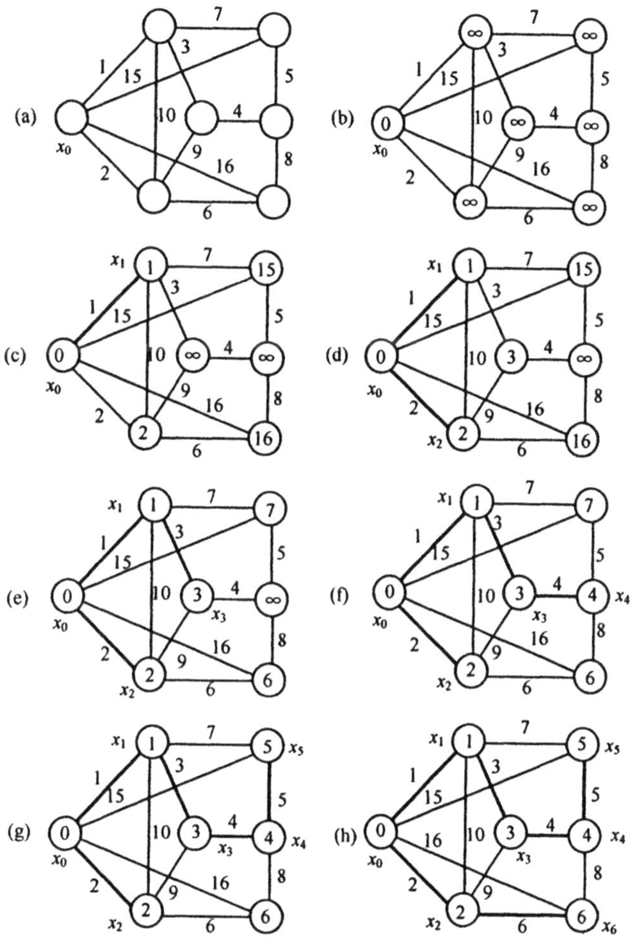

Figure 2.9: An application of Prim's algorithm

Theorem 2.8 Any spanning tree $T_{\upsilon-1}$ constructed by Prim's algorithm is a minimum tree of the weighted connected graph $(G, \mathbf{w})$.

Proof It suffices to prove that T_k is a subgraph of some minimum tree of G. By induction on $k \geq 2$. There is nothing to do for $k = 0$ and suppose that the assertion is true for $k-1$. Consider $T_k = T_{k-1} + e_k$ obtained by Prim's algorithm.

We first prove that T_k is a tree. By the induction hypothesis T_{k-1} is a tree, and by Prim's algorithm $V(T_{k-1}) = V_{k-1}$ and $e_k \in [V_{k-1}, \overline{V}_{k-1}]$ (the exercise 1.5.4 (a) ensures that such an edge exists), thus, T_k is connected. Because $e_k = ux_k$, where $u \in V_{k-1}$ and $x_k \in \overline{V}_{k-1}$, x_k is a vertex of degree one. This implies that T_k contains no cycle, thereby T_k is a tree.

We now prove that T_k is a subgraph of some minimum tree of G. Suppose that T_{k-1} is a subgraph of some minimum tree T^* of G by the induction hypothesis. If $e_k \in T^*$, then $T_k = T_{k-1}+e_k$ is a subgraph of T^* clearly. Suppose that $e_k \notin T^*$ below. Then $T^* + e_k$ contains a unique cycle, C, by Theorem 2.3. Because $e_k \in [V_{k-1}, \overline{V}_{k-1}]$ there is $e'_k \in C$ and $e'_k \in [V_{k-1}, \overline{V}_{k-1}]$, but $e'_k \neq e_k$. Let $T' = (T^* + e_k) - e'_k$. Then $T_k \subseteq T'$. Thus we merely need to prove that T' is a minimum tree of G. Since T' is connected and has edges $v-1$, by Theorem 2.1, T' is a spanning tree other than T^* in G and its weight satisfies

$$\mathbf{w}(T') = \mathbf{w}(T^*) + \mathbf{w}(e_k) - \mathbf{w}(e'_k). \tag{2.13}$$

Note that, in Prim's algorithm, $e_k \in [V_{k-1}, \overline{V}_{k-1}]$ is so chosen that its weight is as small as possible. This means

$$\mathbf{w}(e_k) \leq \mathbf{w}(e'_k). \tag{2.14}$$

Combining (2.13) with (2.14), we have $\mathbf{w}(T') \leq \mathbf{w}(T^*)$. This implies that T' is indeed a minimum tree, and theorem follows. ∎

We now estimate that how many computational times are required for the implementation of Prim's algorithm to construct a minimum tree in any weighted graph $(G, \mathbf{w})$. Step 1, Step 2 and Step 3 are required to execute $v+3$, $v-1$ and $v-1$ times, respectively. The computations mainly involves in Step 2, within which, the execution of "replace $\mathbf{l}(x)$ by $\mathbf{w}(x_{k-1}x)$" requires a total of $|\overline{V}_{k-1}| = v - k + 1$ examinations of the labelling $\mathbf{l}(x)$ of every vertex $x \in V(N_G(x_{k-1}) \cap \overline{V}_{k-1})$; the execution of "choose $x_k \in \overline{V}_{k-1}$ such that $\mathbf{l}(x_k) = \min\{\mathbf{l}(x) : x \in \overline{V}_{k-1}\}$" requires at most $v - k$ comparisons; the execution of "$e_k = ux_k$, $u \in V_{k-1}$ such that $\mathbf{w}(e_k) = \mathbf{l}(x_k)$" requires at most $k - 1$ comparisons of $u \in V(N_G(x_{k-1}) \cap V_{k-1})$. Thus, the total number of computation required for the execution of Step 2 is at most

$$\sum_{k=1}^{v-1}[(v-k+1) + (v-k) + (k-1) + 2] = \frac{3}{2}v^2 + \frac{1}{2}v - 2.$$

It follows that the total number of computation required for this algorithm is at most

$$(v+3)+\left(\frac{3}{2}v^2+\frac{1}{2}v-2\right)+(v-1)=\frac{3}{2}v^2+\frac{5}{2}v.$$

Thus, Prim's algorithm is an efficient algorithm. A graph-theoretic algorithm is *efficient* if the computational times required for its execution on any graph G, called the *complexity* of the algorithm, are bounded above by a polynomial $p(v,\varepsilon)$. A graph-theoretic algorithm is called an $o(g(v,\varepsilon))$-algorithm if there exists a positive constant c such that its complexity is bounded by $c\,g(v,\varepsilon)$, where $g(v,\varepsilon)$ is a polynomial in v and ε. For example, the complexity of Prim's algorithm is bounded by cv^2, and so it is an $o(v^2)$-algorithm.

An algorithm whose complexity is exponential or factorial (such as 2^v or $v\,!$) might be very *inefficient* for some large graphs. There are many problems in graph theory for which no efficient algorithm is known. An important class of these types of problems is called an *NP-hard problem*, the reader is referred to Garey and Johnson [76], and Hochbaum [96] for further details.

Exercises

2.4.1 Constructing a minimum tree in the following weighted graph by using Prim's algorithm.

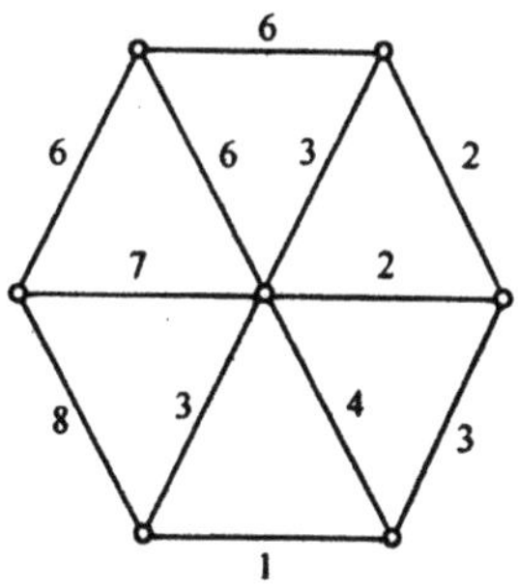

(the exercise 2.4.1 and the exercise 2.4.3)

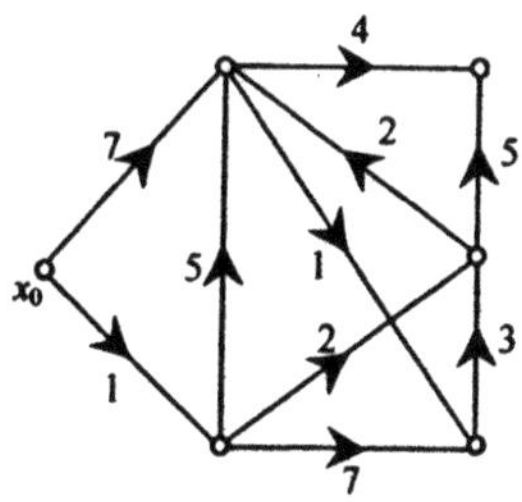

(the exercise 2.4.5)

2.4.2 Prove that if every edge of a connected simple graph G has different weight then a minimum tree constructed by Prim's algorithm is unique.

2.4.3 **Kruskal's algorithm** for finding a minimum tree in a weighted connected graph $(G, \mathbf{w})$ is stated as follows.

1. Choose $e_1 \in E(G)$ such that $\mathbf{w}(e_1)$ is as small as possible.
2. If $e_1, e_2, \cdots, e_i$ have been chosen, then choose $e_{i+1} \in E(G) \setminus \{e_1, e_2, \cdots, e_i\}$ such that $\mathbf{w}(e_{i+1})$ is as small as possible and $G[\{e_1, e_2, \cdots, e_{i+1}\}]$ contains no cycle.
3. Stop when Step 2 can not be implemented further.

(a) Prove that a subgraph of G constructed by Kruskal's algorithm is a minimum tree in G.

(b) Constructing a minimum tree in the above weighted graph by using Kruskal's algorithm.

2.4.4 Can Prim's and Kruskal's algorithms be adapted to find a *maximum-weight spanning tree* in a weighted connected graph? If so, how?

2.4.5 By making a modification to Prim's algorithm, we can obtain the following algorithm for finding a *minimum weight spanning out-tree* rooted at x_0 in a weighted strongly connected digraph $(G, \mathbf{w})$ for a given vertex x_0 in G.

1. Set $\mathbf{l}(x_0) = 0$, $\mathbf{l}(x) = \infty$ $(x \neq x_0)$, $V_0 = \{x_0\}$ and $T_0 = x_0, k = 0$.
2. For any $x \in N_G^+(x_{k-1}) \cap \overline{V}_{k-1}$, if $\mathbf{w}(x_{k-1}, x) < \mathbf{l}(x)$, then replace $\mathbf{l}(x)$ by $\mathbf{w}(x_{k-1}, x)$. So choose $x_k \in \overline{V}_{k-1}$ that $\mathbf{l}(x_k) = \min\{\mathbf{l}(x) : x \in \overline{V}_{k-1}\}$. Let $e_k = (u, x_k)$, $u \in V_{k-1}$ such that $\mathbf{w}(e_k) = \mathbf{l}(x_k)$. Let $V_k = V_{k-1} \cup \{x_k\}$, and $T_k = T_{k-1} + e_k$.
3. Stop when step 2 can not be implemented further.

(a) Prove that a subdigraph constructed by this algorithm is a minimum weight spanning out-tree rooted at x_0 (if such an out-tree exists).

(b) Using the algorithm find a minimum weight spanning out-tree rooted at x_0 in the above digraph.

(c) Prove that this algorithm can be used to verify whether or not a connected digraph with a given vertex x_0 contains a spanning out-tree rooted at x_0.

2.5 The Shortest Path Problem

A railway system connects various cities in some country. Suppose x_0 is the capital of the country. Find a set of shortest paths from x_0 to all of the other cities in the system. This is known as the *shortest path problem.*

Denote this railway system by a weighted undirected simple graph $(G, \mathbf{w})$, where the vertices represent the cities, the edges represent the directly-linked rails and the weight $\mathbf{w}(xy)$ of the edge xy of G represents travelling distance between two cities x and y linked by a rail directly. Thus, the shortest path problem is just that of finding a spanning tree T rooted at x_0 in G such that the weight of a path from x_0 to any other vertex in T is the minimum in $(G, \mathbf{w})$. We will say that such a rooted spanning tree is *optimal.*

At first glance, the shortest path problem and the minimum connector problem, discussed in the preceding section, are similar, but it is really not difficult to see their difference. For example, the spanning tree T of G shown in Figure 2.9 (h) obtained by Prim's algorithm is a unique minimum tree by the exercise 2.4.2. The weight of the x_0x_5-path $P = (x_0, x_1, x_3, x_4, x_5)$ in T is 13, but the weight of the path (x_0, x_1, x_5) in G is 8. This shows that P is not a shortest x_0x_5-path in $(G, \mathbf{w})$. Thus it is necessary to find an efficient algorithm for solving the shortest path problem.

More generally, suppose that $(G, \mathbf{w})$ is a weighted simple digraph with a fixed vertex x_0 and positive weights. The shortest path problem for the this case is to find an *optimal out-tree rooted at* x_0, in which the directed path from x_0 to any other vertex in G is of the minimum weight. It is clear that a necessary condition for the shortest path problem in $(G, \mathbf{w})$ to have a solution is that G contains a spanning out-tree rooted at x_0. Conversely, if G satisfies this necessary condition, then $(G, \mathbf{w})$ certainly contains an optimal out-tree since the number of spanning out-tree rooted at x_0 in G is finite (the exercise 2.3.7). We will assume that G contains a spanning out-tree rooted at x_0 in following discussion. However, such a beforehand assumption is not necessary for a weighted connected undirected graph.

We describe an efficient algorithm for finding an optimal out-tree, proposed by Moore [135], Dijkstra [37], Dantzing [34], Whiting and Hillier [175], independently. This algorithm, based on the exercise 1.5.4(d), can be obtained by making a minor modification to Step 2 in Prim's algorithm. The process of the algorithm attempts to find an increasing sequence $V_0, V_1, \cdots$, $V_{\upsilon-1}$ of subsets of $V(G)$ and a sequence $P_0, P_1, \cdots, P_{\upsilon-1}$ of directed paths with the origin x_0 by a labelling procedure for every vertex of G. Throughout process of the algorithm, each vertex x carries a label $\mathbf{l}(x)$ starting with

$$\mathbf{l}(x) = \begin{cases} 0, & \text{if } x = x_0; \\ \infty, & \text{if } x \neq x_0, \end{cases} \qquad V_0 = \{x_0\} = P_0.$$

The label $\mathbf{l}(x)$ provides an upper bound on the weight of a shortest (x_0, x)-path. V_k and P_k can be recursively constructed by successively correcting the value of $\mathbf{l}(x)$, and vice versa. When the algorithm terminates, the minimum weight of a directed path from x_0 to x is given by the final value of the label $\mathbf{l}(x)$. An outline of the algorithm is stated as follows.

Dijsktra's Algorithm

1. Set $\mathbf{l}(x_0) = 0$, $\mathbf{l}(x) = \infty$ $(x \neq x_0)$, $V_0 = \{x_0\}$ and $P_0 = x_0$ and $k = 0$.

2. For any $x \in N_G^+(x_{k-1}) \cap \overline{V}_{k-1}$, replace $\mathbf{l}(x)$ by $\min\{\mathbf{l}(x), \mathbf{l}(x_{k-1}) + \mathbf{w}(x_{k-1}, x)\}$. Choose $x_k \in N_G^+(V_{k-1}) \cap \overline{V}_{k-1}$ and $x_j \in V_{k-1}$ $(j \leq k-1)$ such that $(x_j, x_k) \in E(G)$ and $\mathbf{l}(x_k) = \min\limits_{x \in \overline{V}_{k-1}} \{\mathbf{l}(x), \mathbf{l}(x_j) + \mathbf{w}(x_j, x_k)\}$.
Let $V_k = V_{k-1} \cup \{x_k\}$ and $P_k = P_j + (x_j, x_k)$.

3. If $k = v - 1$, stop. If $k < v - 1$, replace k by $k + 1$ and go to Step 2.

We give an example to show an application of Dijsktra's algorithm.

Example 2.5.1 We consider the weighted digraph $(G, \mathbf{w})$ shown in Figure 2.10 (a). The process of the execution of Dijsktra's algorithm is as follows.

1. Start with the labels $\mathbf{l}(x_0) = 0$, $\mathbf{l}(y_i) = \infty$ for $i = 1, 2, \cdots, 5$ and set $V_0 = \{x_0\}$, $P_0 = \{x_0\}$ and $k = 0$ (see Figure 2.10 (b)).

2. This step is executed by five iterations. The details are as follows.

1) $N_G^+(x_0) \cap \overline{V}_0 = N_G^+(V_0) \cap \overline{V}_0 = \{y_1, y_2\}$.
$\mathbf{l}(y_1) = \min\{\infty, \mathbf{l}(x_0) + \mathbf{w}(x_0, y_1)\} = \min\{\infty, 0 + 7\} = 7$;
$\mathbf{l}(y_2) = \min\{\infty, \mathbf{l}(x_0) + \mathbf{w}(x_0, y_2)\} = \min\{\infty, 0 + 1\} = 1$.
Set $x_1 = y_2$, $V_1 = V_0 \cup \{x_1\} = \{x_0, x_1\}$, $P_1 = P_0 + (x_0, x_1)$, and set $k = 1$ (see Figure 2.10 (c)).

2) $N_G^+(x_1) \cap \overline{V}_1 = N_G^+(V_1) \cap \overline{V}_1 = \{y_1, y_4, y_5\}$.
$\mathbf{l}(y_1) = \min\{\mathbf{l}(y_1), \mathbf{l}(x_1) + \mathbf{w}(x_1, y_1)\} = \min\{7, 1 + 5\} = 6$;

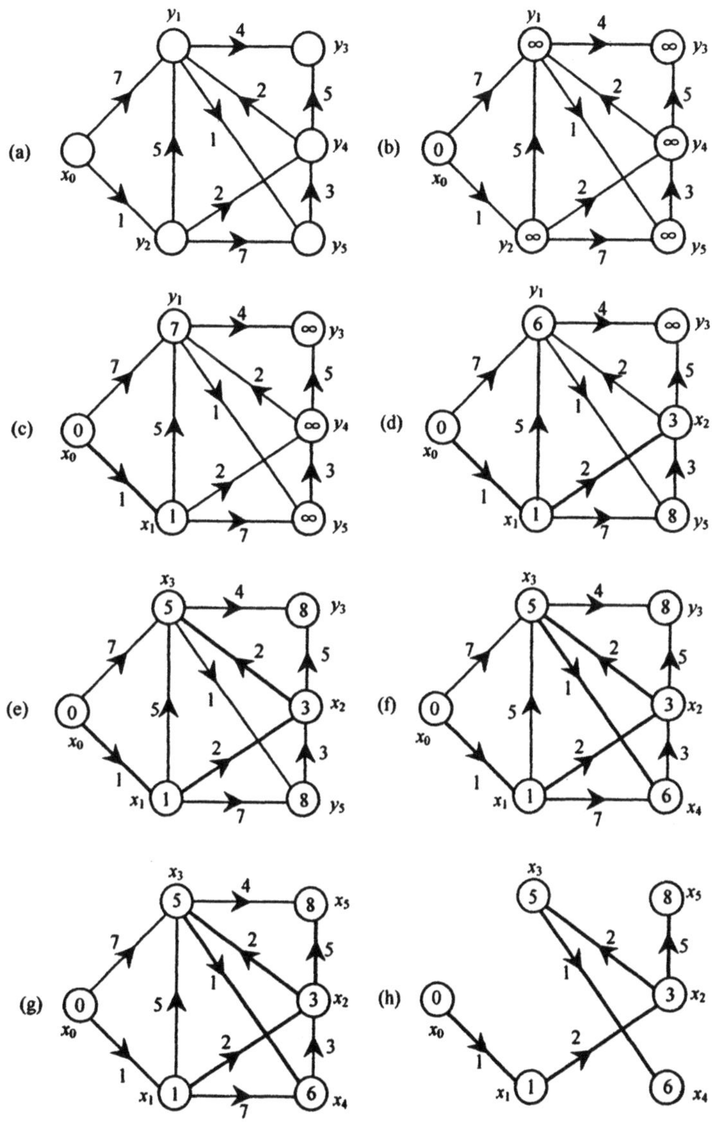

Figure 2.10: An application of Dijsktra's algorithm

$$\mathbf{l}(y_4) = \min\{\mathbf{l}(y_4), \mathbf{l}(x_1) + \mathbf{w}(x_1, y_4)\} = \min\{\infty, 1+2\} = 3;$$
$$\mathbf{l}(y_5) = \min\{\mathbf{l}(y_5), \mathbf{l}(x_1) + \mathbf{w}(x_1, y_5)\} = \min\{\infty, 1+7\} = 8.$$

Set $x_2 = y_2$, $V_2 = V_1 \cup \{x_2\} = \{x_0, x_1, x_2\}$, $P_2 = P_1 + (x_1, x_2)$, and set $k = 2$ (see Figure 2.10 (d)).

3) $N_G^+(x_2) \cap \overline{V}_2 = \{y_1, y_3\}$ and $N_G^+(V_2) \cap \overline{V}_2 = \{y_1, y_3, y_5\}$.
$\mathbf{l}(y_1) = \min\{\mathbf{l}(y_1), \mathbf{l}(x_2) + \mathbf{w}(x_2, y_1)\} = \min\{6, 3+2\} = 5$;
$\mathbf{l}(y_3) = \min\{\mathbf{l}(y_3), \mathbf{l}(x_2) + \mathbf{w}(x_2, y_3)\} = \min\{\infty, 3+5\} = 8$;
$\mathbf{l}(y_5) = 8$.
Set $x_3 = y_1$, $V_3 = V_2 \cup \{x_1\} = \{x_0, x_1, x_2, x_3\}$, $P_3 = P_2 + (x_2, x_3)$, and set $k = 3$ (see Figure 2.10 (e)).

4) $N_G^+(x_3) \cap \overline{V}_3 = \{y_5\}$ and $N_G^+(V_3) \cap \overline{V}_3 = \{y_3, y_5\}$.
$\mathbf{l}(y_5) = \min\{\mathbf{l}(y_5), \mathbf{l}(x_3) + \mathbf{w}(x_3, y_5)\} = \min\{8, 5+1\} = 6$;
$\mathbf{l}(y_3) = 8$.
Set $x_4 = y_5$, $V_4 = V_3 \cup \{x_4\} = \{x_0, x_1, x_2, x_3, x_4\}$, $P_4 = P_3 + (x_3, x_4)$, and set $k = 4$ (see Figure 2.10 (f)).

5) $N_G^+(x_4) \cap \overline{V}_4 = \emptyset$ and $N_G^+(V_4) \cap \overline{V}_4 = \{y_3\}$.
$\mathbf{l}(y_3) = 8$.
Set $x_5 = y_3$, $V_5 = V_4 \cup \{x_5\} = \{x_0, x_1, x_2, x_3, x_4, x_5\}$, $P_5 = P_2 + (x_2, x_5)$, and let $k = 5$ (see Figure 2.10 (g)).

3. Stop as $k = 5 = 6 - 1$.

Figure 2.10(h) illustrates the out-tree T rooted at x_0 obtained when the algorithm terminates, in which the label $\mathbf{l}(x_k)$ gives the minimum weight of the directed path from x_0 to x_k in $(G, \mathbf{w})$ for each $k = 1, 2, \cdots, 5$. ∎

The process of Dijsktra's algorithm describes a growing procedure of an out-tree rooted at x_0 in G, see Figure 2.10. Each iteration of Step 2 obtains a new vertex x_k and a new edge e_k. Therefore, $v - 1$ iterations of Step 2 are required. The digraph T obtained when the algorithm terminates is a connected simple spanning subgraph of G with $v - 1$ edges and $d_T^-(x_0) = 0$ and $d_T^-(x_k) = 1$ for each $k = 1, 2, \cdots, v-1$. Thus, by the exercise 2.1.1, T is a spanning out-tree at rooted x_0. The following theorem ensures that the (x_0, x_k)-path in T is an (x_0, x_k)-path with the minimum weight in $(G, \mathbf{w})$ for each $k = 1, 2, \cdots, v-1$.

Theorem 2.9 Let T be a spanning out-tree at rooted x_0 of $(G, \mathbf{w})$ constructed by Dijsktra's algorithm. Then the (x_0, x_k)-path in T has the minimum weight in $(G, \mathbf{w})$ for each $k = 1, 2, \cdots, v-1$.

Proof In order to prove the theorem, it is sufficient to show that $\mathbf{l}(x_k)$ is the minimum weight over all paths from x_0 to x_k in $(G, \mathbf{w})$, denoted by $d_G(x_0, x_k)$ such a minimum weight.

On the one hand, from the construction of T, the (x_0, x_k)-path P in T has weight $\mathbf{l}(x_k)$. Thus, $d_G(x_0, x_k) \leq \mathbf{w}(P) = \mathbf{l}(x_k)$ and there exist a vertex $x_j \in V_{k-1}$ and an (x_0, x_j)-path P_j of T such that $P = P_j + (x_j, x_k)$. On the other hand, let Q be an (x_0, x_k)-path with minimum weight in $(G, \mathbf{w})$ and let y be the first vertex of Q in $\overline{V}_{k-1}$. Thus y divides Q into two sections $Q(x_0, y)$ and $Q(y, x_k)$ with weights l_1 and l_2, respectively. Since $y, x_k \in \overline{V}_{k-1}$, we have $l_1 \geq \mathbf{l}(y) \geq \mathbf{l}(x_k)$ and $l_2 \geq 0$. Thus, the weight of Q is equal to $d_G(x_0, x_k) = l_1 + l_2 \geq \mathbf{l}(x_k)$. This implies $d_G(x_0, x_k) = \mathbf{l}(x_k)$. ∎

Dijsktra's algorithm is an efficient algorithm with the complexity $o(v^2)$ (the exercise 2.5.1).

Dijsktra's algorithm can find a minimum weight (x_0, x)-path from x_0 to any other vertex x. Thus, when x_0 takes all over the vertices of G, that is, Dijsktra's algorithm is repeatedly executed v times, we will obtain diameter $d(G)$ of a strongly connected digraph G. In other words, Dijsktra's algorithm provides an $o(v^3)$-algorithm for finding weighted diameter of a strongly connected weighted digraph.

Example 2.5.2 Consider the weighted digraph $(G, \mathbf{w})$ shown in Figure 2.11. Dijsktra's algorithm is repeatedly executed 6 times, we obtain 6 spanning out-trees T_i rooted at x_i $(i = 0, 1, \cdots, 5)$, where T_0 is shown in Figure 2.10(h), every label $\mathbf{l}(x_j)$ in T_i indicates weighted distance $d_G(x_i, x_j)$. These weighted distances can be expressed as the following matrix

$$\begin{pmatrix} 0 & 1 & 3 & 5 & 6 & 8 \\ 12 & 0 & 2 & 4 & 7 & 7 \\ 10 & 11 & 0 & 2 & 3 & 5 \\ 8 & 9 & 4 & 0 & 1 & 4 \\ 13 & 14 & 3 & 5 & 0 & 8 \\ 17 & 18 & 13 & 9 & 10 & 0 \end{pmatrix},$$

where the (i, j)-th entry is equal to $d_G(x_i, x_j)$. This matrix is called the *weighted distance matrix* of $(G, \mathbf{w})$. We obtain weighted diameter $d(G)$ of $(G, \mathbf{w})$ by comparing entries in the weighted distance matrix, that is,

$$d(G) = \max\{d_G(x_i, x_j) : 0 \leq i, j \leq 5\} = 18.$$

It should be noted that Dijsktra's algorithm could not be used to find a minimum weight spanning out-tree rooted at x_i. For example, the spanning out-tree T_4 shown in Figure 2.11 has weight 19, however, it is easy to find a spanning out-tree rooted at x_4 with weight 18.

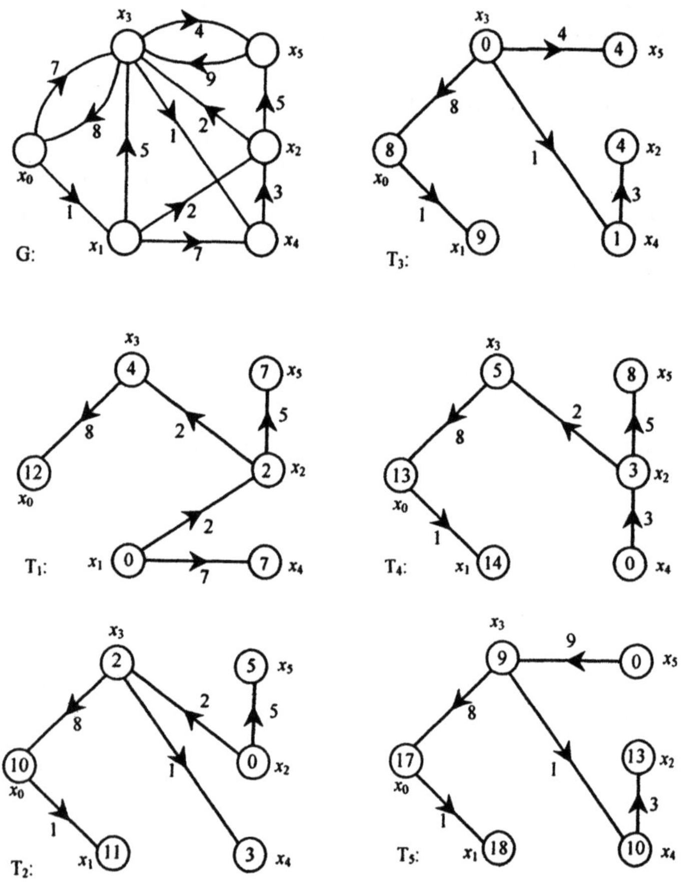

Figure 2.11: Optimal spanning trees rooted at x_i for $i = 1, 2, \cdots, 5$

We conclude this section with a very interesting game, solving by Dijsktra's algorithm.

Example 2.5.3 Suppose that there are a full twelve-gallon jug of wine, and two empty jugs of nine and five gallons capacity, respectively. What is the simplest way to divide the wine equally?

Let A, B, C denote the three jugs of twelve, nine and five gallons capacity, respectively. We divide the wine equally by using two empty jugs. A possible way, to need to pour at least eight times, for dividing the wine in the jug A equally is as follows.

A(12 gallons)	7	7	2	2	11	11	6	6
B(9 gallons)	0	5	5	9	0	1	1	6
C(5 gallons)	5	0	5	1	1	0	5	0

Is eight pours necessary? We present a way to answer and solve this problem by Dijsktra's algorithm.

Denote by an ordered pair (b, c) of integers the volume of the wine in the jugs B and C. For example, $(0, 0)$ denotes the jugs B and C both are empty; $(6, 0)$ six gallon of the wine in the jug B and the jug C is empty. It is clear that $b + c \leq 12$. All possible values of (b, c) are as follows.

$$(0,0),(1,0),(2,0),(3,0),(4,0),(5,0),(6,0),(7,0),(8,0),$$
$$(9,0),(0,1),(0,2),(0,3),(0,4),(0,5),(1,5),(2,5),(3,5),$$
$$(4,5),(5,5),(6,5),(7,5),(8,4),(9,1),(9,2),(9,3).$$

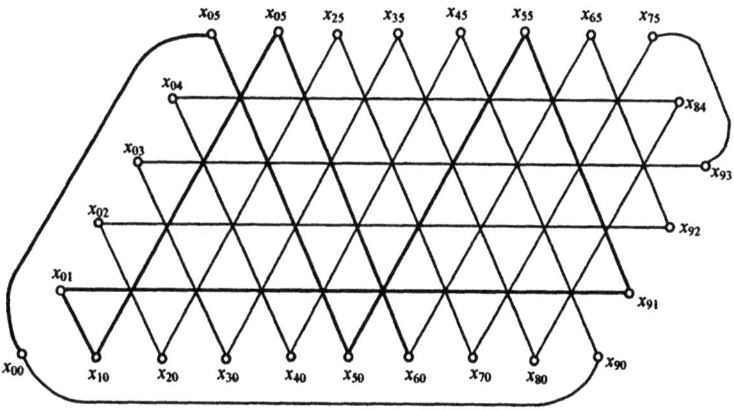

Figure 2.12: A graph G used in Example 2.5.3

Construct a simple undirected graph G as follows. A vertex x_{bc} of G denotes the ordered pair (b, c); two vertices x_{bc} and $x_{b'c'}$ are linked by an undirected edge in G if and only if the situations (b, c) and (b', c') may be obtained from one to the other by pouring only once. For instance, two vertices x_{35} and x_{80} are adjacent for two situations $(3,5)$ and $(8,0)$, the latter may be attained by pouring all of five gallon wine in the jug C into the jug B; the former may be attained by pouring three gallon wine into the jug C from the jug B. The resulting graph G is shown in Figure 2.12.

It follows from the above discussion that our problem is reduced to find a shortest path from the vertex x_{00} to the vertex $_{60}$ in G. Such a path P can be easily found by Dijsktra's algorithm:

$$P = (x_{00}, x_{05}, x_{50}, x_{55}, x_{91}, x_{01}, x_{10}, x_{15}, x_{60}),$$

which consists of the heavy edges in Figure 2.12, and indicates a simplest way in which the wine are poured only eight times to divide the wine equally.

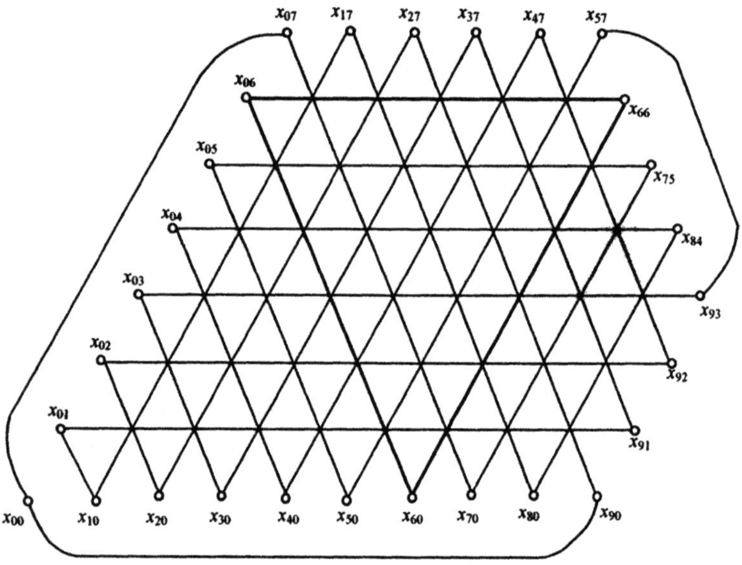

Figure 2.13: Another graph used in Example 2.5.3

In the above problem, if the capacity of the empty jug C is seven gallon, then it is impossible to divide the wine equally since the corresponding graph shown in Figure 2.13 is not connected, the vertex x_{00} and the vertices x_{06}, x_{60}, x_{66} are in different components of the graph. In fact, the component that contains the vertices x_{06}, x_{60}, x_{66} is a triangle, which is separated from x_{00} in the graph. ∎

Exercises

2.5.1 Prove that the complexity of Dijsktra's algorithm is $o(v^2)$.

2.5.2 A company has branches in each of six cities $x_1, \cdots, x_6$. The fare for a direct fight from x_i to x_j is given by the (i, j)th entry in the following matrix:

$$\begin{pmatrix} 0 & 50 & \infty & 40 & 25 & 10 \\ 50 & 0 & 15 & 20 & \infty & 25 \\ \infty & 15 & 0 & 10 & 20 & \infty \\ 40 & 20 & 10 & 0 & 10 & 25 \\ 25 & \infty & 20 & 10 & 0 & 55 \\ 10 & 25 & \infty & 25 & 55 & 0 \end{pmatrix}$$

Design a scheme of cheapest routes between pair of cities.

2.5.3 Find shortest directed paths from x_0 to all other vertices in the following weighted digraph.

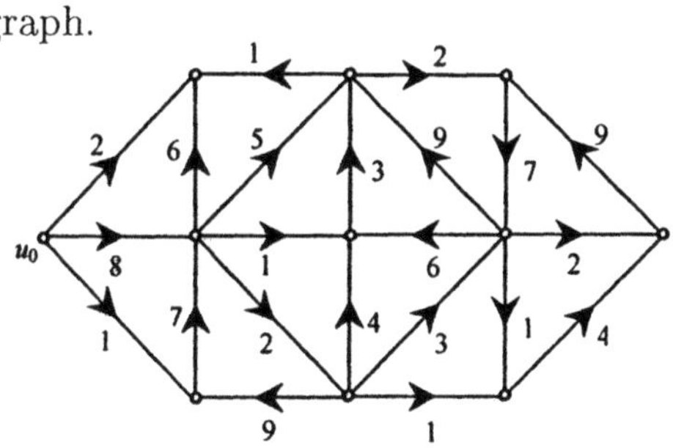

(the exercise 2.5.3)

2.5.4 Suppose that there are a full twelve-gallon jug of eight, and two empty jugs of five and three gallons capacity, respectively. Question:

(a) whether or not it is possible to divide the wine equally. If so, what is the simplest way?

(b) whether or not it is possible to divide the wine equally if replace the jug of three gallons by seven.

2.5.5 A wolf, a goat and a cabbage are on one bank of a river. A ferryman wants to take them across, but, since his boat is small, he can take only one of them at a time. For obvious reason, neither the wolf and the goat nor the goat and the cabbage can be left unguarded. How is the ferryman going to get them across the river?

2.6 The Electrical Network Equations

In 1847, Kirchhoff [106] formulated two important rules which govern the flow of electricity in a network of wires; these rules are now known as Kirchhoff's laws.

Kirchhoff's current law (KCL): The total outflow and inflow of current at any point sum up to 0.

Kirchhoff's voltage law (KVL): The potential differences round any cycle sum up to 0.

Kirchhoff's laws lead to two systems of simultaneous linear equations. These equations are not all independent, and so there arises the question of how many of them are necessary in order to obtain the solution. This question was asked and answered by Kirchhoff himself [106], who abstracted an electrical network as a digraph. It is his great pioneering work that was regarded as an important milestone in history of graph theory.

An electrical network may, in natural way, be thought of as a weighted digraph $(G, \mathbf{w}, \mathbf{u})$, in which each edge e_i with end-vertices x and y is oriented from x to y if an electrical current of $\mathbf{w}(e_i)$ will go through the edge e_i from x to y; and $\mathbf{u}(e_i)$ denotes the potential difference (or voltage) from x to y. It is clear that G is connected. An example is shown in Figure 2.14.

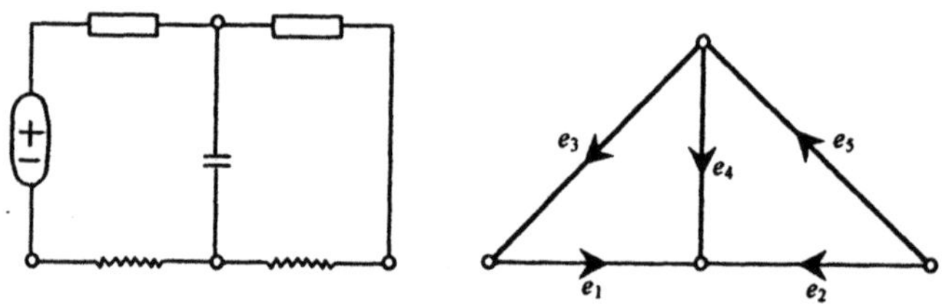

Figure 2.14: An electrical network and its associated digraph

Note that a vector of the electrical currents $\mathbf{w}$ and a vector of the potential differences $\mathbf{u}$ in the electrical network can be thought of as a cycle-vector and a bond-vector in the edge-space of the associated digraph G, respectively. Thus, in the language of graph theory, the KCL and the KVL can be expressed as the following forms:

$$\mathbf{M}\,\mathbf{w} = \mathbf{O}, \tag{2.15}$$

and

$$\mathbf{C}\,\mathbf{u} = \mathbf{O}, \tag{2.16}$$

respectively, where $\mathbf{M}$ is the incidence matrix of G, $\mathbf{w} \in \mathcal{C}(G)$ is a column vector of the currents, $\mathbf{C}$ is a matrix whose rows consist of all vectors in $\mathcal{C}(G)$ and $\mathbf{u} \in \mathcal{B}(G)$ is a column vector of the potential differences.

Choose a spanning tree T of G, and label all edges of G such that the basis matrices of $\mathcal{C}(G)$ and $\mathcal{B}(G)$ corresponding to the spanning tree T have the form

$$\mathbf{C}_T = (\mathbf{I}_{\varepsilon-\upsilon+1} \quad \mathbf{C}_2) \qquad \mathbf{B}_T = (\mathbf{B}_1 \quad \mathbf{I}_{\upsilon-1}),$$

respectively. Let

$$\mathbf{w} = \begin{pmatrix} \mathbf{w}_c \\ \mathbf{w}_t \end{pmatrix}, \qquad \mathbf{u} = \begin{pmatrix} \mathbf{u}_c \\ \mathbf{u}_t \end{pmatrix},$$

where

$$\mathbf{w}_c = \mathbf{w}|\overline{T}, \quad \mathbf{w}_t = \mathbf{w}|T,$$
$$\mathbf{u}_c = \mathbf{u}|\overline{T}, \quad \mathbf{u}_t = \mathbf{u}|T.$$

Since rank $(\mathbf{M})= \upsilon - 1$, only $\upsilon - 1$ equations are independent in (2.15) and have the same solution as the following equations

$$\mathbf{B}_t\mathbf{w} = \mathbf{O}.$$

Thus, we have

$$\mathbf{w}_t = -\mathbf{B}_1\mathbf{w}_c,$$

which shows that the current in each edge in T may be expressed as a linear combination of the currents in the edges in $\overline{T}$.

On the other hand, suppose that $\mathbf{w}$ is a column vector that satisfies (2.15), then its transposition $\mathbf{w}^T$ is a cycle-vector in $\mathcal{E}(G)$, that is $\mathbf{w}^T \in \mathcal{C}(G)$. This shows that $\mathbf{w}^T$ may be expressed as a linear combination of row vectors in $\mathbf{C}_T$ and so, from the exercise 2.3.6,

$$\mathbf{w}^T = (\mathbf{w}_c)^T \mathbf{C}_T,$$

that is,

$$\mathbf{w} = (\mathbf{C}_T)^T \mathbf{w}_c,$$

which shows that the current in each edge in G may be expressed as a linear combination of the currents in the edges in $\overline{T}$.

Similarly, the equations (2.16) have the same solution as the following $\varepsilon - \upsilon + 1$ independent equations

$$\mathbf{C}_T\mathbf{u} = \mathbf{O},$$

from which we have

$$\mathbf{u}_c = -\mathbf{C}_2\mathbf{u}_t.$$

This shows that the potential difference in each edge in $\overline{T}$ may be expressed as a linear combination of the potential differences in the edges in T.

On the other hand, suppose that $\mathbf{u}$ is a column vector that satisfies (2.16), then its transposition $\mathbf{u}^T$ is orthogonal to any vector in $\mathcal{C}(G)$, that is, $\mathbf{u}^T \in \mathcal{B}(G)$. This shows that $\mathbf{u}^T$ may be expressed as a linear combination of row vectors in $\mathbf{B}_T$ and so, from the exercise 2.3.6,

$$\mathbf{u}^T = (\mathbf{u}_t)^T \mathbf{B}_T,$$

that is,

$$\mathbf{u} = (\mathbf{B}_T)^T \mathbf{u}_t,$$

which shows that the potential difference in each edge in G may be expressed as a linear combination of the potential differences in the edges in T.

Example 2.6.1 Consider the electrical network and its associated digraph G shown in Figure 2.14. Choose a spanning tree T, for example, induced by the subset of the edges $\{e_3, e_4, e_5\}$. The basis matrices of $\mathcal{C}(G)$ and $\mathcal{B}(G)$ corresponding to the spanning tree T are, respectively,

$$\mathbf{C}_T = \begin{pmatrix} 1 & 0 & 1 & -1 & 0 \\ 0 & 1 & 0 & -1 & -1 \end{pmatrix}$$

and

$$\mathbf{B}_T = \begin{pmatrix} -1 & 0 & 1 & 0 & 0 \\ 1 & 1 & 0 & 1 & 0 \\ 0 & 1 & 0 & 0 & 1 \end{pmatrix}.$$

Suppose that

$$\mathbf{w} = \begin{pmatrix} w_1 \\ w_2 \\ w_3 \\ w_4 \\ w_5 \end{pmatrix} \quad \text{and} \quad \mathbf{u} = \begin{pmatrix} u_1 \\ u_2 \\ u_3 \\ u_4 \\ u_5 \end{pmatrix}$$

are the vector of currents and the vector of voltages, respectively. Then the independent equations corresponding to KCL and KVL are, respectively,

$$\mathbf{O} = \mathbf{B}_T\,\mathbf{w} = \begin{pmatrix} -1 & 0 & 1 & 0 & 0 \\ 1 & 1 & 0 & 1 & 0 \\ 0 & 1 & 0 & 0 & 1 \end{pmatrix} \begin{pmatrix} w_1 \\ w_2 \\ w_3 \\ w_4 \\ w_5 \end{pmatrix}$$

$$= \begin{pmatrix} -w_1 + w_3 \\ w_1 + w_2 + w_4 \\ w_2 + w_5 \end{pmatrix}$$

and

$$\mathbf{O} = \mathbf{C}_T\,\mathbf{u} = \begin{pmatrix} 1 & 0 & 1 & -1 & 0 \\ 0 & 1 & 0 & -1 & -1 \end{pmatrix} \begin{pmatrix} u_1 \\ u_2 \\ u_3 \\ u_4 \\ u_5 \end{pmatrix}$$

$$= \begin{pmatrix} u_1 + u_3 - u_4 \\ u_2 - u_4 - u_5 \end{pmatrix}.$$

Thus, we have

$$\mathbf{w} = (\mathbf{C}_T)^T \mathbf{w}_c = \begin{pmatrix} 1 & 0 \\ 0 & 1 \\ 1 & 0 \\ -1 & -1 \\ 0 & -1 \end{pmatrix} \begin{pmatrix} w_1 \\ w_2 \end{pmatrix} = \begin{pmatrix} w_1 \\ w_2 \\ w_1 \\ -w_1 - w_2 \\ -w_2 \end{pmatrix}$$

and

$$\mathbf{u} = (\mathbf{B}_T)^T \mathbf{u}_t = \begin{pmatrix} -1 & 1 & 0 \\ 0 & 1 & 1 \\ 1 & 0 & 0 \\ 0 & 1 & 0 \\ 0 & 0 & 1 \end{pmatrix} \begin{pmatrix} u_3 \\ u_4 \\ u_5 \end{pmatrix} = \begin{pmatrix} -u_3 + u_4 \\ u_4 + u_5 \\ u_3 \\ u_4 \\ u_5 \end{pmatrix}.$$

If let $w_1 = 2, w_2 = 2$ and $u_3 = 1, u_4 = 2, u_5 = 1$, then we have $\mathbf{w} = (1, 2, 1, -3, -2)^T$ and $\mathbf{u} = (1, 3, 1, 2, 1)^T$.

Example 2.6.2 We now choose another spanning tree H induced by the subset of the edges $\{e_1, e_3, e_5\}$. The basis matrices of $\mathcal{C}(G)$ and $\mathcal{B}(G)$ corresponding to the spanning tree H are, respectively,

$$\mathbf{C}_H = \begin{pmatrix} -1 & 1 & -1 & 0 & -1 \\ -1 & 0 & -1 & 1 & 0 \end{pmatrix}, \qquad \mathbf{B}_T = \begin{pmatrix} 1 & 1 & 0 & 1 & 0 \\ 0 & 1 & 1 & 1 & 0 \\ 0 & 1 & 0 & 0 & 1 \end{pmatrix}.$$

In this case, $\mathbf{w}_c = (w_2, w_4)^T$ and $\mathbf{u}_h = (u_1, u_3, u_5)^T$. Thus, we have

$$\mathbf{w} = (\mathbf{C}_H)^T \mathbf{w}_c = \begin{pmatrix} -w_2 - w_4 \\ w_2 \\ -w_2 - w_4 \\ w_4 \\ -w_2 \end{pmatrix}, \quad \mathbf{u} = (\mathbf{B}_H)^T \mathbf{u}_h = \begin{pmatrix} u_1 \\ u_1 + u_3 + u_5 \\ u_3 \\ u_1 + u_3 \\ u_5 \end{pmatrix}.$$

If let $w_2 = 2, w_4 = -3$ and $u_1 = u_3 = u_5 = 1$, then we have $\mathbf{w} = (1, 2, 1, -3, -2)^T$ and $\mathbf{u} = (1, 3, 1, 2, 1)^T$, which is identical with the result in Example 2.6.1.

Graph theory has become a fundamental and powerful mathematical tool in analyzing theory of electrical networks. The interested reader is referred to Chan [28].

Exercises

2.6.1 Let $T = \{e_1, e_2, e_4\}$ be a spanning tree in the digraph shown in Figure 2.14. Write out

(a) a system of independent equations by the KCL;

(b) a system of independent equations by the KVL;

(c) the expression of $\mathbf{w}$ in $\mathbf{w}_c$;

(d) the expression of $\mathbf{u}$ in $\mathbf{u}_t$.

2.6.2 Prove that Tellegen's theorem: if two electrical networks N and $\overline{N}$ have the same associated digraph G, then

(a) $\mathbf{u}^{\mathbf{T}}\overline{\mathbf{w}} = \mathbf{O}$ and

(b) $\mathbf{w}^{\mathbf{T}}\overline{\mathbf{u}} = \mathbf{O}$,

where $\mathbf{u}$, $\overline{\mathbf{u}}$ and $\mathbf{w}$, $\overline{\mathbf{w}}$ are the column vectors of the voltages and the currents in N and $\overline{N}$, respectively.

Chapter 3

Plane Graphs and Planar Graphs

As we have seen, a graph can be represented graphically, that is, a graph can be drawn in the plane, and it is this kind of graphical presentation that helps us intuitively understand many of structural properties of graphs. In many real-world problems, for example, layout of printed circuits, one wish to draw a graph in the plane such that its edges intersect only at their end-vertices, such a graph is called to be planar. A natural and important question is how to decide whether a given graph is planar or not. We will, in this chapter, discuss and answer this problem.

First, we will introduce the well-known Euler's formula on a connected plane graph, from which we deduce some of the basic properties of planar graphs. Then, present a remarkably simple characterization of a planar graph, that is, the well-known Kuratowski's theorem. With theory of graphic spaces, we introduce the concept of combinatorial dual of a graph, and making use of Kuratowski's theorem, describe another characterization of a planar graph, that is, Whitney's theorem.

Lastly, as applications of planar graphs, we will show that there are only five regular polyhedra, which were known by the ancient Greeks over two thousand years ago; and describe an efficient algorithm for testing the planarity of a given graph and, thus, for solving the problem of layout of printed circuits.

3.1 Plane Graphs and Euler's Formula

Let S be a given surface such as the plane, the sphere, the torus and so on. If a graph G can be drawn in S such that its edges intersect only at their end-vertices, then G is said to be *embeddable on the surface* S. Such a drawing of G in S is called an *embedding* of G in S, denoted by $\widetilde{G}$. Since the concept of embedding of a graph has no relation to orientations of edges, in the following discussions, we will restrict ourselves to undirected graphs. Furthermore, we consider the surface S as the plane or the sphere in this chapter. In fact, we have the following result.

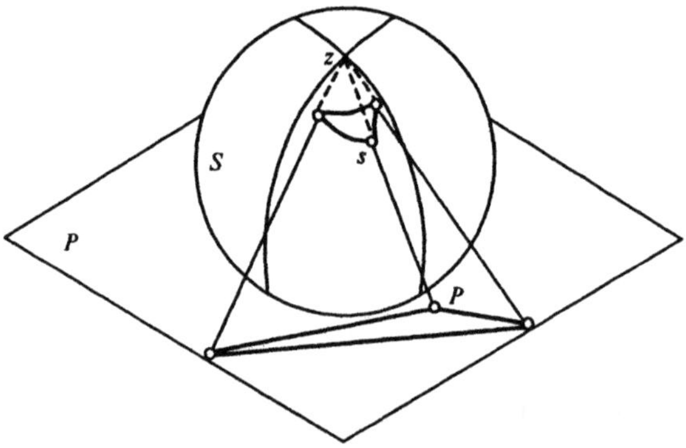

Figure 3.1: Stereographic projection

Theorem 3.1 A graph G is embeddable on the sphere S if and only if it is embeddable on the plane P.

Proof To show this theorem we make use of a mapping known as stereographic projection. Consider a sphere S resting on a plane P, and denoted by z the point of S that is diagonally opposite the point of contact of S and P. The mapping $\phi: S \to P$, defined by $\phi(z) = \infty$ and $\phi(s) = p \in P \setminus \{\infty\}$ for any $s \in S \setminus \{z\}$ if and only if the points z, s and p are collinear; see Figure 3.1. The mapping ϕ is bijective clearly.

Suppose that G is embeddable on the sphere S and $\widetilde{G}$ is its embedding in P. Then $\phi^{-1}(\widetilde{G})$ is an embedding of G in the sphere S. Conversely, suppose that $\widetilde{G'}$ is an embedding of G in S. Without loss of generality, suppose that z is not in $\widetilde{G'}$. Then $\phi(\widetilde{G'})$ is an embedding of G in P. Thus, G is embeddable in S if and only if it is embeddable on P. ∎

If a graph G is embeddable on the plane (or the sphere), G is called a *planar graph*; otherwise G is called a *non-planar graph.* If G is a planar graph, then any embedding $\widetilde{G}$ of G on the plane can itself be regarded as a graph isomorphic to G. Therefore, we refer to an embedding $\widetilde{G}$ of G as a *plane graph.*

Figure 3.2 shows a planar graph $K_{3,3}^-$, obtained from $K_{3,3}$ by deleting any one edge, and its embedding on the boundary of a tetrahedron. Such an embedding of $K_{3,3}^-$ will be useful in the proof of Theorem 3.6 in the next section.

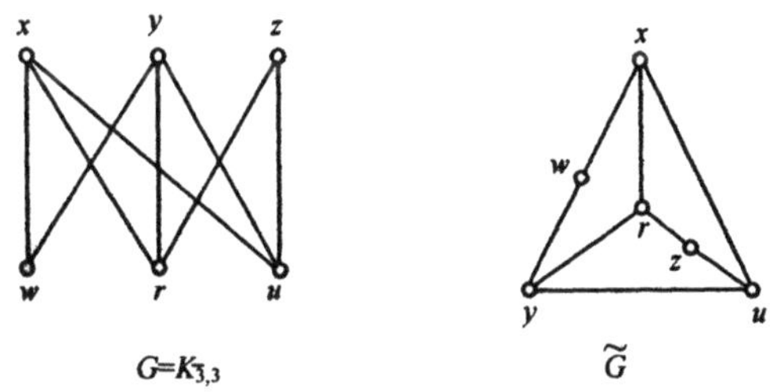

Figure 3.2: $K_{3,3}^-$ and its planar embedding

Let G be a nonempty plane graph. It can partition the plane into several connected regions, which are called *faces.* We use $F(G)$ and $\phi(G)$ to denote the set and the number of faces of G, respectively. For G shown in Figure 3.3, for example, we have

$$F(G) = \{f_0, f_1, f_2, f_3, f_4, f_5\} \quad \text{and} \quad \phi(G) = |F(G)| = 6.$$

It is clear that $\phi(G) \geq 1$ for any plane graph G, and $\phi(G) = 1$ if and only if G is a forest.

Denote by $B_G(f)$ the *boundary* of the face $f \in F(G)$, in general, which consists of several edge-disjoint closed walks. For example, the face f_0 of G shown in Figure 3.3 has boundary

$$B_G(f_0) = x_1\, e_2\, x_2\, e_4\, x_6\, e_7\, x_3\, e_8\, x_5\, e_8\, x_5\, e_8\, x_3\, e_1\, x_1.$$

The number of edges in $B_G(f)$ is the *degree* of f, denoted by $d_G(f)$. For G shown in Figure 3.3, for example, we have $d_G(f_0) = 7$, $d_G(f_1) = 1$.

Any planar embedding of a planar graph has exactly one unbounded face, called the *exterior face*; in the plane graph of Figure 3.3, f_0 is the exterior face. For any vertex x or any edge e of a planar graph G, it is easy to prove that G can be embedded in the plane in such a way that x or e is on the boundary of the exterior face of the embedding (the exercise 3.1.2).

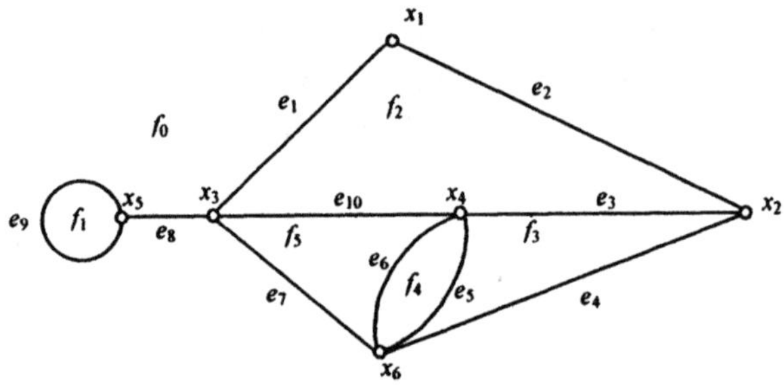

Figure 3.3: The faces of a plane graph

For a given plane graph G, there is the following relations between the face degrees and $\varepsilon(G)$, similar to one between the vertex degrees and $\varepsilon(G)$ (see Corollary 1.1.1).

Theorem 3.2 For any plane graph G,

$$\sum_{f \in F(G)} d_G(f) = 2\varepsilon(G).$$

Proof If G is empty, then the conclusion holds clearly. Suppose now that G is nonempty, and e is any edge of G. Then e either is on a common boundary of two distinct faces (for example, the edge e_1 of the graph shown in Figure 3.3 is on the boundary of f_0 and f_2) or appears in a boundary of some face twice (for example, the edge e_8 of the graph shown in Figure 3.3 appears on the boundary of f_0 twice). Thus the conclusion follows. ∎

There is a simple formula relating to the numbers of vertices v, the number of edges ε and the number of faces ϕ of a connected plane graph. It is the well-known *Euler's formula.*

Theorem 3.3 (Euler [58]) If G is a connected plane graph, then

$$v - \varepsilon + \phi = 2.$$

Proof Let G be a connected plane graph and T be a spanning tree of G. Then $\phi(T) = 1$ and $\varepsilon(\overline{T}) = \varepsilon - v + 1$. On the one hand, addition of each edge of $\overline{T}$ to T, the number of faces increases by at least one by Theorem 2.3, which implies $\phi(G) \geq \phi(T) + \varepsilon - v + 1$. On the other hand, to obtain a new face, one edge of $\overline{T}$ must be added to T, which implies $\phi(G) \leq \phi(T)+\varepsilon-v+1$. Thus

$$\phi(G) = \phi(T) + \varepsilon - v + 1 = \varepsilon - v + 2,$$

as required. ∎

Corollary 3.3.1 If G is a plane graph, then $v - \varepsilon + \phi = 1 + \omega$. ∎

Corollary 3.3.2 All planar embeddings of a given connected planar graph have the same number of faces. ∎

Corollary 3.3.3 If G is a simple connected planar bipartite graph of order v (≥ 3), then $\varepsilon \leq 2v - 4$.

Proof Let $\widetilde{G}$ be a planar embedding of G. If $\widetilde{G}$ is a tree, then by Theorem 2.3, $\varepsilon = v - 1 \leq 2v - 4$ for $v \geq 3$. Suppose that $\widetilde{G}$ contains a cycle below. Since G is a simple bipartite graph, then by Corollary 1.6.2, G contains no odd cycle and so $d_{\widetilde{G}}(f) \geq 4$ for each face f of $\widetilde{G}$. It follows from Theorem 3.2 that

$$4\phi \leq \sum_{f \in F(\widetilde{G})} d_{\widetilde{G}}(f) = 2\varepsilon,$$

that is, $\varepsilon \geq 2\phi$. It follows from Euler's formula that $\varepsilon \leq 2v - 4$. ∎

Corollary 3.3.4 $K_{3,3}$ is non-planar.

Proof Since $K_{3,3}$ is simple and bipartite, $\varepsilon(K_{3,3}) = 9$ and $v(K_{3,3}) = 6$. Suppose to the contrary that $K_{3,3}$ is planar, then we can deduce from Corollary 3.3.3 a contradiction as follows.

$$9 = \varepsilon(K_{3,3}) \leq 2\,v(K_{3,3}) - 4 = 8.$$

Therefore, $K_{3,3}$ is non-planar. ∎

A simple planar graph G is called to be *maximal* if $G + xy$ is non-planar for any two nonadjacent vertices x and y of G. It is clear that each face of any planar embedding of a maximal planar graph is a triangle. A planar embedding of a maximal planar graph is called a *plane triangulation*.

Theorem 3.4 Let G be a simple planar graph of order $v \geq 3$. Then G is maximal if and only if $\varepsilon = 3v - 6$.

Proof Let G be a simple planar graph of order $v \geq 3$ and $\widetilde{G}$ be a planar embedding of G. Then it is clear that $\widetilde{G}$ is maximal if and only if $d_{\widetilde{G}}(f) = 3$

for any $f \in F(\widetilde{G})$. It follows from Theorem 3.2 that

$$2\varepsilon = \sum_{f \in F(\widetilde{G})} d_{\widetilde{G}}(f) = 3\,\phi.$$

By Euler's formula, we have

$$v - \varepsilon + \tfrac{2}{3}\,\varepsilon = 2,$$

as desired. ∎

Corollary 3.4.1 If G is a simple planar graph of order $v \geq 3$, then $\varepsilon \leq 3v - 6$. ∎

Corollary 3.4.2 K_5 is non-planar.

Proof If K_5 is planar, then, by Corollary 3.4.1, we should have

$$10 = \varepsilon(K_5) \leq 3\,v(K_5) - 6 = 9.$$

But this is impossible. Thus, K_5 is non-planar. ∎

Corollary 3.4.3 If G is a simple planar graph, then $\delta \leq 5$.

Proof The conclusion is clearly true for $v = 1$ or 2. For $v \geq 3$, by Corollary 1.1 and Corollary 3.4.1, we have

$$\delta\, v \leq \sum_{x \in V} d_G(x) = 2\varepsilon \leq 6v - 12,$$

which implies that $\delta \leq 5$. ∎

The following feature of planar graphs is found by Wagner [169] and, rediscovered by Fáry [60].

Theorem 3.5 Any simple planar graph can be embedded in the plane so that each edge is a straight line segment.

Proof It is sufficient to prove the conclusion for maximal planar graphs. By induction on v. The theorem holds clearly for $v \leq 3$. Suppose that it is true for all maximal planar graphs with fewer than v vertices, and let G be a maximal planar graph of order $v \geq 4$. We can, without loss of generality, suppose that G is a plane triangulation. Let x be a vertex of G that it is not on the boundary of the exterior face of G, and $y \in N_G(x)$. Then the edge xy is the common boundary of two triangles, say, (z_1, x, y, z_1) and (z_2, x, y, z_2)

(see Figure 3.4 (b)). Consider the graph $G \cdot xy$, and let G' be the resulting graph by deleting parallel edges of $G \cdot xy$ (see Figure 3.4 (c) in which p is a new vertex obtained by identifying x and y of G). It is clear that G' is planar and, by Theorem 3.4,

$$\varepsilon(G') = \varepsilon(G) - 3 = 3(v(G) - 1) - 6 = 3v(G') - 6.$$

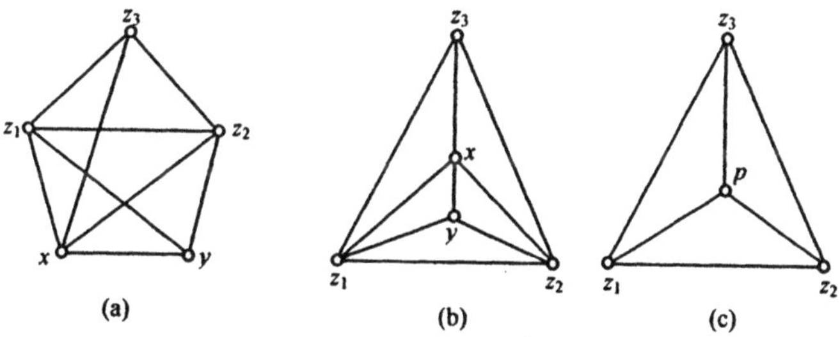

Figure 3.4: An illustration of the proof of Theorem 3.5

Also by Theorem 3.4, G' is a maximal planar graph of order $(v - 1)$. By the induction hypothesis, G' has a planar embedding $\widetilde{G'}$ in which each edge is a straight line segment.

We can obtain a planar embedding $\widetilde{G}$ in which each edge is a straight line segment from $\widetilde{G'}$ by returning the vertex p of $\widetilde{G'}$ to two vertices x and y of G, adding the edge xy, and replacing two edges pz_1, pz_2 by four edges xz_1, yz_1, xz_2, yz_2. The theorem follows by the principle of induction. ∎

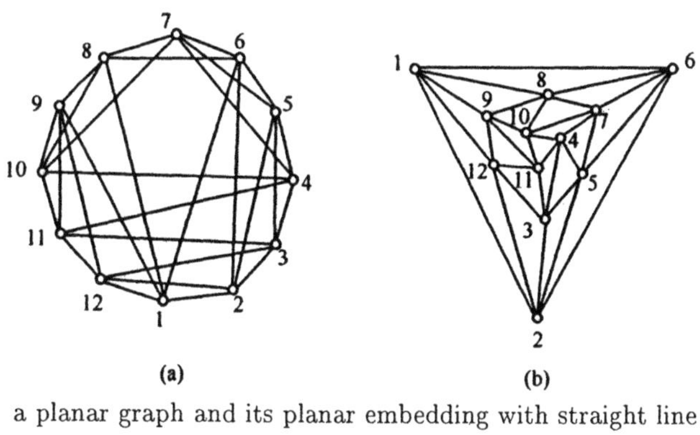

Figure 3.5: a planar graph and its planar embedding with straight line segments

Figure 3.5 shows a planar graph and its planar embedding with straight line segments.

Exercises

3.1.1 Prove that

(a) a graph G is planar if and only if each of its components is planar;

(b) if G is a plane graph, then $\upsilon - \varepsilon + \phi = \omega + 1$.

3.1.2 Let G be a graph. Prove that

(a) if G is planar, $x \in V(G)$ and $e \in E(G)$ then G can be embedded in the plane in such a way that x (resp. e) is on the boundary of the exterior face of the embedding;

(b) G is planar if and only if each block of G is planar.

3.1.3 Prove that each simple planar graph of order $\upsilon \geq 3$ is a spanning subgraph of some plane triangulation.

3.1.4 Let G be a graph of order $\upsilon \geq 4$, and υ_i be the number of i-degree vertices of G. Prove that

(a) if G is a plane triangulation, then

$$3\upsilon_3 + 2\upsilon_4 + \upsilon_5 = \upsilon_7 + 2\upsilon_8 + \cdots + (\Delta - 6)\upsilon_\Delta + 12;$$

(b) if $\delta(G) = 5$, then G contains at least 12 vertices of degree 5;

(c) if G is a tree, then

$$\upsilon_1 = \upsilon_3 + 2\upsilon_4 + 3\upsilon_5 + \cdots + (\Delta - 2)\upsilon_\Delta + 2.$$

3.1.5 Prove that, if G is a connected plane graph and each of its faces has degree four, then $\varepsilon = 2\upsilon - 4$.

3.1.6 Let G be a connected 3-regular plane graph, and ϕ_i be the number of the faces of i-degree of G. Prove that

(a) $12 = 5\phi_1 + 4\phi_2 + 3\phi_3 + 2\phi_4 + \phi_5 - \phi_7 - 2\phi_8 - \cdots$;

(b) G contains a face of degree fewer than 6.

3.1.7 Prove that

(a) if G is a connected planar graph with girth $g \geq 3$, then

$$\varepsilon \leq g(\upsilon - 2)/(g - 2);$$

(b) Petersen graph is non-planar.

3.1.8 (a) Prove that, if G is a simple planar graph of order $v \geq 11$, then G^c is non-planar. (W. T. Tutte (1973) showed that the assertion is true for $v \geq 9$.)

(b) Construct a simple planar graph G of order 8 such that G^c is also planar.

3.1.9 Let G be a simple planar graph. Prove that

(a) G contains at least 4 vertices of degree fewer than 6 if $v \geq 4$;

(b) there exists exactly one 4-regular plane triangulation.

3.1.10 Let $S = \{x_1, x_2, \cdots, x_n\}$ be a set of $n\,(\geq 3)$ points in the plane such that the distance between any two points is at least one. Prove that there are at most $3n-6$ pairs of points at distance exactly one.

3.1.11 Prove that if G is a simple planar graph with $v \geq 5$ and $\Delta = v-1$ then G contains two nonadjacent vertices of degree at most 3.

(N.Alon and Y.Caro, 1984)

3.1.12 Find a planar embedding of the following graphs in which each edge is a straight line segment.

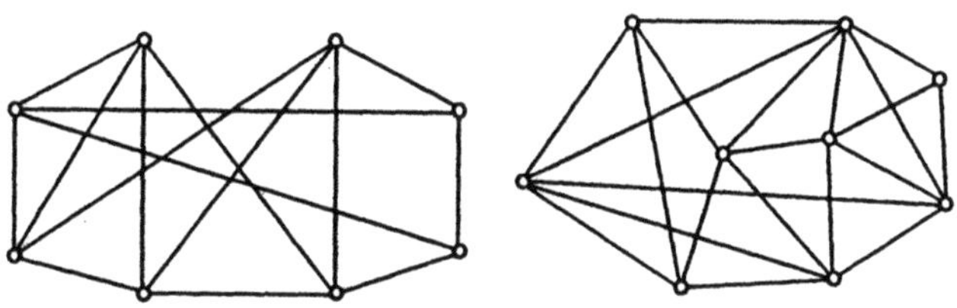

(the exercise 3.1.12)

3.1.13 A graph G is called a *minimal non-planar graph* if G is non-planar and each subgraph of G is planar. Prove that

(a) both K_5 and $K_{3,3}$ are minimal non-planar;

(b) each minimal non-planar graph is a block.

3.1.14 Let G be a simple planar graph of order v. Prove that

(a) $\sum\limits_{x \in V} d_G(x)^2 \leq 2(v+3)^2 - 62$ if $v \geq 4$;

(b) $\sum_{x \in V} d_G(x)^2 < 2(v+3)^2 - 62$ if $\delta \geq 4$.

3.1.15 If a graph G can be drawn in the 3-dimensional space $\mathcal{R}^3$, then G is said to be *embeddable in* $\mathcal{R}^3$ so that its edges intersect only at their end-vertices. Prove that

(a) all graphs are embeddable in $\mathcal{R}^3$;

(b) all simple graphs are embeddable in $\mathcal{R}^3$ so that each edge is a straight line segment.

3.1.16 The *thickness* of G, $\vartheta(G)$, is the minimum number of planar graphs into which the edges of G can be partitioned. It is clear that $\vartheta(G) = 0$ if and only if G is planar. Prove that

(a) if G is a simple graph of order $v \geq 3$ then $\vartheta(G) \geq \left\lceil \dfrac{\varepsilon}{3v-6} \right\rceil$, and the equality holds for all K_v's with $3 \leq v \leq 8$;

(b) $\vartheta(K_9) = \vartheta(K_{10}) = 3$;

(c) $\vartheta(K_v) \geq \left\lfloor \dfrac{v+7}{6} \right\rfloor$.

(It has been shown that the equality holds for all K_v's with $v \geq 3$ and $v \neq 9, 10$ by L. W. Beineke and F. Harary (1965), V. B. Alekeev and V. S. Gonchakov (1976).)

3.1.17 The *crossing number* of G, $r(G)$, is the minimum number of pairwise intersections of its edges when G is drawn in the plane. Obviously, $r(G) = 0$ if and only if G is planar. Prove that $r(K_5) = r(K_{3,3}) = 1$ and $r(K_6) = 3$.

(It has been prove that

$$r(K_{m,n}) = \left\lfloor \frac{m}{2} \right\rfloor \left\lfloor \frac{m-1}{2} \right\rfloor \left\lfloor \frac{n}{2} \right\rfloor \left\lfloor \frac{n-1}{2} \right\rfloor$$

for $\min\{m, n\} \leq 6$ by Kleitman [109] and

$$r(K_v) = \frac{1}{4} \left\lfloor \frac{v}{2} \right\rfloor \left\lfloor \frac{v-1}{2} \right\rfloor \left\lfloor \frac{v-2}{2} \right\rfloor \left\lfloor \frac{v-3}{2} \right\rfloor$$

for $v \leq 10$ by Guy [81]. Moreover, it has been conjectured that the two equalities hold for any positive integers m, n and v.)

3.2 Kuratowski's Theorem

It is clearly of importance to know which graphs are planar and which are not. In the preceding section we obtain some necessary conditions for a graph to be planar. Making use of these conditions we have already shown that, in particular, both K_5 and $K_{3,3}$ are non-planar. We will, in this section, see that these two non-planar graphs play an important role in the characterization of planarity of a graph.

A remarkably simple, useful criteria for graphs to be planar was found in 1930 by Kuratowski [116] and, independently, Frink and Smith [68]. This criteria is called *Kuratowski's theorem* in the literature and textbooks on graph theory. For this classical theorem, there are many simpler proofs than the original. The first relatively simple proof was given by Dirac and Schuster [42], and some of other proofs have been given in Thomassen's paper [159], Klotz [110] and Makarychev [125]. A discussion of its history, the reader is referred to Kennedy, Quintas and Syslo [105]. The proof presented here is due to Tverberg [165].

Before stating and proving Kuratowski's theorem, we need to describe other concepts on graphs.

An edge e is said to be *subdivided* when it is deleted and replaced by a single path of length two connecting its end-vertices of e, the internal vertex of this single path being a new vertex. This is illustrated in Figure 3.6.

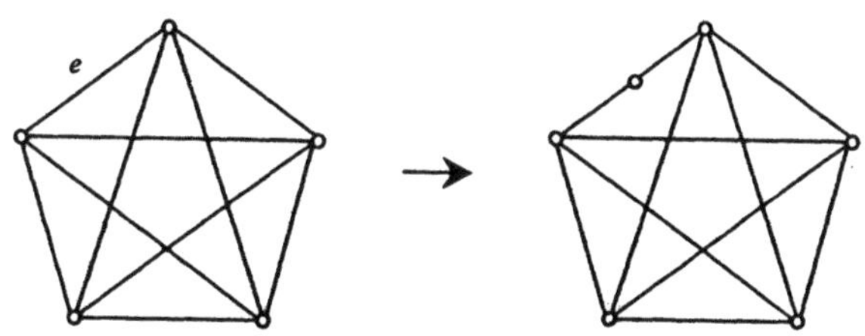

Figure 3.6: Subdivision of an edge e of K_5

A *subdivision* of a graph G is a graph obtained from G by a sequence of edge subdivisions. Figure 3.7 illustrates two subdivisions of $K_{3,3}$.

Let H be a subgraph of G. Define a relation "$\sim$" on $E(G) \setminus E(H)$ by the condition that $e_1 \sim e_2$ if there exists a walk W such that

(i) the first and last edges of W are e_1 and e_2, respectively, and
(ii) W is internally disjoint from H.

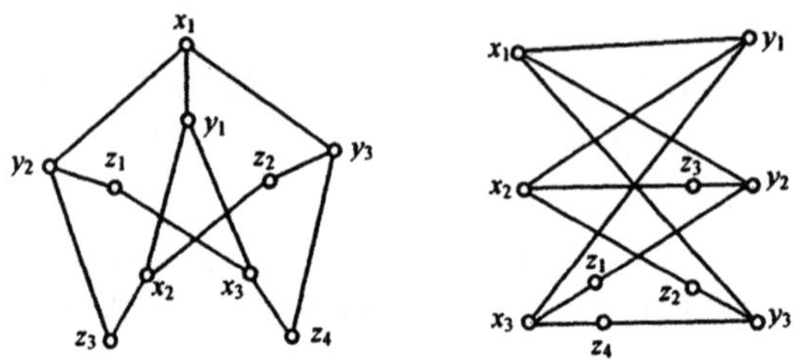

Figure 3.7: Two subdivisions of $K_{3,3}$

It is easy to verify that "$\sim$" is an equivalence relation on $E(G) \setminus E(H)$. The subgraph B induced by the edges in a resulting equivalence class is called a *bridge relative to* H, for short, H-*bridge.*

It follows immediately from the definition that each H-bridge B is connected and any two H-bridges are edge-disjoint. For an H-bridge B of G, we write $V_G(B, H) = V(B) \cap V(H)$, and called the vertices in this set the *vertices of attachment* of B to H in G. Figure 3.8 illustrates all H-bridges B_1, B_2, B_3, B_4 of G, where $H = (x_1, x_2, \cdots, x_9, x_1)$ is a cycle, $V_G(B_1, H) = V_G(B_2, H) = \{x_1, x_2, x_3\}$, $V_G(B_3, H) = \{x_4, x_6, x_7, x_9\}$, $V_G(B_4, H) = \{x_5, x_8\}$.

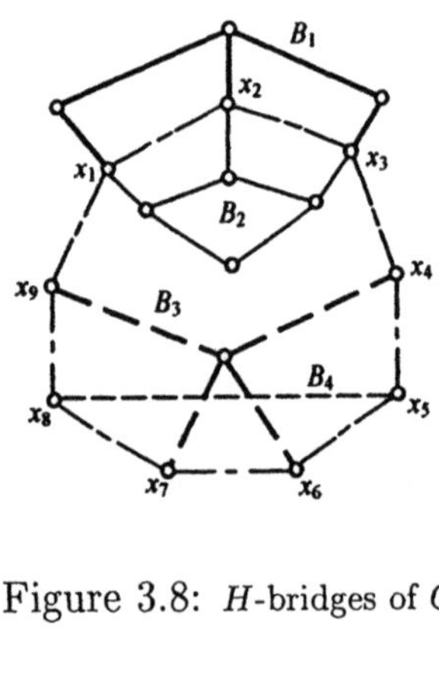

Figure 3.8: H-bridges of G

Theorem 3.6 A graph is planar if and only if it contains no subdivision of K_5 or $K_{3,3}$ as its subgraph.

Proof The necessity is a simple observation, and we leave the proof to the reader as an exercise (the exercise 3.2.1). We will prove the sufficiency for a simple and connected graph of order $v \geq 6$ by contradiction.

Let G be a minimal non-planar simple graph containing no subdivision of K_5 or $K_{3,3}$ and having as few vertices as possible. Then G contains no cut-vertex by the exercise 3.1.13. Moreover, we can, without loss of generality, suppose that G contains a subdivision H of K_5^- or $K_{3,3}^-$ as a subgraph, where K_5^- (resp. $K_{3,3}^-$) denotes K_5 (resp. $K_{3,3}$) minus one edge.

Suppose that H is a subdivision of K_5^-. Since H is planar and G is not, there is an H-bridge B of G such that $|V_G(B,H)| \geq 2$ as G contains no cut-vertex. Choose $x, y \in V_G(B,H)$ and let P be an xy-path in B (see Figure 3.9 (a)). Then $xy \in E(H)$ as G contains no subdivision of K_5. Since G is simple, $v(P) \geq 3$ and $H \cup P$ contains a subdivision of $K_{3,3}^-$ (see Figure 3.9 (b)).

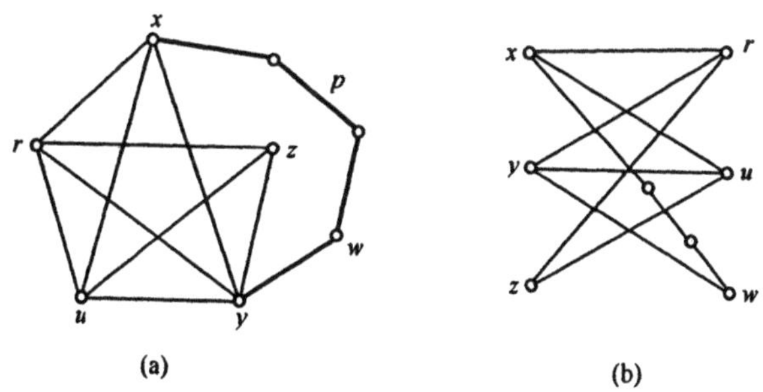

Figure 3.9: $H \cup P$ and a subdivision of $K_{3,3}^-$ as a subgraph

Thus, we can suppose that H is a subdivision of $K_{3,3}^-$. Let $\{X, Y\}$ be a "bipartition" of $K_{3,3}^-$, where $X = \{x, y, z\}$, $Y = \{r, u, w\}$ and $K_{3,3}^-$ does not contain "the edge" P_{zw}, which denotes an edge zw or a zw-path obtained from $K_{3,3}$ by a sequence of subdividing the edge zw.

In view of the equivalence between spherical and planar embedding, we will attempt embedding G on a tetrahedron T with vertices x, y, r, u. We start by embedding all vertices of H in edges of T except for x, y, r, u. Thus there are at least two vertices of H, say w and z, that are, respectively, embedded in two different edges which are not incident with a common end-vertex. Let w and z be embedded in the edges xy and ru, respectively (see Figure 3.10). Because of the non-planarity and minimality of G, $d_G(w) \geq 3$ and $d_G(z) \geq 3$, moreover, w and z are vertices of attachment of some H-bridges in G.

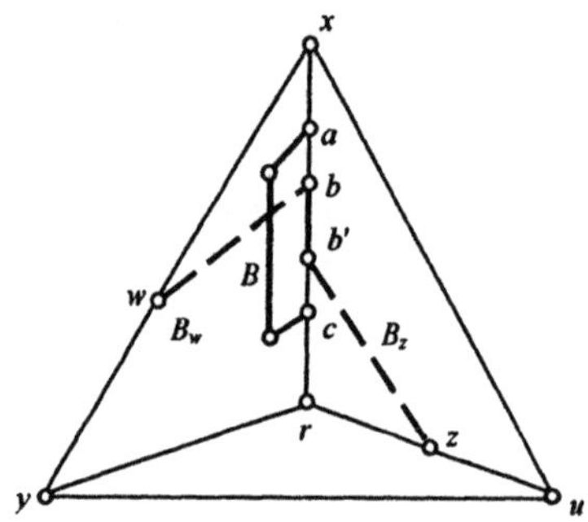

Figure 3.10: An illustration of the proof of Theorem 3.6

Let B be an H-bridge of G. Since G contains no subdivision of $K_{3,3}$, then, all vertices of attachment of B to H are contained in either an edge of T (such a B is called a small bridge) or the union of two or three of the boundary of a face of T (such a B is called a large bridge). This means that B is embedded in a face of T.

Let B_w and B_z be H-bridges of G containing w and z, respectively. By vertex-minimality, $G - w$ and $G - z$ can be embedded in T. If B_w is a small bridge, then it does not provide any obstacle to embedding of other bridges on T. Hence B_w must be large since G can not be embedded on T. Similarly, B_z is also large. Since B_w and B_z do not hamper embedding of any large bridge, they must hamper embedding of some small bridge B. Without loss of generality, we suppose that the vertices of attachment of B to H are in the edge xr of T. Thus there are vertices a, b, b', c in the edge xr, where b and b' (maybe $b' = b$), which are vertices of attachment of B_w and B_z to H, respectively, are between a and c, which are vertices of attachment of B to H (see Figure 3.10).

Let H' be a graph obtained from H by replacing ac-path on the edge xr by ac-path in B, and let P be a wz-path in G, which is the union of three paths: wb-path in B_w, bb'-path on xr and $b'z$-path in B_z. Then $H' \cup P$ is a subdivision of $K_{3,3}$, a contradiction. ∎

As a direct consequence of Theorem 3.6, we have immediately that Petersen graph is non-planar since it contains the subdivision of $K_{3,3}^{-}$ shown in Figure 3.7.

There are several other characterizations of planar graphs. For example, in 1937, Wagner [170] proved that a graph is planar if and only if it contains no subgraph contractible to K_5 or $K_{3,3}$; McLane [129] proved that a graph is planar if and only if it has a fundamental cycles together with one additional cycle such that this collection of cycles contains each edge of the graph exactly twice (the exercise 3.2.4).

Another well-known characterization of planar graphs, due to Whitney [174], concerns with the concept of dual graphs, which will be presented in the next section.

Exercises

3.2.1 Prove that (a) any subdivision of a non-planar graph is also non-planar; (b) any subgraph of a planar graph is also planar.

3.2.2 Prove that G is planar if either $\varepsilon < 9$ or $v < 5$.

3.2.3 Prove that for any three vertices x, y, x of a simple planar graph G,

$$d_G(x) + d_G(y) + d_G(z) \leq 2v + 2.$$

3.2.4 Prove that if G is a minimal planar graph, then G has a basic cycles together with one additional cycle such that this collection of cycles contains each edge of G exactly twice.

3.2.5 Prove that if C is a cycle of a planar graph G, then G has a planar embedding $\widetilde{G}$ such that C partitions all faces of $\widetilde{G}$ into two parts, Int C and Ext C, one part is contained in Int C and the other in Ext C.

3.2.6 Prove that if G is a plane graph of odd order and contains a Hamilton cycle, then G has even (≥ 2) faces of odd-degree.

3.2.7 Prove that

(a) if G is a loopless plane graph with a Hamilton cycle C, then

$$\sum_{i=1}^{v}(i-2)(\phi_i' - \phi_i'') = 0$$

where ϕ_i' and ϕ_i'' are the numbers of faces of degree i contained in Int C and Ext C, respectively; (È. Ja. Grinberg [78])

(b) Grinberg graph is non-planar;

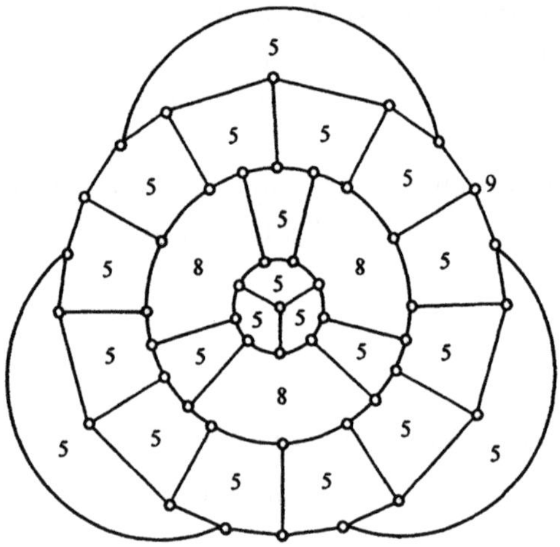

(the exercise 3.2.7 (b) Grinberg graph)

(c) the following graph contains no xy-Hamilton path;

(d) Tutte graph is non-planar.

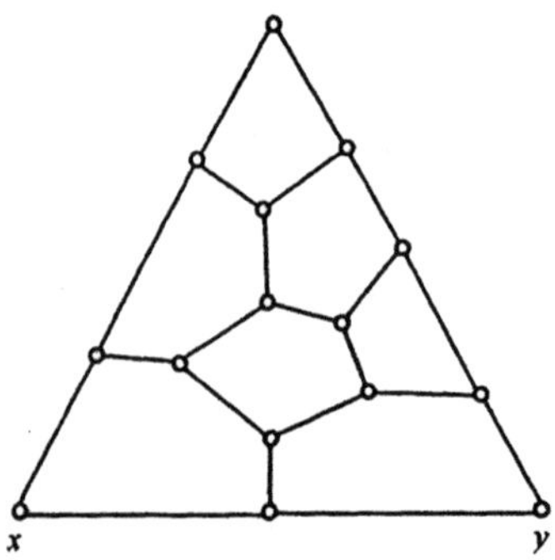

(the exercise 3.2.7 (c))

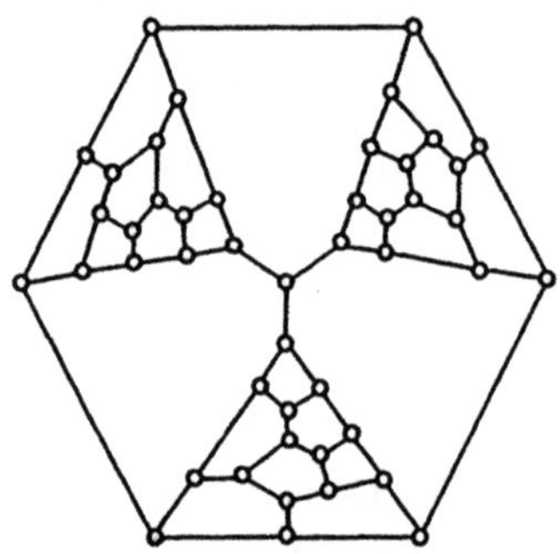

(the exercise 3.2.7 (d) Tutte graph)

3.3 Dual Graphs

Let G be a plane graph with the edge-set $\{e_1, e_2, \cdots, e_\varepsilon\}$ and the face-set $F(G) = \{f_1, f_2, \cdots, f_\phi\}$. We can define a graph G^* with vertex-set $V(G^*) = \{f_1^*, f_2^*, \cdots, f_\phi^*\}$ and the edge-set $\{e_1^*, e_2^*, \cdots, e_\varepsilon^*\}$, and two vertices f_i^* and f_j^* are linked by an undirected edge e_i^* if and only if e_i is on a common boundary of two faces f_i and f_j of G. The graph G^* is called the *geometric dual* of G.

A plane graph G and its geometric dual G^* are shown in Figure 3.11, where G is depicted by the light lines and G^* by the heavy lines.

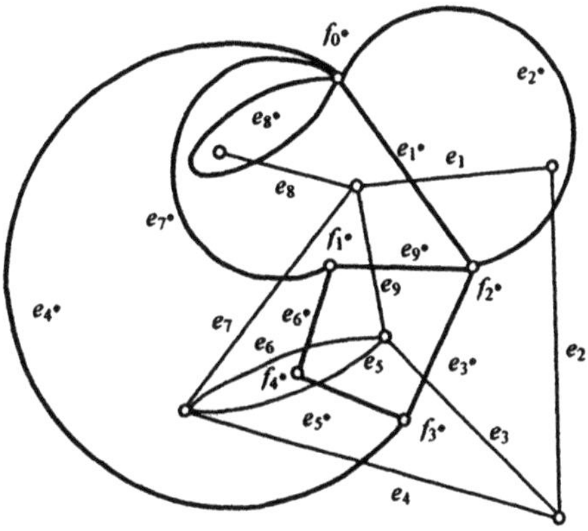

Figure 3.11: A plane graph and its geometric dual

It is a simple observation that the geometric dual G^* of a plane graph G is planar and satisfies the following relations:

$$\begin{cases} v(G^*) = & \phi(G), \\ \varepsilon(G^*) = & \varepsilon(G) \\ d_{G^*}(f^*) = & d_G(f), \quad \forall\, f \in F(G). \end{cases} \tag{3.1}$$

Making use of (3.1) we can easily give another proof of Theorem 3.2, left to the reader as an exercise for the details (the exercise 3.3.3 (a)).

It should be noted that isomorphic plane graphs may have non-isomorphic geometric duals (the exercise 3.3.1 (d)).

Theorem 3.7 Let G be a plane graph and G^* the geometric dual of G, $B \subseteq E(G)$ and $B^* = \{e^* \in E(G^*) : e \in B\}$. Then

(a) $G[B]$ is a cycle of G if and only if B^* is a bond of G^*;
(b) $G[B]$ is a bond of G if and only if $G^*[B^*]$ is a cycle of G^*.

Proof Suppose that B^* is a bond of G^*. Then there is a bipartition $\{F_1^*, F_2^*\}$ of $V(G^*)$ such that $B^* = [F_1^*, F_2^*]$. Let F_1 and F_2 be the sets of the faces of G corresponding to F_1^* and F_2^*, respectively. Then B separates all faces of G into two parts. Since G^* is connected by the exercise 3.3.1 (a), thus $G^*[F_i^*]$ is connected for $i = 1, 2$ by the exercise 2.1.11, and so $G[B]$ is connected. Note that e^* joints two distinct vertices f_i^* and f_j^* in G^* if and only if e is a common edge of two distinct face f_i and f_j in G. This implies that $G[B]$ is 2-regular, connected and, hence, a cycle by the exercise 1.7.2 (b).

Conversely, let $G[B]$ is a cycle of G. Then $G[B]$ divides the plane into two regions and so the vertices of G^* are divided into two nonempty subsets. Removal of the set of edges B^* of G^* clearly separates G^* into two connected components. Hence B^* is a bond of G^* by the exercise 2.1.11.

A similar argument yields a proof of (b), the detail is left to the reader as an exercise (the exercise 3.3.3 (a)). ∎

Corollary 3.7.1 Let T be a spanning tree of a connected plane graph G and $E^* = \{e^* \in E(G^*) : e \in E(\overline{T})\}$. Then $T^* = G^*[E^*]$ is a spanning tree of G^*.

Proof If $\phi(G) = 1$, then $G = T$, $E^* = \emptyset$ and, hence, the conclusion is true. Suppose that $\phi(G) \geq 2$ below. Because each edge of $\overline{T}$ is contained in a common boundary of two distinct faces of G, each vertex of T^* is incident with some element in E^*, that is, T^* is a spanning subgraph of G^*. If T^* contains a cycle, then $E(\overline{T})$ contains a bond of G by Theorem 3.7, which contradicts Theorem 2.4. Thus T^* contains no cycle. Since $\phi(G - e) = \phi(G) - 1$ for any $e \in E(\overline{T})$, it follows that

$$1 = \phi(G - E(\overline{T})) = \phi(G) - \varepsilon(\overline{T}) = v(G^*) - \varepsilon(T^*),$$

that is, $\varepsilon(T^*) = v(G^*) - 1$. Note that G^* is connected. Thus T^* is a spanning tree of G^* by Theorem 2.1 (c). ∎

Corollary 3.7.2 Let G be a connected plane graph. Then
(a) $\dim \mathcal{C}(G^*) = \dim \mathcal{B}(G)$;
(b) $\dim \mathcal{B}(G^*) = \dim \mathcal{C}(G)$. ∎

Motivated by the facts in Theorem 3.7, Whitney [173, 174] formulated an abstract notion of duality for general graphs, combinatorial dual of a graph.

Let G and G' be two graphs. If there is a bijective mapping $\varphi : E(G) \to E(G')$ such that for any $B \subseteq E(G)$, $G[B]$ is a cycle of G if and only if $\varphi(B) = \{e' \in E(G') : \varphi(e) = e', e \in B\}$ is a bond of G', then G' is called the *combinatorial dual* of G.

Figure 3.12 shows a graph G and its combinatorial dual G', where

$$\begin{aligned} \varphi : E(G) &\to E(G') \\ e_i &\mapsto \varphi(e_i) = e'_i,\ i = 1, 2, \cdots, 9. \end{aligned}$$

If $B = \{e_3, e_5, e_6, e_8\}$, then $G[B]$ is a cycle of G, and $\varphi(B) = \{e'_3, e'_5, e'_6, e'_8\}$ is a bond of G'. Notice that the definition of combinatorial dual makes no allusion to G and G' being planar. Although, in general, it is difficult to find the combinatorial dual of a given graph, the combinatorial definition coincides with the geometric definition for plane graphs.

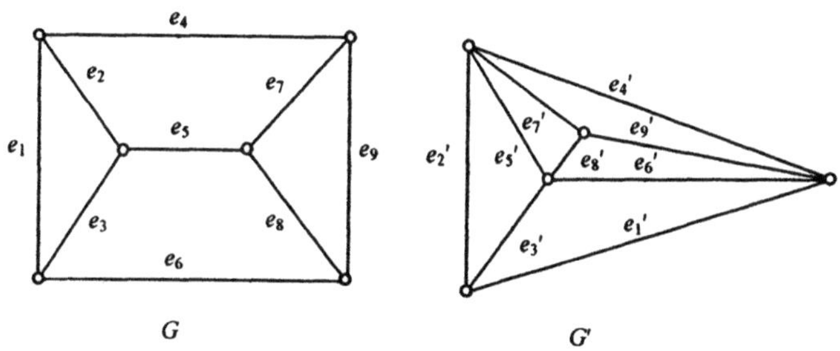

Figure 3.12: A graph and its combinatorial dual

Theorem 3.8 Let G be a plane graph and G^* is its geometric dual. Then G^* is the combinatorial dual of G. Moreover, G is the combinatorial dual of G^*.

Proof Define the mapping

$$\begin{aligned} \varphi : E(G) &\to E(G^*) \\ e_i &\mapsto \varphi(e) = e^*. \end{aligned}$$

It is easy to verify that φ is bijective. The theorem follows immediately by the definition of combinatorial dual and Theorem 3.7. ∎

We have noticed the definition of combinatorial dual makes no reference to planarity of a graph. With this concept, however, Whitney [174] obtained another characterization of planar graphs.

Theorem 3.9 (Whitney's theorem) A graph G is planar if and only if it has combinatorial dual.

Proof Let G be a planar graph and $\widetilde{G}$ be its planar embedding. By Theorem 3.8 the geometric dual $\widetilde{G}^*$ of $\widetilde{G}$ is the combinatorial dual of G and, thus, the necessity follows.

To prove the sufficiency, we need to only show that a non-planar graph has no combinatorial dual. From the definition it is easy to prove that G has a combinatorial dual if and only if every connected component or subdivision of G has a combinatorial dual (the exercise 3.3.2). By Kuratowski's theorem (3.6), we need, therefore, to only prove that K_5 and $K_{3,3}$ have no combinatorial dual. We do this in (i) and (ii) below.

(i) Suppose that K_5 has a combinatorial dual, K_5'. We will deduce a contradiction. Observe that K_5 has 10 edges and two bonds with 4 and 6 edges, respectively, but no cycle of length 2, no bond with 2 edges. These, respectively, lead to the following consequences: K_5' has 10 edges and cycles of length 4 and 6 only, no vertex of degree less than 3, no cycle of length 2. It follows from Corollary 1.6.2 that K_5' is a simple bipartite graph. Notice that any simple bipartite graph of order fewer than 7 has at most 9 edges. Therefore, $v(K_5') \geq 7$. So we get a contradiction:

$$10 = \varepsilon(K_5) = \varepsilon(K_5') \geq \tfrac{1}{2} \cdot 7 \cdot 3 > 10.$$

(ii) Suppose now that $K_{3,3}$ has a combinatorial dual, $K_{3,3}'$. We can similarly deduce a contradiction. $K_{3,3}$ has 9 edges, no bond consisting of 2 edges and so $K_{3,3}'$ has no cycle of length 2. Also $K_{3,3}$ has cycles of length 4 and 6 only, therefore $K_{3,3}'$ has no minimal cut with less than 4 edges. It follows that the degree of every vertex in $K_{3,3}'$ is at least 4, that is, $\delta(K_{3,3}') \geq 4$, and so $v(K_{3,3}') \geq 5$. Thus

$$9 = \varepsilon(K_{3,3}) = \varepsilon(K_{3,3}') \geq \tfrac{1}{2} \cdot 5 \cdot 4 = 10,$$

a contradiction. ∎

Exercises

3.3.1 Let G be a plane graph and G^* be the geometric dual of G.

(a) Prove that G^* is a connected plane graph.

(b) Prove that G is isomorphic to the geometric dual G^{**} of G^* if and only if G is connected.

(c) Construct a plane graph G such that G is not isomorphic to the geometric dual G^{**} of G^*.

(d) Construct an example to show that isomorphic plane graphs may have non-isomorphic geometric duals.

3.3.2 Prove that

(a) G has the geometric (combinatorial) dual if and only if every connected component of G has the geometric (combinatorial) dual;

(b) G has the geometric (combinatorial) dual if and only if every subdivision of G has the geometric (combinatorial) dual.

3.3.3 Prove that

(a) Theorem 3.2 using (3.1) and Theorem 3.7 (b);

(b) the geometric dual of any plane graph without odd-vertices is bipartite;

(c) any plane graph has even faces of odd degree.

3.3.4 Prove that if G is a connected plane graph then $\varsigma(G) = \varsigma(G^*)$, where G^* is the geometric dual of G.

3.3.5 A plane graph is *self-dual* if it is isomorphic to its geometric dual.

(a) Prove that if G is self-dual plane graph then $\varepsilon = 2v - 2$.

(b) Find a self-dual plane graph of order v for each $v \geq 4$.

(c) Prove that a *wheel* W_n $(= K_1 \vee C_{n-1})$ is self-dual.

3.3.6 A bond B of a connected graph G is called a *Hamilton cut* if two connected components of $G - B$ both are trees. Prove that

(a) a connected plane graph G contains a Hamilton cycle if and only if the geometric dual G^* contains a Hamilton cut;

(b) if G contains a Hamilton cut B, then

$$\sum_{i=1}^{\Delta}(i-2)(v_i' - v_i'') = 0$$

where v_i' and v_i'' are the numbers of i-degree vertices of G in two connected components G_1 and G_2 of $G - B$, respectively.

Applications

3.4 Regular Polyhedra

The theory of planar graphs is closely allied to study of convex polyhedra. In fact, every convex polyhedron P is associated with a connected plane graph $G(P)$ whose vertices and edges are the vertices and the edges of P. Necessarily, then, $\delta(G(P)) \geq 3$. Moreover, the faces of P are faces of $G(P)$ and every edge of $G(P)$ is on the boundary of two faces. Figure 3.13 shows such convex polyhedra and its associated plane graphs.

It is customary to denote the numbers of vertices, edges and faces of a convex polyhedron P by V, E and F, respectively. However, these are the numbers of vertices, edges and faces of $G(P)$, respectively. Then Euler's formula (Theorem 3.3) can be written as

$$V - E + F = 2. \tag{3.2}$$

This is the well-known Euler's Polyhedra Formula.

For the sake of convenience, we use V_n and F_n to denote the numbers of vertices and faces of degree n in a convex polyhedron, respectively. Then $n \geq 3$ and it follows from Corollary 1.1 and Theorem 3.2 that

$$2E = \sum_{n\geq 3} n\, V_n = \sum_{n\geq 3} n\, F_n. \tag{3.3}$$

Theorem 3.10 Every convex polyhedron has at least one face of degree n with $3 \leq n \leq 5$.

Proof Suppose to the contrary that $F_3 = F_4 = F_5 = 0$. It follows from (3.3) that

$$2E = \sum_{n\geq 6} n\, F_n \geq \sum_{n\geq 6} 6\, V_n = 6 \sum_{n\geq 6} F_n = 6F,$$

that is,

$$F \leq \frac{1}{3} E. \tag{3.4}$$

On the other hand, from (3.3) we have

$$2E = \sum_{n\geq 3} n\, V_n \geq 3 \sum_{n\geq 6} V_n = 3V,$$

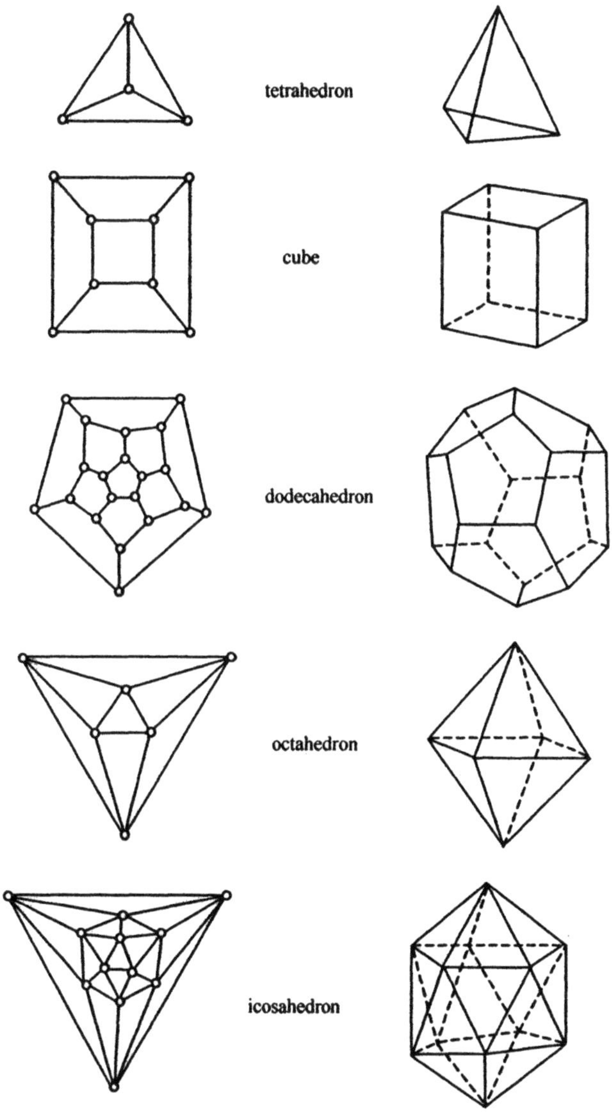

Figure 3.13: The regular polyhedra and the corresponding plane graphs

that is,

$$V \leq \frac{2}{3} E. \tag{3.5}$$

It follows from (3.3), (3.4) and (3.5) that

$$E = V + F - 2 \leq \frac{2}{3} E + \frac{1}{3} E - 2 = E - 2,$$

a contradiction. ∎

A *regular polyhedron* is a convex polyhedron whose faces are bounded by congruent regular polygons and whose polyhedral angles are congruent. There are exactly five regular polyhedra (see Figure 3.13), as was proved by Euclid in the last proposition of *The Elements.* These five regular polyhedra were known to the ancient Greeks, and were described by Plato in his *Timaeus ca.* 350 BC, and so are called *platonic solids.*

Theorem 3.11 There are exactly five regular polyhedra.

Proof Let P be a regular polyhedron and $G(P)$ be an associated plane graph. By (3.2) and (3.3), we have

$$\begin{aligned} -8 = & \; 4E - 4V - 4F = 2E + 2E - 4V - 4F \\ = & \sum_{n \geq 3} n\, F_n + \sum_{n \geq 3} n\, V_n - 4 \sum_{n \geq 3} V_n - 4 \sum_{n \geq 3} F_n \\ = & \sum_{n \geq 3} (n-4)\, F_n + \sum_{n \geq 3} (n-4)\, V_n. \end{aligned} \tag{3.6}$$

Since P is regular, there exist integers $h\,(\geq 3)$ and $k\,(\geq 3)$ such that $F = F_h$ and $V = V_k$. It follows from (3.6) and (3.3) that

$$-8 = (h-4)F_h + (k-4)V_k \tag{3.7}$$

and, respectively,

$$hF_h = 2E = kV_k. \tag{3.8}$$

According to Theorem 3.10, we have $3 \leq h \leq 5$. This gives us three cases to consider.

Case 1 $h = 3$. From (3.7) and (3.8), we have

$$12 = (6-k)V_k. \tag{3.9}$$

Thus $3 \le k \le 5$. From from (3.9) and (3.8), we have
$V_3 = 4, F_3 = 4$, implying P is the tetrahedron;
$V_4 = 6, F_3 = 8$, implying P is the octahedron;
$V_5 = 12, F_3 = 20$, implying P is the icosahedron.

Case 2 $h = 4$. From (3.7), we have $8 = (4 - k)V_k$. Thus, $k = 3$ and $V_3 = 8, F_4 = 6$, implying P is the cube.

Case 3 $h = 5$. From (3.7) and (3.8), we have $20 = (10 - 3k)V_k$. Thus, $k = 3$ and $V_3 = 20, F_5 = 12$, implying P is the dodecahedron. ∎

Exercises

3.4.1 Prove that there is no such convex polyhedron that has odd faces of odd degree.

3.4.2 Prove that any convex polyhedron contains at least 6 edges.

3.4.3 Prove that there is no such convex polyhedron that has 7 edges.

3.4.4 Prove that excepting the tetrahedron there is no such convex polyhedron that each of its vertices is adjacent to all of others.

3.4.5 Prove that any convex polyhedron contains either a face of degree three or a vertex of degree three.

3.5 Layout of Printed Circuits

There are many practical situations in which it is important to decide whether a given graph is planar, and if so, to then find a planar embedding of the graph. For example, a VLSI (very large scale integrated)-designer has to place the cells on printed circuit boards according to several designing requirements. One of these requirements is to avoids crossings since crossings lead to undesirable signals. One is, therefore, interested in knowing if the graph corresponding to a given electrical network is planar, where the vertices correspond to electrical cells and the edges correspond to the conductor wires connecting the cells.

Several different $o(v)$-algorithms for solving this problem have been proposed by different authors, for example, Hopcroft, Tarjan [99] and Liu [119] who used different techniques. These algorithms require lengthy explanations and verification. We therefore in this section describe a much simpler but nevertheless fairly efficient algorithm due to Demoucron, Malgrange and Pertuiset [36], *DMP algorithm* for short. The presentation stated here is due to Bondy and Murty [18], which is rather simpler than the original text.

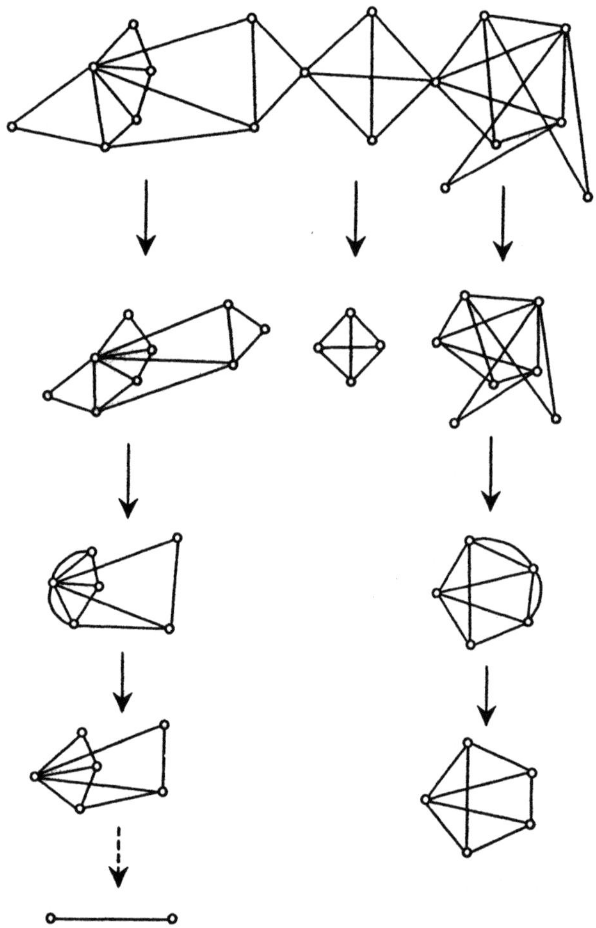

Figure 3.14: The preprocessing of a graph

Before presenting this algorithm, some preprocessing are necessary since they may considerably simplify the task. In this connection we note the following facts (see Gibbons [74]):

(a) Consider each component separately if the graph is not connected.

(b) Consider each block separately if the graph has cut-vertices.

(c) Removed all loops.

(d) Replace each 2 degree vertex plus its incident edges by a single edge.

(e) Identify parallel edges with one edge.

The last two simplifying steps ought to be applied repeatedly and alternately until neither can be applied further. Following these simplifications two elementary tests can be applied:

(f) The graph must be planar if $\varepsilon < 9$ or $v < 5$.

(g) The graph must be non-planar if $\varepsilon > 3v - 6$.

Figure 3.14 shows a graph with three blocks subjected to this processing which resolves that the graphs is planar.

Before describing DMP algorithm we need one further definition.

Let H be a planar subgraph of a graph G and let $\widetilde{H}$ be a planar embedding of H. We say that $\widetilde{H}$ is *G-admissible* if G is planar and there is a planar embedding $\widetilde{G}$ of G such that $\widetilde{H} \subseteq \widetilde{G}$.

For example, Figure 3.15 shows a planar graph G and two subgraphs $G - xy$, of which one is G-admissible and the other is not.

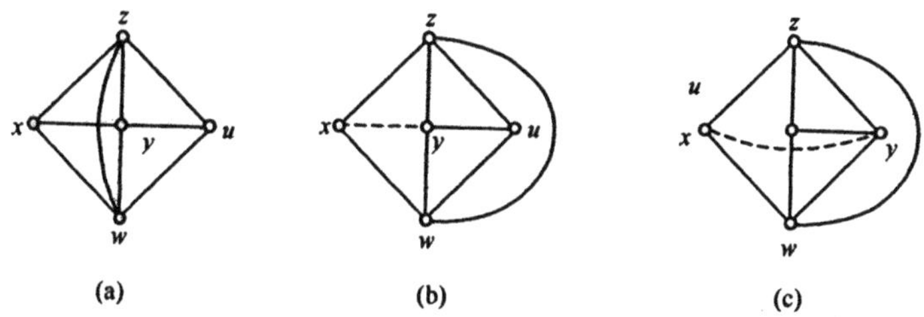

Figure 3.15: (a) G; (b) G-admissible; (c) G-inadmissible

An H-bridge B in G is said to be *drawable* in a face f of $\widetilde{H}$ if the vertices of attachment of B to H are contained in the boundary of f. We write $F(B, \widetilde{H})$ for the set of the faces of $\widetilde{H}$ in which B is drawable, that is,

$$F(B, \widetilde{H}) = \{f \in F(\widetilde{H}) : B \text{ is drawable in } f\}.$$

Theorem 3.12 If $\widetilde{H}$ is G-admissible then $F(B, \widetilde{H}) \neq \emptyset$ for every H-bridge B in G.

Proof If $\widetilde{H}$ is G-admissible then, by definition, there exists a planar embedding $\widetilde{G}$ of G such that $\widetilde{H} \subseteq \widetilde{G}$. Clearly, the subgraph of $\widetilde{G}$ which corresponds to an H-bridge B in G must be confined to one face $\widetilde{H}$. Hence $F(B, \widetilde{H}) \neq \emptyset$. ∎

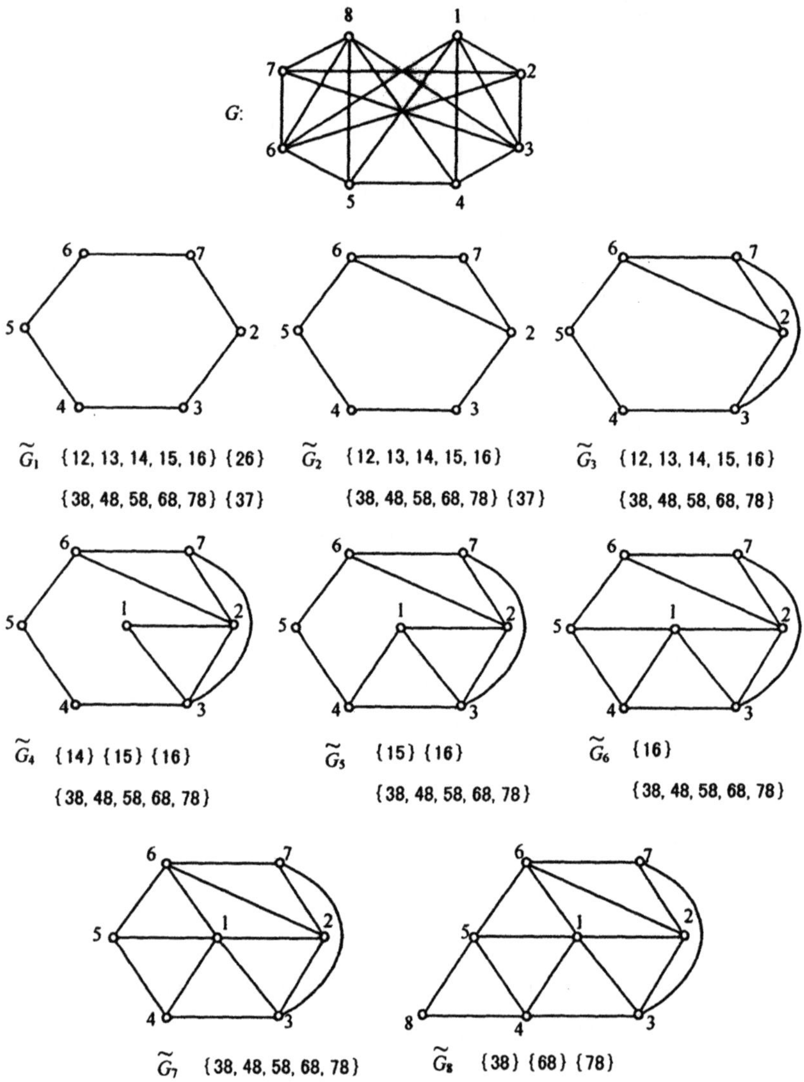

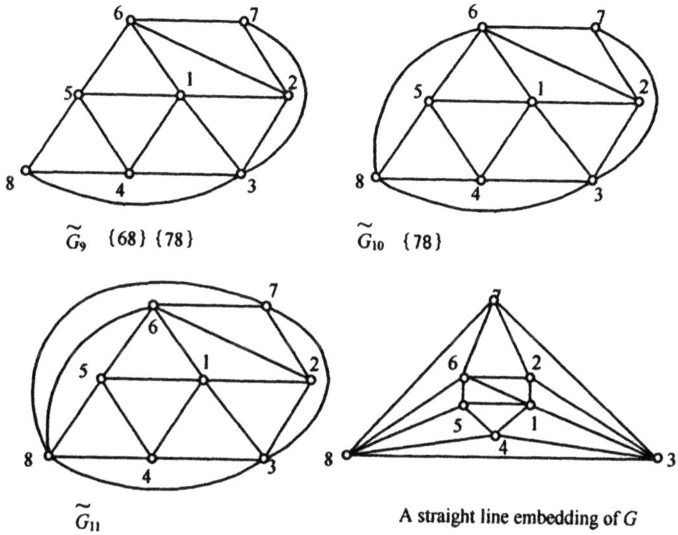

Figure 3.16: An illustration of DMP algorithm

We now describe the DMP planarity testing algorithm, of course, what is subjected to the algorithm is a block following any preprocessing, as follows:

DMP Algorithm:

1. Let G_1 be a cycle in G. Find a planar embedding $\widetilde{G}_1$ of G_1. Set $i = 1$.

2. If $E(G) \setminus E(G_i) = \emptyset$, then stop. Otherwise, determine all G_i-bridges in G; for each such bridge B find the set $F(B, \widetilde{G}_i)$.

3. If there exists a G_i-bridge B such that $F(B, \widetilde{G}_i) = \emptyset$, then stop; by Theorem 3.12, G is non-planar. If there exists a G_i-bridge B such that $F(B, \widetilde{G}_i) = 1$, let $F(B, \widetilde{G}_i) = \{f\}$. Otherwise, let B be any G_i-bridge and f any face such that $f \in F(B, \widetilde{G}_i)$.

4. Choose a path $P_i \subseteq B$ connecting two vertices of attachment of B to G_i. Set $G_{i+1} = G_i \cup P_i$ and obtain a planar embedding $\widetilde{G}_{i+1}$ of G_{i+1} by drawing P_i in the face f of $\widetilde{G}_i$. Replace i by $i + 1$ and go to step 2.

Example 3.5.1 To illustrate this algorithm, we consider the graph G in Figure 3.16. We start the algorithm with the cycle $\widetilde{G}_1 = 2345672$. By 10 iterations of the second step of the algorithm, we obtain $\widetilde{G}_i$'s ($2 \le i \le 11$) and and a list of $\widetilde{G}_i$-bridges (denoted, for brevity, by their edge sets), shown in Figure 3.16, at each stage, the $\widetilde{G}_i$-bridges B ($1 \le i \le 10$) for which $|F(B, \widetilde{G}_i)| = 1$ are indicated in bold face. The algorithm determines with a planar embedding $\widetilde{G}_{11}$ of G. Thus G is planar. ∎

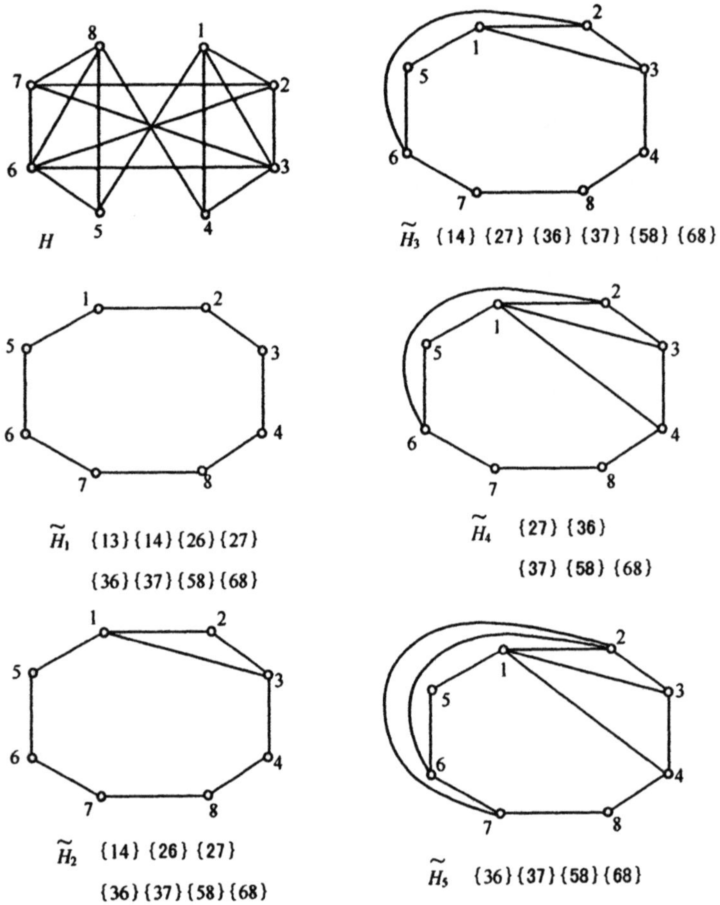

Figure 3.17: An illustration of DMP algorithm

Example 3.5.2 Consider the graph H in Figure 3.17. Apply the DMP algorithm to H starting with the cycle $\widetilde{H}_1 = 123487651$. We proceed as shown in Figure 3.17. It can be seen that, having constructed $\widetilde{H}_5$, we find an

$\widetilde{H}_5$-bridge $B = \{36\}$ such that $|F(B, \widetilde{H}_5)| = \emptyset$. At this point the algorithm stops (Step 3), and we conclude that H is non-planar. ∎

Theorem 3.13 The DMP algorithm is valid.

Proof If G is a connected planar graph, then, by Euler's formula, $\phi = \varepsilon - v + 2$ for any planar embedding $\widetilde{G}$. The DMP algorithm starts with a cycle $\widetilde{G}_1$, and $\phi(\widetilde{G}_1) = 2$. In process of the algorithm, at each stage, replacing i by $i+1$, the number of faces increases by one, that is, $\phi(\widetilde{G}_{i+1}) = \phi(\widetilde{G}_i)+1 = i+2$. It follows that if G is a connected planar graph then the sequence of graphs $\widetilde{G}_1, \widetilde{G}_2, \cdots, \widetilde{G}_{\varepsilon-v+1}$ must be obtained when the algorithm terminates. Thus, in order to prove the theorem, it is sufficient to show that each term of the sequence of graphs $\widetilde{G}_1, \widetilde{G}_2, \cdots, \widetilde{G}_{\varepsilon-v+1}$ is G-admissible if G is planar.

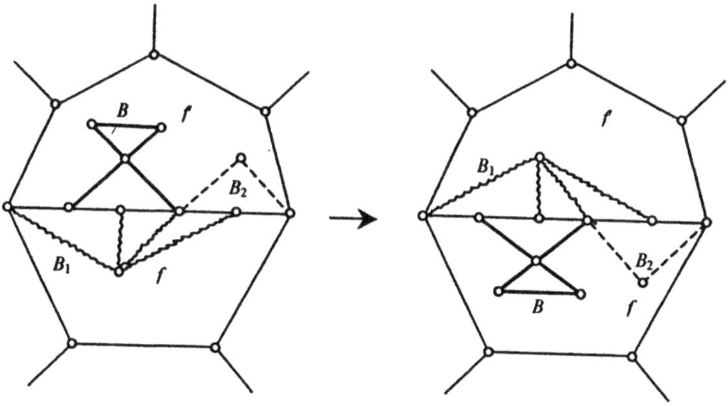

Figure 3.18: An illustration of the proof of Theorem 3.13

By induction on $k = \varepsilon - v + 1$. If $k = 1$, for G is planar, $\widetilde{G}_1$ is G-admissible clearly. Suppose that $\widetilde{G}_i$ is G-admissible for $1 \leq i \leq k < \varepsilon - v + 1$. By definition, there is planar embedding $\widetilde{G}$ of G such that $\widetilde{G}_k \subseteq \widetilde{G}$. We wish to show that $\widetilde{G}_{k+1}$ is G-admissible. Let B and f be a G_k-bridge and a face in $F_G(B, \widetilde{G}_k)$ as defined in step 3 of the algorithm. If $|F_G(B, \widetilde{G}_k)| = \{f\}$, then $\widetilde{G}_k \subseteq \widetilde{G}_{k+1} \subseteq \widetilde{G}$ by the construction of $\widetilde{G}_{k+1}$ clearly, that is, $\widetilde{G}_{k+1}$ is G-admissible.

So suppose that $|F_G(B, \widetilde{G}_k)| > 1$ and suppose that B can be drawn in some other face f' of $F_G(B, \widetilde{G}_k)$. Since G contains no cut-vertex, $|V_G(B', \widetilde{G}_k)| > 1$ for any G_k-bridge B' of G, that is, B' can be exactly drawn in two faces. Thus, each G_k-bridge B' of G whose vertices of attachment are restricted to the common boundary of f and f' can be drawn in either f and f'. For such a B', we can interchange it across the common boundary of f and f' and thereby obtain other planar embedding $\widetilde{G}'$ of G in which B is in the face f of $\widetilde{G}'$ and B' is in the face f' of $\widetilde{G}'$ (see Figure 3.18). Thus, $\widetilde{G}_{k+1} \subseteq \widetilde{G}'$ and $\widetilde{G}_{k+1}$ is G-admissible. The theorem holds by the principle of induction. ∎

The DMP algorithm is efficient, we leave the details to the reader as an exercise (the exercise 3.5.4).

Exercises

3.5.1 Preprocess the following graph by applying the way described in the beginning of this section.

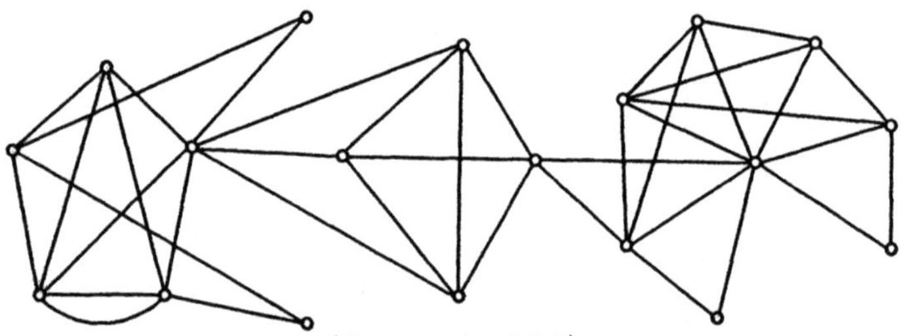

(the exercise 3.5.1)

3.5.2 Test planarity of the following graph by applying the DMP algorithm.

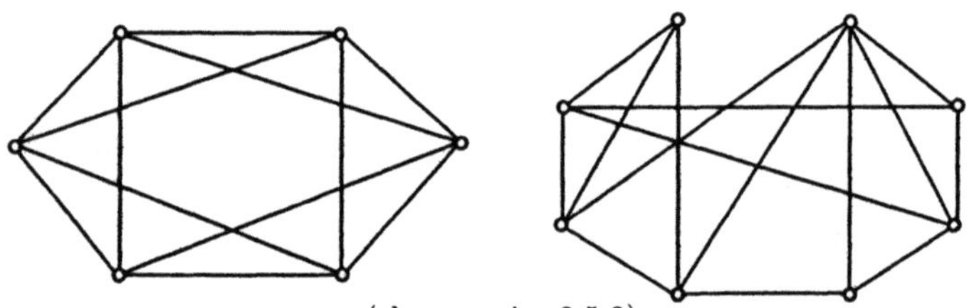

(the exercise 3.5.2)

3.5.3 Prove that the Petersen graph, K_5 and $K_{3,3}$ are non-planar by applying the DMP algorithm.

3.5.4 (a) Describe the main operations involved in the DMP algorithm.

(b) Show that the DMP algorithm is efficient.

Chapter 4

Flows and Connectivity

As we have seen in the preceding chapters, the most basic property a graph possesses is that of being connected. A most basic concept to study connectedness of a connected graph is connectivity of the graph. The most sensitive parameters measuring the connectivity of a graph are (vertex-) connectivity and edge-connectivity of the graph. We will, in this chapter, discuss these two parameters in two its local and global aspects.

Menger's theorem is the best-known, the most classical and fundamental result on local connectivity of a graph, which is the basis of almost every proof of connectivity properties, and thus is one of the cornerstones of graph theory. We will present Menger's theorem and its edge version, deduced from the max-flow min-cut theorem of Ford and Fulkerson. We will point out the equivalence of these theorems. Then we discuss the relations among the global (vertex-)connectivity, the global edge-connectivity and the minimum degree of a graph as well, and establish a sufficient and necessary condition for a graph to be k-connected (or k-edge-connected), due to Menger and Whitney.

The analytical method of network flow is an important one in graph theory. In application part of this chapter, we will make use of theory of flows and graphic spaces to describe three efficient algorithms, labelling method, Klein's algorithm and Edmonds-Johnson's algorithm, for solving the designs of transport schemes, optimal transport schemes, and the Chinese Postman Problem, respectively. Lastly, we will introduce a famous combinatorial problem, squared rectangles, whose solution needs the help of theory of flows.

4.1 Network Flows

We say what is called a *network* N is a connected weighted loopless graph $(G, \mathbf{w})$ with two specified vertices x and y, called the *source* and the *sink*, respectively, denoted by $N = (G_{xy}, \mathbf{w})$. We can, without loss of generality, suppose that G is a simple digraph since we may identify parallel edges with one edge whose weight is the sum of all weights on these edges if parallel edges exist. If $\mathbf{w}$ is a nonnegative *capacity function* $\mathbf{c}$, then the network $N = (G_{xy}, \mathbf{c})$ is called a *capacity network*, and the value $\mathbf{c}(a)$ is the *capacity* of the edge a. If $\mathbf{c}(a)$ is an integer for any $a \in E(G)$, then N is called an *integral capacity network.* Figure 4.1 (a) shows an integral capacity network.

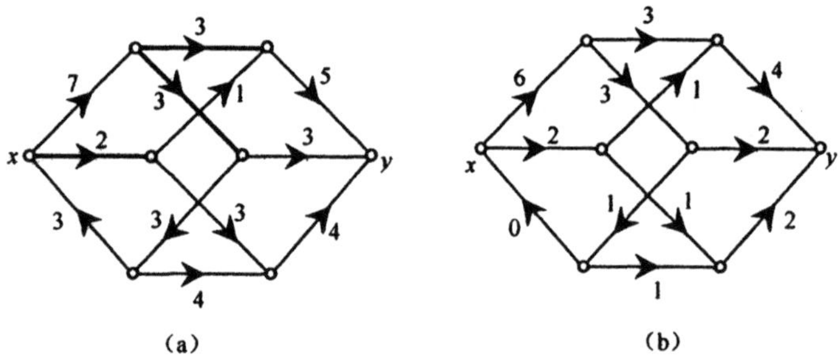

Figure 4.1: (a) An integral capacity network N; (b) an (x, y)-flow in N

Intuitively, the capacity of an edge $a = (u, z)$ may be thought of as the maximum amount of some materials that can be transported along a from u to z per unit of time. For example, the capacity of the edge a may present the number of seats available on a direct flight from city u to city z in some airline system. One the other hand, this capacity might be the capacity of a pipeline from city u to city z in an oil network, or perhaps the maximum weight of items that can be transported by truck along a highway from city u to city z. The problem in general, then, is to maximize the "flow" from the source x to the sink y without exceeding the capacity of the edges. We now introduce the notion of the flow.

Let $N = (G_{xy}, \mathbf{w})$ be a capacity network. A function $\mathbf{f} \in \mathcal{E}(G)$ is called a flow in N from x to y, in short (x, y)-flow, if it satisfies the capacity constraint condition

$$0 \le \mathbf{f}(a) \le \mathbf{c}(a), \quad \forall\, a \in E(G), \tag{4.1}$$

and the conservation condition

$$\mathbf{f}^+(u) = \mathbf{f}^-(u), \quad \forall\, u \in V(G) \setminus \{x, y\}. \tag{4.2}$$

It is clear that there is at least one (x, y)-flow in every capacity network, since the function $\mathbf{f}$ defined by $\mathbf{f}(a) = 0$ for all $a \in E(G)$, i.e., the *zero flow*, satisfies both (4.1) and (4.2) clearly. Figure 4.1 (b) gives a nontrivial example of a flow in the capacity network shown in Figure 4.1 (a).

According to the condition (4.2), it is easily verified that any (x, y)-flow $\mathbf{f}$ satisfies

$$\mathbf{f}^+(x) - \mathbf{f}^-(x) = \mathbf{f}^-(y) - \mathbf{f}^+(y). \tag{4.3}$$

This common quantity is called the *value* of $\mathbf{f}$, denoted by val $\mathbf{f}$. For example, the value of the flow $\mathbf{f}$ indicated in Figure 4.1 (b) is 8, that is, val $\mathbf{f} = 8$.

An (x, y)-flow $\mathbf{f}$ in N is *maximum* if there is no (x, y)-flow $\mathbf{f}'$ in N such that val $\mathbf{f}' >$ val $\mathbf{f}$.

An (x, y)-*cut* in N is a set of edges of the form $(S, \overline{S})$, where $x \in S$ and $y \in \overline{S}$. The *capacity* of an (x, y)-cut B, denoted by cap B, is the sum of the capacities of edges in B, that is,

$$\text{cap}B = \mathbf{c}(B) = \sum_{a \in B} \mathbf{c}(a).$$

An (x, y)-cut B in N is called to be *minimum*, if there exists no (x, y)-cut B' in N such that cap $B' <$ cap B.

For example, in the network of Figure 4.1 (a), an (x, y)-cut B is indicated by the heavy lines, and its capacity cap $B = 8$.

The maximum flow and the minimum cut are of obvious importance in the context of transportation networks. We will, in Section 4.4, present an efficient algorithm for finding such flows and cuts in a given network. We now show an important and the best-known theorem on relationship between the maximum flow and the minimum cut, due to Ford and Fulkerson [65].

Theorem 4.1 (max-flow min-cut theorem) In any capacity network, the value of a maximum flow is equal to the capacity of a minimum cut.

Proof Let $N = (G_{xy}, \mathbf{c})$ be a capacity network, $\mathbf{f}$ be a maximum (x, y)-flow and $B = (S, \overline{S})$ be a minimum (x, y)-cut in N. From the definition, for any $u \in S$, we have

$$\mathbf{f}^+(u) - \mathbf{f}^-(u) = \begin{cases} \text{val}\, \mathbf{f}, & \text{if } u = x; \\ 0, & \text{if } u \in S \setminus \{x\}. \end{cases}$$

It is not difficult to show (the exercise 4.1.2) that

$$\begin{aligned} \operatorname{val}\mathbf{f} &= \mathbf{f}^+(x) - \mathbf{f}^-(x) = \sum_{u\in S}(\mathbf{f}^+(u) - \mathbf{f}^-(u)) \\ &= \mathbf{f}^+(S) - \mathbf{f}^-(S) \le \mathbf{f}^+(S) \le \mathbf{c}^+(S) = \operatorname{cap} B. \end{aligned}$$

Thus, it is sufficient to show that

$$\operatorname{val}\mathbf{f} \ge \operatorname{cap} B. \tag{4.4}$$

To the end, we define a new digraph G' obtained from G by adding a symmetric edge $\overleftarrow{a}$ for each edge a of G, where a is said an old edge and $\overleftarrow{a}$ a new edge. Define a function $\widetilde{\mathbf{f}} \in \mathcal{E}(G')$ as follows:

$$\widetilde{\mathbf{f}} = \begin{cases} \mathbf{c}(a) - \mathbf{f}(a), & \text{if } a \text{ is old;} \\ \mathbf{f}(a), & \text{if } a \text{ is new.} \end{cases}$$

For example, for the digraph G shown in Figure 4.1, the resulting digraph G' according to this way is shown in Figure 4.2 (a), where edges indicated by curves are new, the digit nearby the edge a is $\widetilde{\mathbf{f}}(a)$.

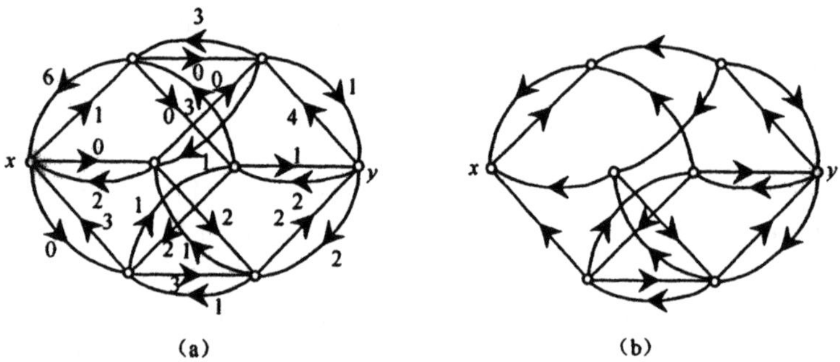

Figure 4.2: (a) G' and $\widetilde{\mathbf{f}}$; (b) $H = G'_{\widetilde{\mathbf{f}}}$

Let $H = G'_{\widetilde{\mathbf{f}}}$, the support of $\widetilde{\mathbf{f}}$, for example, in Figure 4.2, the digraph in (b) is H corresponding to G' and $\widetilde{\mathbf{f}}$ in (a).

We now claim that there exists no (x, y)-path in H. Suppose to the contrary that P is an (x, y)-path in H. Let

$$\sigma = \min\{\, \widetilde{\mathbf{f}}(a) : \ a \in E(P)\}.$$

Then $\sigma > 0$. Define a function $\mathbf{f}' \in \mathcal{E}(G)$ as follows: For each $a \in E(G)$,

$$\mathbf{f}'(a) = \begin{cases} \mathbf{f}(a) + \sigma, & \text{if } a \in E(P); \\ \mathbf{f}(a) - \sigma, & \text{if } \overleftarrow{a} \in E(P); \\ \mathbf{f}(a), & \text{otherwise.} \end{cases}$$

It is not difficult to show that $\mathbf{f}'$ is an (x, y)-flow and val $\mathbf{f}' =$ val $\mathbf{f} + \sigma >$ val $\mathbf{f}$ (the exercise 4.1.3), which contradicts the maximality of $\mathbf{f}$. Therefore, there exists no (x, y)-path in H. Let

$$S' = \{u \in V(H) : \text{ there exists an } (x, u)\text{-path in } H\}.$$

Then $x \in S'$ and $y \notin S'$ since there exists no (x, y)-path in H. This implies that $B' = (S', \overline{S}')$ is an (x, y)-cut in G. Thus, in G' we have (the exercise 4.1.3)

$$\widetilde{\mathbf{f}}(a) = \begin{cases} 0, & \text{for } a \in (S', \overline{S}'); \\ \mathbf{c}(a), & \text{for } a \in (\overline{S}', S'), \end{cases} \tag{4.5}$$

that is, in G we have

$$\mathbf{f}(a) = \begin{cases} \mathbf{c}(a), & \text{for } a \in (S', \overline{S}'); \\ 0, & \text{for } a \in (\overline{S}', S'). \end{cases}$$

This implies that

$$\text{val}\,\mathbf{f} = \mathbf{f}^+(S') - \mathbf{f}^-(S') = \mathbf{f}^+(S') = \text{cap}B' \geq \text{cap}B.$$

The inequality (4.4) is proved, and so the theorem follows. ∎

Corollary 4.1 In any integral capacity network, there must be an integral maximum flow, and its value is equal to the capacity of a minimum cut.

The proof is left to the reader as an exercise (the exercise 4.1.4).

The analytical method of network flow is an important one in graph theory. The max-flow min-cut theorem, that is, Theorem 4.1 is not only the base of network analysis by flows, but also of central importance in graph theory. We will later find that many well-known theorems on graph theory, in particular, Menger's theorem stated in the next section, can be deduced from it. In application part of this chapter, we will make use of network flow techniques to describe efficient algorithms for solving three classes of real-world problems.

The classical book on network flow of Ford and Fulkerson [67] is even today a refreshing reading. Ahuja, Magnanti and Orlin [1] provided a comprehensive book on network flow techniques.

Exercises

4.1.1 Find all (x, y)-cuts and an (x, y)-flow $\mathbf{f}$ with value 2 in the following integral capacity network, and prove that $\mathbf{f}$ is maximum.

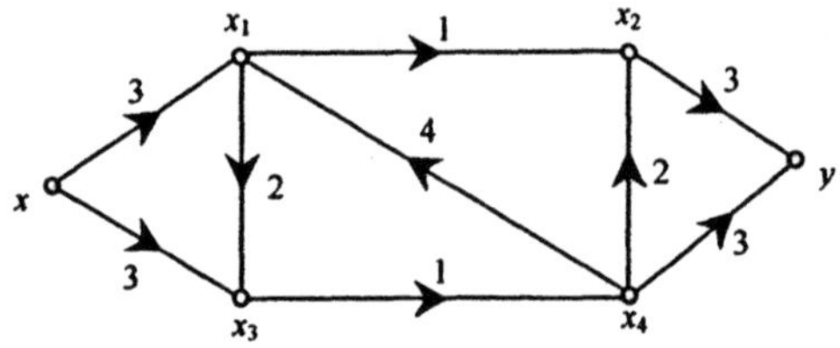

(the exercise 4.1.1)

4.1.2 (a) Prove that the equality (4.3).

(b) Prove that for any (x, y)-flow $\mathbf{f}$ in a capacity network $N = (G_{xy}, \mathbf{c})$ and $\emptyset \neq S \subset V(G)$,

$$\sum_{u \in S} (\mathbf{f}^+(u) - \mathbf{f}^-(u)) = \mathbf{f}^+(S) - \mathbf{f}^-(S).$$

(c) Construct an example to show

$$\sum_{u \in S} \mathbf{f}^+(u) \neq \mathbf{f}^+(S), \qquad \sum_{u \in S} \mathbf{f}^-(u) \neq \mathbf{f}^-(S).$$

4.1.3 Prove that (a) the assertion in the proof of Theorem 4.1: $\mathbf{f}'$ is an (x, y)-flow and val $\mathbf{f}' =$ val $\mathbf{f} + \sigma$; (b) the equality (4.5).

4.1.4 Prove that (a) the corollary 4.1;

(b) for any nonnegative real capacity network, there must be a maximum flow.

4.1.5 Let $\mathbf{f}$ be an (x, y)-flow in N and $B = (S, \overline{S})$ an (x, y)-cut. Prove that

$$\text{val } \mathbf{f} = \text{cap } B \Longleftrightarrow \mathbf{f}(a) = \begin{cases} \mathbf{c}(a), & a \in (S, \overline{S}); \\ 0, & a \in (\overline{S}, S) \end{cases}$$

$$\Longleftrightarrow \mathbf{f} \text{ is maximum and } B \text{ is minimum.}$$

4.1.6 Let $(S, \overline{S})$ and $(T, \overline{T})$ be two minimum (x, y)-cuts in N. Prove that both $(S \cup T, \overline{S \cup T})$ and $(S \cap T, \overline{S \cap T})$ are minimum (x, y)-cuts in N.

4.2 Menger's Theorem

In this section, we will present the best-known Menger's theorem in graph theory. To state this theorem we need some notation.

Let x and y be two distinct vertices of a graph G. We said (x,y)-paths $P_1, P_2, \cdots, P_n$ in G to be *internally disjoint* if $V(P_i) \cap V(P_j) = \{x,y\}$, and *edge-disjoint* if $E(P_i) \cap E(P_j) = \emptyset$ for any i and j with $1 \le i \ne j \le n$. We denote by $\zeta_G(x,y)$ and $\eta_G(x,y)$ the maximum numbers of internally disjoint and edge-disjoint (x,y)-paths in G, respectively. Denote by $\lambda_G(x,y)$ the minimum number of edges in an (x,y)-cut in G, which is call the *local edge-connectivity* of G.

The following inequality holds clearly

$$\eta_G(x,y) \le \lambda_G(x,y) \tag{4.6}$$

as to destroy all edge-disjoint (x,y)-paths in G we need to delete at least one edge from each of $\eta_G(x,y)$ (x,y)-paths. We will show that the equality in (4.6) always holds by making use of Corollary 4.1, due to Ford and Fulkerson [65], Elians, Feinstein and Shannon [52].

Theorem 4.2 Let x and y be two distinct vertices in a graph G. Then

$$\eta_G(x,y) = \lambda_G(x,y).$$

Proof By the inequality (4.6), it is sufficient to prove the following inequality:

$$\eta_G(x,y) \ge \lambda_G(x,y). \tag{4.7}$$

To the end, we consider a capacity network $N = (G_{xy}, \mathbf{c})$ with a unit of capacity of each edge of G. By Corollary 4.1, there exist a maximum (x,y)-integral flow $\mathbf{f}$ and a minimum (x,y)-cut $B = (S, \overline{S})$ such that val $\mathbf{f}$ = cap B (see Figure 4.3 (a), where B is depicted by the heavy edges). Thus, it is clear that

$$\lambda_G(x,y) \le |B| = \operatorname{cap} B = \operatorname{val} \mathbf{f}.$$

In order to prove (4.7), therefore, we need to only prove $\eta_G(x,y) \ge \operatorname{val} \mathbf{f}$.

Let $H = G_{\mathbf{f}}$, the support of $\mathbf{f}$ (see Figure 4.3 (b)). Since $\mathbf{c}(a) \equiv 1$ for any $a \in E(G), \mathbf{f}(a) \equiv 1$ for any $a \in E(H)$. This implies that

$$\begin{cases} d_H^+(x) - d_H^-(x) = \operatorname{val} \mathbf{f} = d_H^-(y) - d_H^+(y), \\ d_H^+(u) = d_H^-(u), \quad \forall\, u \in V(H) \setminus \{x,y\}. \end{cases}$$

Therefore there are val $\mathbf{f}$ edge-disjoint (x, y)-paths in H (the exercise 1.8.3). This means val $\mathbf{f} \le \eta_G(x, y)$ and, thus, the theorem follows. ∎

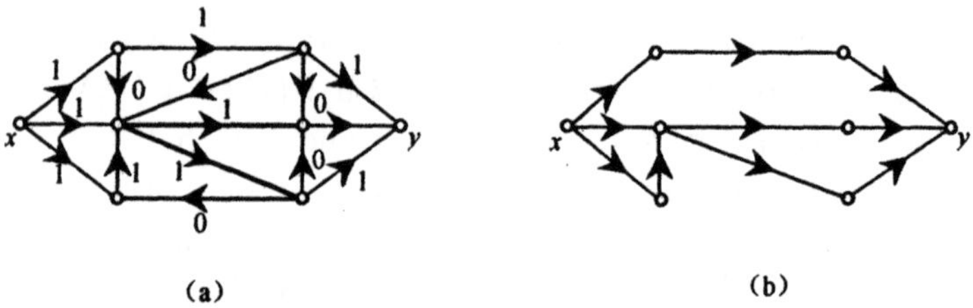

Figure 4.3: (a) A maximum (x, y)-flow $\mathbf{f}$ and a minimum (x, y)-cut B; (b) $H = G_{\mathbf{f}}$

We note that $\eta_G(x, y) = \eta_G(y, x)$ may be not, in general, always true for a digraph (see the exercise 4.2.2 (b)). However, Lováz [120] obtained the following result, which gives another characterization of an eulerian digraph.

Corollary 4.2 A connected digraph G is eulerian if and only if

$$\eta_G(x, y) = \eta_G(y, x), \quad \forall\ x, y \in V(G).$$

Proof Let x and y be two distinct vertices of an eulerian digraph G. By Theorem 4.2, there is an (x, y)-cut $(S, \overline{S})$ such that

$$|(S, \overline{S})| = \eta_G(x, y), \quad x \in S,\ y \in \overline{S}.$$

Since G is eulerian, G is balanced by Theorem 1.7. It follows from Example 1.4.1 that

$$\eta_G(y, x) \le |(\overline{S}, S)| = |(S, \overline{S})| = \eta_G(x, y).$$

Similarly, we can prove that $\eta_G(x, y) \le \eta_G(y, x)$, and the necessity follows.

To prove the sufficiency, it is sufficient to show $d_G^+(x) = d_G^-(x)$ for any $x \in V(G)$. To the end, we arbitrarily choose a vertex x of G and an element $y \in N_G^+(x)$. Then, by Theorem 4.2, there is a subset $Y_y \subseteq V \setminus \{x\}$ such that $y \in Y_y$ and $d_G^-(Y_y) = \eta_G(x, y)$. Let $Y_1, Y_2, \cdots, Y_k$ be distinct such sets that $y_i \in Y_i \cap N_G^+(x)$ and contains as many vertices as possible.

We first claim that $y_i \notin Y_j$ for any i and j with $j \ne i$. Indeed, suppose to the contrary that there exist some i and some j with $j \ne i$ such that $y_i \notin Y_i \cap Y_j$. Then the definitions of Y_i and Y_j give

$$d_G^-(Y_i) = \eta_G(x, y_i), \qquad d_G^-(Y_j) = \eta_G(x, y_j).$$

Note that the set of edges $E_G^-(Y_i \cap Y_j)$ is an (x, y_i)-cut of G. By Theorem 4.2, we have

$$d_G^-(Y_i \cap Y_j) \geq \lambda_G(x, y_i) = \eta_G(x, y_i).$$

It follows from the exercise 1.3.7 that

$$\begin{aligned} d_G^-(Y_i \cup Y_j) &\leq d_G^-(Y_i) + d_G^-(Y_j) - d_G^-(Y_i \cap Y_j) \\ &\leq d_G^-(Y_i) + d_G^-(Y_j) - \eta_G(x, y_i) \\ &= d_G^-(Y_j) = \eta_G(x, y_j). \end{aligned}$$

On the other hand, also note that the set of edges $E_G^-(Y_i \cup Y_j)$ is an (x, y_j)-cut of G. By Theorem 4.2, we have

$$d_G^-(Y_i \cup Y_j) \geq \lambda_G(x, y_j) = \eta_G(x, y_j).$$

It follows that

$$d_G^-(Y_i \cup Y_j) = \eta_G(x, y_j).$$

This implies that $Y_i \cup Y_j \subseteq Y_j$, which contradicts the choice of Y_i and Y_j. Therefore, $y_i \notin Y_j$ for any $i \neq j$. Let

$$V_i = Y_i \setminus \sum_{j \neq i} Y_j, \quad i = 1, 2, \cdots, k.$$

Then $y_i \in V_i$ for each $i = 1, 2, \cdots, k$. By Theorem 4.2 and our hypothesis, we have

$$d_G^+(V_i) \geq \lambda_G(y_i, x) = \eta_G(y_i, x) = \eta_G(x, y_i) = d_G^-(Y_i).$$

It follows that

$$\sum_{i=1}^k d_G^+(V_i) \geq \sum_{i=1}^k d_G^-(Y_i).$$

Note that the edges counted in the left of the above inequality are l incoming edges of x and all edges in (V_i, Y_j) $(j \neq i)$, moreover each of these edges is counted only once; the edges counted in the right are all out-going edges of x and of Y_j, each of the latter is counted at least once. Thus,

$$0 \leq \sum_{i=1}^k d_G^+(V_i) - \sum_{i=1}^k d_G^-(Y_i) \leq l - d_G^+(x) \leq d_G^-(x) - d_G^+(x).$$

By consider vertices in $N_G^-(x)$, similarly, we can obtain $d_G^-(x) \leq d_G^+(x)$. Thus the sufficiency is proved and the theorem follows. ∎

Theorem 4.2 is often known as the edge version of Menger's theorem in the literature. We now turn to Menger's theorem, that is, the vertex version of Theorem 4.2. In order to state Menger's theorem, we need the following notion.

A nonempty set $S \subseteq V(G) \setminus \{x, y\}$ is said to be an (x, y)-separating set in G if there exists no (x, y)-path in $G - S$. We denote by $\kappa_G(x, y)$ the minimum cardinality of an (x, y)-separating set in G, which is call the *local (vertex-) connectivity* of G.

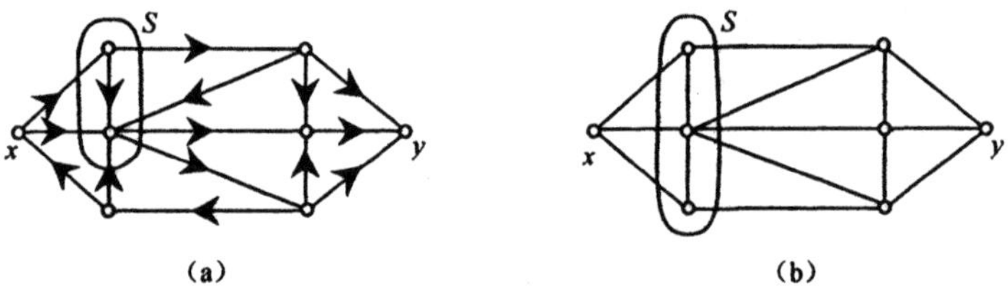

Figure 4.4: An (x, y)-separating set in a digraph or an undirected graph

Figure 4.4 illustrates an (x, y)-separating set in a digraph and an (x, y)-separating in an undirected graph.

The following inequality holds clearly

$$\zeta_G(x, y) \leq \kappa_G(x, y) \tag{4.8}$$

as to destroy all internally disjoint (x, y)-paths in G we need to delete at least one vertex from each of $\zeta_G(x, y)$ (x, y)-paths. Menger [130] found that the equality in (4.8) always holds in 1927. We will deduce it from Theorem 4.2.

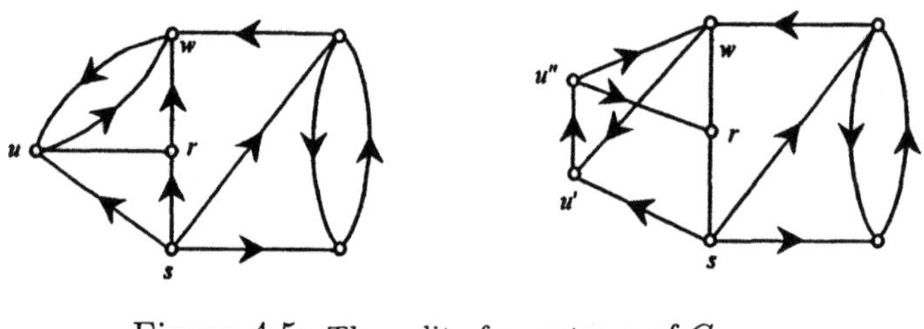

Figure 4.5: The split of a vertex u of G

The proof of this result needs a new operation on graphs, *split of a vertex*. Let $u \in V(G)$. The split of u is such an operation. First replace u by two new vertices u' and u'', and join them by an edge (u', u''), then replace each edge of G with head u by a new edge with head u', and each edge of G with tail u by a new edge with tail u''. This operation is illustrated in Figure 4.5.

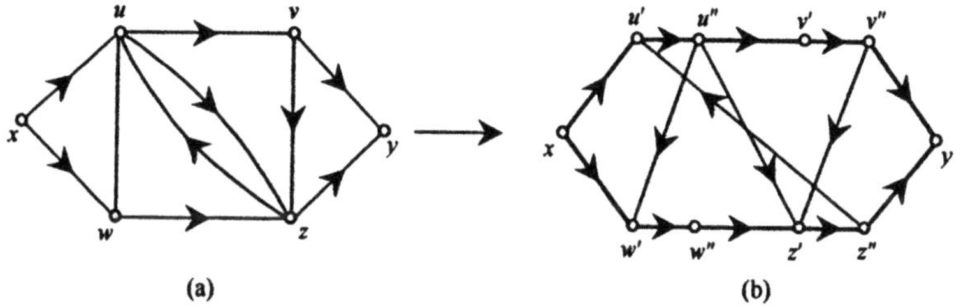

Figure 4.6: (a) G; (b) H obtained from G by splitting all vertices

Theorem 4.3 (Menger's theorem) Let x and y be two distinct vertices of G without edges from x to y. Then $\zeta_G(x, y) = \kappa_G(x, y)$.

Proof By the inequality (4.8), we need only to prove the inequality

$$\zeta_G(x, y) \geq \kappa_G(x, y). \tag{4.9}$$

To the end, let H be the graph obtained from G by splitting all $u \in V(G) \setminus \{x, y\}$ (see Figure 4.6). By Theorem 4.2, we have

$$\eta_H(x, y) = \lambda_H(x, y).$$

By construction of H, it is not difficult to observe that $\eta_H(x, y)$ edge-disjoint (x, y)-paths in H there correspond $\eta_H(x, y)$ internally disjoint (x, y)-paths in G obtained by contracting all edges of type (u', u'') (see Figure 4.6). It follows that

$$\zeta_G(x, y) \geq \eta_H(x, y) = \lambda_H(x, y).$$

Thus, we need to only prove the inequality $\lambda_H(x, y) \geq \kappa_G(x, y)$.

Let B be an (x, y)-cut in H with $|B| = \lambda_H(x, y)$. Then there exists $\emptyset \neq S \subset V(H)$ such that $B = (S, \overline{S})$, and $x \in S$ and $y \in \overline{S}$. Let S' be the set of the tails of edges in B. Then $|S'| \leq |B|$, and there is no (x, y)-path in $H - S'$. Let S'' be the set of the vertices in G obtained by contracting all edges of type (u', u''). Then $|S''| \leq |S'|$, and there is no (x, y)-path in $G - S''$. It follows that

$$\kappa_G(x, y) \leq |S''| \leq |S'| \leq |B| = \lambda_H(x, y)$$

as desired and so the theorem follows. ∎

We conclude this section with presenting direct proofs of Menger's theorem (that is, Theorem 4.3, Theorem 4.2) and the equivalence of the max-flow min-cut theorem (Corollary 4.1) with Menger's theorem, at the first reading some readers may wish to skip them.

The original proof of Menger's theorem for an undirected graph is due to Menger [130] by topological considerations, for the detail background, the reader is referred to Menger [131]. The proof of Menger's theorem in the language of graph theory is given and generalized to a digraph first by Dirac [40, 41], for more proofs, see Nash-Williams and Tutte [140], a short and self-contained proof, stated below, due to McCuaig [128].

Direct proof of Theorem 4.3 By (4.8) we need to only prove $\zeta_G(x,y) \geq \kappa_G(x,y)$ for any pair (x,y) of vertices in G, by induction on $\kappa_G(x,y) = n$.

For $n = 1$, the conclusion holds obviously. Suppose $\kappa_G(x,y) = n+1$ with $n \geq 1$. By the induction hypothesis, there are n internally disjoint (x,y)-paths in G, say, $P_1, P_2, \cdots, P_n$. Since $d^+_G(x) \geq n+1$, there is an (x,y)-path P whose the first edge is not contained in any P_i. Let u be the first vertex other than x in P and some P_i in common, and let $P_{n+1} = P(x,u)$, a section of P from x to u. We choose such $P_1, P_2, \cdots, P_n, P_{n+1}$ and u that the distance $d_{G-x}(u,y)$ is as small as possible.

If $u = y$, then the conclusion holds obviously. Suppose $u \neq y$ below. By the induction hypothesis, there are n internally disjoint (x,y)-paths in $G-u$. We choose such (x,y)-paths $Q_1, Q_2, \cdots, Q_n$ in $G-u$ that the number m of edges contained in $B = E(G) \setminus E(P_1 \cup P_2 \cup \cdots \cup P_n \cup P_{n+1})$ as few as possible.

Let $H = G[V(Q_1 \cup Q_2 \cup \cdots \cup Q_n) \cup \{u\}]$. Choose some P_k $(1 \leq k \leq n+1)$ such that the first of its edges is not in $E(H)$. Let w be the first vertex other than x in P_k and $V(H)$ in common.

If $w = y$, then $Q_1, Q_2, \cdots, Q_n, P_k$ are $(n+1)$ internally disjoint (x,y)-paths in G. We show $w = y$ below by contradiction. Suppose $w \neq y$.

If $w = u$, let R be a shortest (u,y)-path in $G-x$, and z the first vertex in R and some Q_j in common, then $d_{G-x}(z,y) < d_{G-x}(u,y)$ since u is not in any Q_i. Replacing $P_1, P_2, \cdots, P_n$ by $Q_1, Q_2, \cdots, Q_n$, P_{n+1} by $P_k(x,u) \cup R(u,z)$ and u by z, we obtain a contradiction to the choice of $P_1, P_2, \cdots, P_n, P_{n+1}$ and u.

If $w \neq u$, then w must be in some Q_i and none of edges in $Q_i(x,w)$ is in B, for otherwise, w is a common vertex of some two of $\{P_1, P_2, \cdots, P_n, P_{n+1}\}$.

Replacing $Q_i(x,w)$ by $P_k(x,w)$, we obtain n internally disjoint (x,y)-paths in $G-u$ whose edges contained in B is fewer than m. This is a contradiction to the choice of $Q_1, Q_2, \cdots, Q_n$. ∎

Direct proof of Theorem 4.2 By (4.6) we need to only prove $\eta_G(x,y) \geq \lambda_G(x,y)$ for any pair of vertices x and y in G.

We first show that if $\lambda_G(x,y) \geq 1$ then there exists a shortest (x,y)-path in G such that $\lambda_{G_P}(x,y) \geq \lambda_G(x,y) - 1$, where $G_P = G - E(P)$. The proof proceeds by induction on $n = d_G(x,y)$.

For $n = 1$, every edge of $E_G(x,y)$ must be in any (x,y)-cut of G. Thus, $\lambda_{G_P}(x,y) \geq \lambda_G(x,y) - 1$, where P is any edge in $E_G(x,y)$. Assume that the conclusion holds for any graph H and any pair (x,y) of vertices in H as long as $d_H(x,y) \leq n$.

Let x any y be two vertices of a graph G with $d_G(x,y) = n+1 \geq 2$, and let P' be a shortest (x,y)-path in G. Choose $e \in E(P')$ and $x' \in N_G^+(x)$ such that $\psi_G(e) = (x,x')$. Let $H = G \cdot e$, a graph obtained from G by contracting e. Then $d_H(x,y) = n$. By the induction hypothesis, there exists a shortest (x,y)-path P'' in H such that $\lambda_{H_{P''}}(x,y) \geq \lambda_H(x,y) - 1$. Let $P'' = xe_1x_1e_2x_2 \cdots e_{n-1}x_{n-1}e_ny$. Then $P = xex'e_1x_1e_2x_2 \cdots e_{n-1}x_{n-1}e_ny$ is an (x,y)-path of length $(n+1)$ in G.

Assume that B is a minimum (x,y)-cut of G_P. If B is an (x,y)-cut of $H_{P''}$, then

$$\lambda_{G_P}(x,y) = |B| \geq \lambda_{H_{P''}}(x,y) \geq \lambda_H(x,y) - 1 \geq \lambda_G(x,y) - 1.$$

If B is not an (x,y)-cut of $H_{P''}$, then each of (x,y)-paths in $G-B$ must contain the edge e. Therefore, $B \cup \{e\}$ is an (x,y)-cut in G. It follows that

$$\lambda_{G_P}(x,y) + 1 = |B \cup \{e\}| \geq \lambda_G(x,y).$$

By the induction principle, the conclusion follows.

We now prove Theorem 4.2 by applying the above conclusion repeatedly. Let $\lambda_G(x,y) = m \geq 1$. By the above conclusion, there is a shortest (x,y)-path P_1 in G such that $\lambda_{G_{P_1}}(x,y) \geq m-1$. If $m-1 \geq 1$, then there is a shortest (x,y)-path P_2 in G_{P_1} such that $\lambda_{G_{P_1P_2}}(x,y) \geq m-2$, where $G_{P_1P_2} = G - E(P_1 \cup P_2)$. Clearly, P_1 and P_2 are edge-disjoint in G.

In general, if $\lambda_{G_{P_1P_2\cdots P_{l-1}}}(x,y) \geq 1$, then then there is a shortest (x,y)-path P_l in $G_{P_1P_2\cdots P_{l-1}}$ such that $\lambda_{G_{P_1P_2\cdots P_l}}(x,y) \geq m-l$, where $G_{P_1P_2\cdots P_l} =$

$G - E(P_1 \cup P_2 \cup \cdots \cup P_l)$. Set $l = m$. Clearly, then, $P_1, P_2, \cdots, P_m$ are m edge-disjoint (x, y)-paths G. This gives $\eta_G(x, y) \geq \lambda_G(x, y)$. ∎

Equivalence of Theorem 4.2 and Theorem 4.3 Theorem 4.3 has been deduced from Theorem 4.2. We now deduce Theorem 4.2 from Theorem 4.3. It suffices to show the inequality $\eta_G(x, y) \geq \lambda_G(x, y)$.

Let H be the graph obtained G by splitting x and y, $a = (x', x'')$ and $b = (y', y'')$ be two edges in H, and let L be the line graph of H. By Theorem 4.3, $\zeta_L(a, b) = \kappa_L(a, b)$. Thus we need to only show

$$\eta_G(x, y) \geq \zeta_L(a, b) \quad \text{and} \quad \kappa_L(a, b) \geq \lambda_G(x, y).$$

It is easily observed that ζ internally disjoint (a, b)-paths in L correspond to ζ edge-disjoint (x, y)-paths in G. This gives the inequality $\eta_G(x, y) \geq \zeta_L(a, b)$. On the other hand, it is also easily observed that an (a, b)-separating set in L corresponds to an (x, y)-cut in G. This gives the inequality $\kappa_L(a, b) \geq \lambda_G(x, y)$. ∎

Equivalence of Theorem 4.2 and Corollary 4.1 Theorem 4.2 has been deduced from the latter part of Corollary 4.1 (the integral max-flow min-cut theorem). We now deduce the integral max-flow min-cut theorem from Theorem 4.2.

Suppose that $N = (G_{xy}, \mathbf{c})$ is an integral capacity networks with source x and sink y. Let $\mathbf{f}$ be a maximum (x, y)-flow and B be an (x, y)-cut with minimum capacity in N. We need only show the inequality $\operatorname{val} \mathbf{f} \geq \operatorname{cap} B$.

To the end, we construct a graph H obtained from G by replacing every edge e by $\mathbf{c}(e)$ parallel edges. By Theorem 4.2, $\eta_H(x, y) = \lambda_H(x, y)$. Thus, we need to only prove

$$\operatorname{val} \mathbf{f} \geq \eta_H(x, y) \quad \text{and} \quad \lambda_H(x, y) \geq \operatorname{cap} B.$$

It is clear that an (x, y)-path in H gives a unite of flow in N and so $\eta_H(x, y)$ edge-disjoint (x, y)-paths in H yield an (x, y)-flow of $\eta_H(x, y)$ units in N. Thus, $\operatorname{val} \mathbf{f} \geq \eta_H(x, y)$. On the other hand, let B' be a minimum (x, y)-cut of H, then it is also clear that B' is an (x, y)-cut of G. Thus, $\lambda_H(x, y) = |B'| \geq \operatorname{cap} B$. ∎

Exercises

4.2.1 (Menger's theorem) Let G be an undirected graph with two distinct vertices x and y. Prove that

(a) $\eta_G(x,y) = \lambda_G(x,y)$;

(b) $\zeta_G(x,y) = \kappa_G(x,y)$ if x and y are not adjacent in G.

4.2.2 Construct digraphs with two distinct x and y to show that the following statements are not true.

(a) If any two (x,y)-path and (y,x)-path have a common edge, then there is an edge which is contained in all of these paths.

(b) If $\eta_G(x,y) \geq k\ (\geq 1)$, then $\eta_G(y,x) \geq k$.

(c) If $\eta_G(x,y) \geq k\ (\geq 1)$ and $\eta_G(y,x) \geq k$, then there are k edge-disjoint (x,y)-paths $P_1, P_2, \cdots, P_k$ and k edge-disjoint (y,x)-paths $Q_1, Q_2, \cdots, Q_k$ which are pairwise-edge-disjoint.

4.2.3 Let G be a digraph with two distinct vertices x and y. Prove that

(a) if G is connected and each vertex of G other than x and y is balanced, and $d^+_G(x) - d^-_G(y) = k$, then $\eta_G(x,y) \geq k$;

(b) if G is balanced, then three statements in the exercise 4.2.2 are true.

4.2.4 Let G be undirected graph with two distinct vertices x and y. Prove that if G has diameter two, then there are $\min\{d_G(x), d_G(y)\}$ edge-disjoint xy-paths of length at most 4. (Peyrat [144])

4.2.5 Let G be a k-regular graph. Prove that if there are k internally disjoint paths between any two vertices in G, then there are two vertices x and y such that at least one is of length at least $d(G)+1$ among any k internally disjoint (x,y)-paths.

4.2.6 Let G be a k-regular bipartite graph with bipartition $\{X, Y\}$. Prove that

(a) if G is undirected, then for any $x \in X$ there is $y \in Y$ such that $\eta_G(x,y) \geq k$; (Y.O.Hamidoune and M.Las Vergnas (1988))

(b) if G is directed, then for any $x \in X$ there is $y \in Y$ and k edge-disjoint (x,y)-paths and k edge-disjoint (y,x)-paths such that all $2k$ of them are pairwise edge-disjoint. (Xu [179])

4.3 Connectivity

In preceding section, we have defined two parameters of a graph G, $\kappa_G(x,y)$ and $\lambda_G(x,y)$, called *local connectivity* and *local edge-connectivity*, respectively. We, in this section, consider the global connectivity of G.

Let G be a strongly connected digraph. A nonempty proper subset S of $V(G)$ is said to be a *separating set* if $G-S$ is not strongly connected. It is clear that every strongly connected digraph contains a separating set provided it contains no complete graph as a spanning subgraph. The parameter

$$\kappa(G) = \begin{cases} 0, & \text{if } G \text{ is not strongly connected;} \\ v-1, & \text{if } G \text{ contains a complete spanning subgraph;} \\ \min\{|S| : S \text{ is a separating set of } G\}, & \text{otherwise} \end{cases}$$

is defined as the *(vertex-)connectivity* of G. Clearly, two parameters $\kappa(G)$ and $\kappa_G(x,y)$ satisfy the following relation.

$$\kappa(G) = \min\{\kappa_G(x,y) : \forall\ x,y \in V(G),\ E_G(x,y) = \emptyset\}.$$

For example, $\kappa(K_n) = n-1$, $\kappa(C_n) = 1$ if C_n is directed and $\kappa(C_n) = 2$ if C_n is undirected for $n \geq 3$.

A separating set S of G is a *κ-separating set* if $|S| = \kappa(G)$. A graph G is said to be *k-connected* if $\kappa(G) \geq k$. All nontrivial connected undirected graphs and strongly connected digraphs are 1-connected.

A nonempty proper subset B of $E(G)$ is said to be a *directed cut* if $G-B$ is not strongly connected. Clearly, every nontrivial strongly connected digraph must contain a directed cut. The parameter

$$\lambda(G) = \begin{cases} 0, & \text{if } G \text{ is trivial or not strongly connected;} \\ \min\{|B| : B \text{ is a dircetd cut of } G\}, & \text{otherwise} \end{cases}$$

is defined as the *edge-connectivity* of G. It is also clear that

$$\lambda(G) = \min\{\lambda_G(x,y) : \forall\ x,y \in V(G)\}.$$

For example, $\lambda(K_n) = n-1$, $\lambda(C_n) = 1$ if C_n is directed and $\lambda(C_n) = 2$ if C_n is undirected for $n \geq 3$.

A directed cut B of G is a *λ-cut* if $|B| = \lambda(G)$. G is said to be *k-edge-connected* if $\lambda(G) \geq k$. All nontrivial connected undirected graphs and strongly connected digraphs are 1-edge-connected.

By definition, if B is a directed cut of G, then there exists a nonempty proper subset $S \subset V(G)$ such that $(S, \overline{S}) = B$. Recall that, in Section 2.2, a cut of G is a subset of $E(G)$ of the form $[S, \overline{S}]$. These two concepts are identical if G is undirected. However, if G is directed, then its cut contains certainly a directed cut, but the converse may be not always true.

The following fundamental inequality relating to $\kappa(G)$, $\lambda(G)$, and $\delta(G)$, is often referred to as *Whitney's inequality*, found first by Whitney [176] for undirected graphs, and generalized to digraphs by Geller and Harary [72].

Theorem 4.4 $\kappa(G) \leq \lambda(G) \leq \delta(G)$ for any graph G.

Proof Without loos of generality, we need to only prove this theorem for a nontrivial, nonempty and loopless digraph G with $\delta(G) = \delta^+(G)$. Let $x \in V(G)$ such that $d_G^+(x) = \delta(G)$. Since the set of out-going edges of x, $E_G^+(x)$, is a directed cut of G, it follows that $\lambda(G) \leq |E_G^+(x)| = \delta(G)$. We now prove $\kappa(G) \leq \lambda(G)$ by induction on $\lambda(G) \geq 0$.

Let $\lambda = \lambda(G)$. If $\lambda = 0$, then G is not strongly connected, i.e., $\kappa(G) = 0$ and, thus, the equality holds. Suppose $\kappa(H) \leq \lambda(H)$ for any digraph H with $\lambda(H) < \lambda$, and suppose $\lambda \geq 1$. There, thus, exists a directed cut B of G such that $|B| = \lambda$. Let $a \in B$ and $H = G - a$. Then $\lambda(H) \leq \lambda - 1$. By the induction hypothesis, we have

$$\kappa(H) \leq \lambda(H) \leq \lambda - 1.$$

If H contains a complete spanning subgraph, then so does G. Thus

$$\kappa(G) = v - 1 = \kappa(H) \leq \lambda(H) \leq \lambda - 1 < \lambda(G).$$

If H does not contain a complete graph as a spanning subgraph, then there exists a κ-separating set S in H. If $G - S$ is not strongly connected, then

$$\kappa(G) \leq |S| = \kappa(H) \leq \lambda(H) < \lambda(G).$$

Suppose that $G - S$ is strongly connected below. If $v(G - S) = 2$, then

$$\begin{aligned} \kappa(G) \leq & \; v - 1 = v(G - S) + |S| - 1 = |S| + 1 \\ = & \; \kappa(H) + 1 \leq \lambda(H) + 1 \leq \lambda(G). \end{aligned}$$

If $v(G - S) > 2$, let $\psi_G(a) = (x, y)$, then $S \cup \{x\}$ or $S \cup \{y\}$ is a separating set of G. Thus, we have

$$\kappa(G) \leq |S| + 1 = \kappa(H) + 1 \leq \lambda(H) + 1 \leq \lambda(G).$$

Thus in each case we have $\kappa(G) \le \lambda(G)$. The theorem follows by the principle of induction. ∎

The inequality in Theorem 4.4 are often strict. For example, the graph G in Figure 4.7 has $\kappa(G) = 2$, $\lambda(G) = 3$ and $\delta(G) = 4$. In fact, it has been proved that for all integers κ, λ, δ with $0 < \kappa \le \lambda \le \delta$, there exists a graph G with $\kappa(G) = \kappa$, $\lambda(G) = \lambda$ and $\delta(G) = \delta$ by Chartrand and Harary [29] for the undirected case and by Geller and Harary [72] for the directed case (see the exercise 4.3.15).

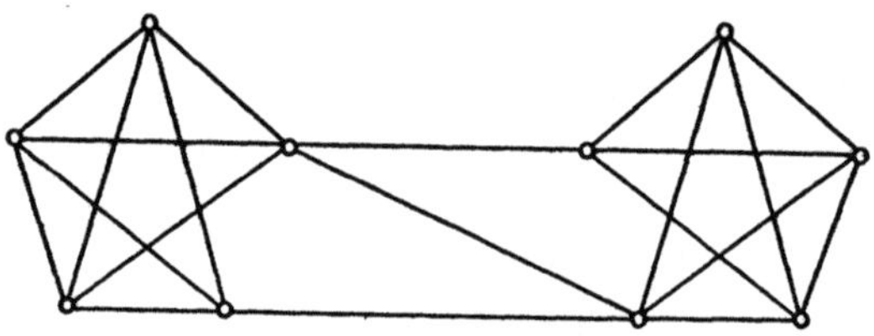

Figure 4.7: An illustration of Whitney's inequality

It is clearly of importance to know which conditions can ensure the equalities in Theorem 4.4. We present a sufficient condition for the latter equality to hold, due to Jolivet [101] for the undirected case, and Plesnik [145] for the directed case.

Example 4.3.1 Let G be a simple graph with diameter $d(G) \le 2$. Then $\lambda(G) = \delta(G)$.

Proof We prove the equality for a digraph G. If $d(G) = 0$ or 1, then G is trivial or a complete graph since G is simple, and so the equality holds clearly. We suppose $d(G) = 2$ below. By Whitney's inequality, we need to only prove $\lambda(G) \ge \delta(G)$ for any simple digraph G.

Let $B = (S, \overline{S})$ be a λ-cut of G, and let $X \subseteq S$ and $Y \subseteq \overline{S}$ be sets of the tails and the heads of the edges in B, respectively. We first note $|S \setminus X| \cdot |\overline{S} \setminus Y| = 0$. Otherwise, for any $x \in S \setminus X$ and $y \in \overline{S} \setminus Y$, any (x, y)-path in G must contains an edge in B, that is, $d_G(x, y) \ge 3$, which contradicts the hypothesis of $d(G) = 2$.

Without loss of generality, we suppose that $\overline{S} \setminus Y$ is an empty set. Let $Y = \{y_1, y_2, \cdots, y_b\}$ and let

$$s_i = |N_G^-(y_i) \cap X|, \quad i = 1, 2, \cdots, b.$$

Since G is simple, it follows that

$$\sum_{i=1}^{b} s_i = \lambda(G) \geq b \geq 1.$$

Let $\delta = \delta(G)$ and $\lambda = \lambda(G)$. Noting

$$\delta \leq d_G^-(y_i) \leq s_i + (b-1) \quad \text{for all } i = 1, 2, \cdots, b,$$

we have

$$b\delta \leq \sum_{i=1}^{b} S_i + b(b-1) = \lambda + b(b-1),$$

that is,

$$\lambda \geq b\lambda - b(b-1) = \delta + (\delta - b)(b-1) \geq \delta$$

as desired. ∎

The following example gives a sufficient condition for a undirected graph to be k-connected, due to Bondy [14] and, independently, Boesch [13].

Example 4.3.2 Let G be a simple undirected graph of order $v \geq 2$, degrees d_i of vertices satisfy $d_1 \leq d_2 \leq \cdots \leq d_v$, and k be an integer with $1 \leq k \leq v-1$. If

$$d_i \leq k + i - 2 \Longrightarrow d_{v-k+1} \geq v - i \quad \text{for } i = 1, 2, \cdots, \left\lfloor \frac{1}{2}(v-k+1) \right\rfloor,$$

then $\kappa(G) \geq k$.

Proof To the contrary suppose $\kappa(G) < k$. Then G is not complete, and there exists a separating set S in G such that $\kappa(G) \leq |S| = k-1$. Let H be a component of $G - S$ with the minimum number of vertices i. Then $i \leq \left\lfloor \frac{1}{2}(v-k+1) \right\rfloor$, that is, $k \leq v - 2i + 1$. Thus, for each $x \in V(H)$, we have

$$d_G(x) \leq v(H) - 1 + |S| = i + k - 2 \leq v - i - 1.$$

On the other hand, $d_G(y) \leq v - i - 1$ for any $y \in V(G) \setminus (V(H) \cup S)$ clearly. This implies that all of vertices of degree at least $v - i$ are in S. Since

$$d_v \geq d_{v-1} \geq \cdots \geq d_{v-k+1} \geq v - i,$$

it follows that $|S| \geq v - (v-k+1) + 1 = k$, a contradiction. ∎

With the aid of Menger's theorem, we now present a sufficient and necessary condition for a graph to be k-connected, its version of the undirected case was first found by Whitney [176] in 1932.

Theorem 4.5 Let G be a graph of order at least $k+1$. Then
(a) $\kappa(G) \geq k \Longleftrightarrow \zeta_G(x,y) \geq k,\ \forall\ x,y \in V(G)$;
(b) $\lambda(G) \geq k \Longleftrightarrow \eta_G(x,y) \geq k,\ \forall\ x,y \in V(G)$.

Proof We prove the conclusion (a) for a digraph G. For $k = 1$, the conclusion holds clearly. Suppose $k \geq 2$ below.

($\Longrightarrow$) Let x and y be two distinct vertices of G, and let $\mu = |E_G(x,y)|$. If $\mu = 0$, then, by Menger's theorem, we have

$$\zeta_G(x,y) = \kappa_G(x,y) \geq \kappa(G) \geq k.$$

Suppose now that $\mu \geq 1$. If $\mu \geq k$, then $\zeta_G(x,y) \geq \mu \geq k$. Thus, we suppose $\mu < k$ below, and let $H = G - E_G(x,y)$. We need to only prove $\zeta_H(x,y) \geq k - \mu$.

To the contrary suppose $\zeta_H(x,y) < k - \mu$. By Menger's theorem, there exists an (x,y)-separating set S in H such that $|S| = \zeta_H(x,y) \leq k - \mu - 1$. Since

$$v(H - S) = v(H) - |S| \geq k + 1 - (k - \mu - 1) = \mu + 2,$$

it follows that there exists $z \in V(H - S)$ other than x and y.

We now prove $\zeta_{H-S}(x,z) \geq 1$. If $E_H(x,z) \neq \emptyset$, then the assertion holds clearly. Suppose $E_H(x,z) = \emptyset$ below. Then $E_G(x,z) = \emptyset$. By the hypothesis and Menger's theorem, we have $\zeta_G(x,z) \geq k$. Thus, $\zeta_H(x,z) \geq k - \mu$. Since $|S| \leq k - \mu - 1$, we have $\zeta_{H-S}(x,z) \geq 1$.

Similarly, we can prove $\zeta_{H-S}(z,y) \geq 1$. Thus, we have $\zeta_{H-S}(x,y) \geq 1$, which contradicts the choice of S. The necessity follows.

($\Longleftarrow$) If G contains a complete spanning subgraph, then $\kappa(G) = v - 1 \geq k$. Suppose now that G contains no complete spanning subgraph. Let S be a κ-separating set of G. Then $G - S$ is not strongly connected. Thus, there are two distinct vertices x and y in $G - S$ such that there is no (x,y)-path in $G - S$. This implies that S is an (x,y)-separating set in G. It follows that $|S| \geq \kappa_G(x,y)$. By the hypothesis $\zeta_G(x,y) \geq k$, and by Menger's theorem, we have

$$\kappa(G) = |S| \geq \kappa_G(x,y) = \zeta_G(x,y) \geq k$$

as desired.

Similarly, we can prove the conclusion (b), left to the reader as an exercise for the details (the exercise 4.3.1). ∎

As application of Theorem 4.5, we now present a result on connectivity of the cartesian product.

Theorem 4.6 If the connectivity $\kappa(G_i) > 0$ for each $i = 1, 2, \cdots, n$, then connectivity

$$\kappa(G_1 \times G_2 \times \cdots \times G_n) \geq \kappa(G_1) + \kappa(G_2) + \cdots + \kappa(G_n).$$

Furthermore, if $\kappa(G_i) = \delta(G_i) > 0$ for each $i = 1, 2, \cdots, n$, then

$$\kappa(G_1 \times G_2 \times \cdots \times G_n) = \kappa(G_1) + \kappa(G_2) + \cdots + \kappa(G_n).$$

Particularly, $\kappa(Q_n) = n$.

Proof Use the associative law and induction on $n \geq 2$, it suffices to prove $\kappa(G_1 \times G_2) \geq \kappa(G_1) + \kappa(G_2)$.

Let $\kappa(G_i) = k_i$ for $i = 1, 2$. Assume that x and y are two distinct vertices in $G_1 \times G_2$. By Theorem 4.5, we need only show there are $k_1 + k_2$ internally disjoint (x, y)-paths in $G_1 \times G_2$. To this purpose, Let $x = x_1x_2$ and $y = y_1y_2$, where $x_1, y_1 \in V(G_1)$ and $x_2, y_2 \in V(G_2)$.

If $x_1 \neq y_1$, then by Theorem 4.5 there must exist k_1 internally disjoint (x_1, y_1)-paths $P_1, P_2, \cdots, P_{k_1}$ in G_1 since G_1 is k_1-connected. Let P be a shortest (x_1, y_1)-path in G_1. We can, without loss of generality, suppose that P_1 has the first edge in common with P. Thus $\varepsilon(P_i) \geq 2$ for each $i = 2, 3, \cdots, k_1$. Let v_i be the first internal vertex in P_i ($v_1 = y_1$ if $(x_1, y_1) \in E(G_1)$). Then v_i cuts the path P_i into two subpaths a_i and P_i', where a_i is the first edge (x_1, v_i) in P_i and P_i' is the subpath of P_i from v_i to y_1. And so the (x_1, y_1)-path P_i can be expressed as

$$P_i = x_1 \stackrel{a_i}{\rightarrow} v_i \stackrel{P_i'}{\rightarrow} y_1, \quad i = 2, 3, \cdots, k_1.$$

Similarly, if $x_2 \neq y_2$ then there are k_2 internally disjoint (x_2, y_2)-paths $W_1, W_2, \cdots, W_{k_2}$ in G_2. Let W be a shortest (x_2, y_2)-path in G_2. Assume that W has the first edge in common with W_1. Let u_j be the first internal vertex in W_j for each $j = 1, 2, \cdots, k_2$ ($u_1 = y_2$ if $(x_2, y_2) \in E(G_2)$). Then W_j can be represented as

$$W_j = x_2 \stackrel{b_j}{\rightarrow} u_j \stackrel{W_j'}{\rightarrow} y_2, \; j = 2, 3, \cdots, k_2,$$

where b_j is the first edge (x_2, u_j) in W_j and W_j' is the subpath of W_j from u_j to y_2.

Since G_i is k_i-connected for $i = 1, 2$, there must be k_1 internally disjoint (v_i, x_1)-paths $T_1, T_2, \cdots, T_{k_1}$ in G_1 and k_2 internally disjoint (u_i, x_2)-paths $U_1, U_2, \cdots, U_{k_2}$ in G_2 (the exercise 4.3.10).

Using the above notation, we can construct $k_1 + k_2$ internally disjoint (x_1x_2, y_1y_2)-paths $R_1, R_2, \cdots, R_{k_1+k_2}$ in $G_1 \times G_2$ as follows.

If $y_1 \neq x_1$ and $y_2 \neq x_2$, then let

$$\begin{cases} R_1 = x_1x_2 \overset{Px_2}{\to} y_1x_2 \overset{y_1W_1}{\to} y_1y_2, \\ R_i = x_1x_2 \overset{a_ix_2}{\to} v_ix_2 \overset{v_iW}{\to} v_iy_2 \overset{P_i'y_2}{\to} y_1y_2, \ i = 2, \cdots, k_1, \\ R_{k_1+1} = x_1x_2 \overset{x_1W}{\to} x_1y_2 \overset{P_1y_2}{\to} y_1y_2, \\ R_{k_1+j} = x_1x_2 \overset{x_1b_j}{\to} x_1u_j \overset{Pu_j}{\to} y_1u_j \overset{y_1W_j'}{\to} y_1y_2, \ j = 2, \cdots, k_2. \end{cases}$$

If $y_1 = x_1$ and $y_2 \neq x_2$, then let

$$\begin{cases} R_i = x_1x_2 \overset{a_ix_2}{\to} v_ix_2 \overset{v_iW}{\to} v_iy_2 \overset{T_iy_2}{\to} x_1y_2 = y_1y_2, \\ \qquad\qquad\qquad\qquad i = 1, 2, \cdots, k_1 \\ R_{k_1+j} = x_1x_2 \overset{x_1W_j}{\to} x_1y_2 = y_1y_2, \ j = 1, 2, \cdots, k_2. \end{cases}$$

If $y_1 \neq x_1$ and $y_2 = x_2$, then let

$$\begin{cases} R_i = x_1x_2 \overset{P_ix_2}{\to} y_1x_2 = y_1y_2, \ i = 1, 2, \cdots, k_1, \\ R_{k_1+j} = x_1x_2 \overset{x_1b_j}{\to} x_1u_j \overset{Pu_j}{\to} y_1u_j \overset{x_1U_j}{\to} y_1x_2 = y_1y_2, \\ \qquad\qquad\qquad\qquad j = 1, 2, \cdots, k_2. \end{cases}$$

It is easy to check that these (x, y)-paths $R_1, R_2, \cdots, R_{k_1+k_2}$ constructed above are internally disjoint in $G_1 \times G_2$ whichever case occurs.

Note that (the exercise 1.4.6)

$$\delta(G) = \delta(G_1) + \delta(G_2) + \cdots + \delta(G_n).$$

If $\kappa(G_i) = \delta(G_i) > 0$ for each $i = 1, 2, \cdots, n$, then by Whitney's inequality and the above result just proved, we have

$$\delta(G) \geq \kappa(G) \geq \kappa(G_1) + \kappa(G_2) + \cdots + \kappa(G_n) = \delta(G).$$

Immediately, $\kappa(Q_n) = n$ since $\kappa(K_2) = 1 = \delta(K_2)$. ∎

The hypercube Q_n has many other nice properties, some of which will be further explored in Section 7.4.

Exercises

4.3.1 Prove the conclusion (b) in Theorem 4.5.

4.3.2 Let G be a graph with $\lambda(G) = \lambda > 0$, and B be a λ-cut of G.

(a) Prove that there exists $\emptyset \neq S \subset V(G)$ such that $B = [S, \overline{S}]$ and two subgraph $G[S]$ and $G[\overline{S}]$ of G are connected if G is undirected.

(b) Construct a digraph to show that (a) is not always true.

4.3.3 Let k be an integer with $1 \le k \le v - 1$.

(a) Prove that, if G is a simple undirected graph of order v and $\delta(G) \ge \left\lceil \frac{1}{2}(v + k - 2) \right\rceil$, then $\kappa(G) \ge k$.

(b) Find a simple undirected graph G of order v such that $\delta(G) = \left\lceil \frac{1}{2}(v + k - 3) \right\rceil$ and $\kappa(G) < k$.

4.3.4 (a) prove that, if G is a simple undirected graph and $\delta(G) \ge v - 2$, then $\kappa(G) = \delta(G)$.

(b) Find a simple undirected graph G of order $v \ge 4$ such that $\delta(G) = v - 3$ and $\kappa(G) < \delta(G)$.

4.3.5 (a) Prove that, if G is a simple undirected graph with order v and $\delta(G) \ge \lfloor \frac{v}{2} \rfloor$, then $\lambda(G) = \delta(G)$.

(b) Find a simple undirected graph G of order v such that $\delta(G) = \lfloor \frac{v}{2} \rfloor - 1$ and $\kappa(G) < \delta(G)$.

4.3.6 Prove that (a) if L is the line graph $L(G)$ of G, then $\kappa(L) \ge \lambda(G)$;

(b) $\kappa(B(d, n)) = d - 1$ and $\kappa(K(d, n)) = d$.

4.3.7 Prove that, if G is a 2-connected undirected graph with order at least 3, then there exist two adjacent vertices x and y in G such that $G - \{x, y\}$ is connected.

4.3.8 Prove that, if G is a plane triangulation of order at least 4, then the geometric dual G^* is simple, 3-regular and 2-edge-connected.

4.3.9 Prove that, if G is a 3-connected undirected graph with order at least 5, then there exists $e \in E(G)$ such that $\kappa(G \cdot e) \ge 3$. (W.T.Tutte, 1961)

4.3.10 Prove that, if G is a k-connected graph, $\{x, x_1, x_2, \cdots, x_k\}$ is any $k+1$ vertices of G, then there exist k internally disjoint (x, x_i)-paths ($i = 1, 2, \cdots, k$). (Such a set of paths is called an *x-fan*.)

4.3.11 Prove that, if G is a k (≥ 2)-connected undirected graph, then any k vertices of G is contained in a cycle of G. (G.A.Dirac 1960)

4.3.12 Prove that the girth $g(G) \leq \lceil \frac{v}{k} \rceil$ for a k (≥ 1)-connected digraph G of order v. (This result is due to Hamidoune [85], who solved a special case of a conjecture of Behzad *et al.* [7] that $g(G) \leq \lceil \frac{v}{k} \rceil$ for any k-regular digraph G of order v.)

4.3.13 Prove that, if G is a strongly connected simple digraph with connectivity κ and diameter d, then $v \geq \kappa(d-3) + \delta^+ + \delta^- + 2$.

4.3.14 Let G be an undirected graph. Prove that

(a) if D is a k-connected oriented graph of G, then $\lambda(G) \geq 2k$;

(b) if $\lambda(G) \geq 2$, then G has a strongly connected orientation;

(c) if G eulerian and $\lambda(G) \geq 2k$ for some $k \geq 1$, then G has an orientation D such that $\lambda(D) \geq k$. (The conclusion (b) is due to Robbins [152]; and generalized by Nash-Williams [138] who showed that, if $\lambda(G) \geq 2k$, then G has an orientation D such that $\lambda(D) \geq k$.)

4.3.15 Let v, δ, κ and λ be given nonnegative integers. Prove that there exists a simple graph G of order v such that $\delta(G) = \delta, \kappa(G) = \kappa, \lambda(G) = \lambda$ if and only if one of the following condition holds:

(a) $0 \leq \kappa \leq \lambda \leq \delta \leq \lfloor \frac{v}{2} \rfloor$;

(b) $1 \leq 2\delta + 2 - v \leq \kappa \leq \lambda \leq \delta < v - 1$;

(c) $\kappa = \lambda = \delta = v - 1$.

4.3.16 The *k-diameter* of a k-connected graph G, denoted by $d_k(G)$, is the maximum integer d for which for any two vertices x and y in G, there exist k internally disjoint (x, y)-paths of length at most d. Prove that

(a) $d_k(G) \leq v - k + 1$ for any k-connected undirected graph G;

(b) $d_k(G) \geq d(G) + 1$ for any k (≥ 2)-regular k-connected graph G;

(c) $d_{d-1}(B(d,n)) = n + 1$ and $d_d(K(d,n)) = n + 2$ by making use of Example 1.8.2 and the exercises 1.8.9.

Applications

4.4 Design of Transport Schemes

Commodities are shipped from their production centers to their markets by a given transport system. Subject to the carrying capacity in the system, one attempts to design a transport scheme by which freight volume is as large as possible. Furthermore, which ways and means must be improved to raise the freight volume.

If the transport system is viewed as a capacity network $N = (G_{xy}, \mathbf{c})$, where the digraph G denotes a structure of the system; the source x and the sink y correspond to the production center and the market, respectively; $\mathbf{c}$ is the freight volume of the system. As a result, the design of transport schemes can be reduced to find a maximum (x, y)-flow and a minimum (x, y)-cut in $N = (G_{xy}, \mathbf{c})$. In this section, we will present an efficient algorithm for solving this problem, which is known as the *labelling method*, proposed first by Ford and Fulkerson [66] in 1957.

To describe the algorithm, we need some necessary theoretic preparations. Let $\mathbf{f} \in \mathcal{E}(G)$ be an (x, y)-flow in $N = (G_{xy}, \mathbf{c})$, u be a vertex in G other than x. For an xu-path P in G, we write P^+ and P^- for the sets of forward and reverse edges of P, respectively. Define

$$\sigma(a) = \begin{cases} \mathbf{c}(a) - \mathbf{f}(a), & \text{for } a \in P^+; \\ \mathbf{f}(a), & \text{for } a \in P^-, \end{cases}$$

and

$$\sigma_P(u) = \min\{\sigma(a) : \ a \in E(P)\}.$$

An xu-path P is called to be **f**-*saturated* if $\sigma_P(u) = 0$; and **f**-*unsaturated* if $\sigma_P(u) > 0$. A **f**-unsaturated xy-path is often called **f**-*incrementing path.*

The existence of an **f**-incrementing path P implies that **f** is not a maximum (x, y)-flow; in fact, by sending an additional flow of $\sigma_P(y)$ along P, one obtains a new (x, y)-flow $\widetilde{\mathbf{f}}$ defined by

$$\widetilde{\mathbf{f}} = \begin{cases} \mathbf{f}(a) + \sigma_P(y), & \text{for } a \in P^+; \\ \mathbf{f}(a) - \sigma_P(y), & \text{for } a \in P^-; \\ \mathbf{f}(a), & \text{otherwise} \end{cases} \tag{4.10}$$

for which val $\widetilde{\mathbf{f}} = \text{val } \mathbf{f} + \sigma_P(y)$ (the exercise 4.4.1). We will refer to $\widetilde{\mathbf{f}}$ as the *revised flow based on P*.

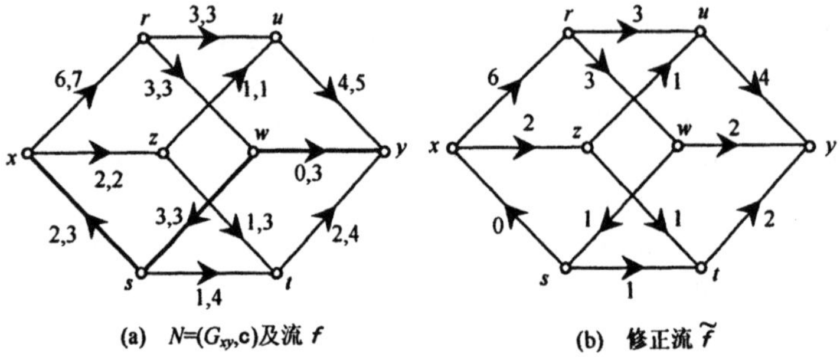

Figure 4.8: (a) A flow $\mathbf{f}$ in $N = (G_{xy}, \mathbf{c})$; (b) a revised flow based on P

Example 4.4.1 Consider the capacity network $N = (G_{xy}, \mathbf{c})$ in Figure 4.8 (a), the two digits nearby edges denote an (x, y)-flow $\mathbf{f}$ and a capacity function $\mathbf{c}$, respectively. For an xy-path $P = x\,(s, x)\,s\,(w, s)\,w\,(w, y)\,y$,

$$\begin{array}{l} P^+ = \{(w, y)\}, \quad P^- = \{(s, x), (w, s)\}; \\ \sigma(w, y) = \mathbf{c}(w, y) - \mathbf{f}(w, y) = 3 - 0 = 3; \\ \sigma(s, x) = \mathbf{f}(s, x) = 2, \quad \sigma(w, s) = \mathbf{f}(w, s) = 3; \\ \sigma_P(y) = \min\{\sigma(w, y), \sigma(s, x), \sigma(w, s)\} = \min\{3, 2, 3\} = 2 > 0. \end{array}$$

It follows that P is an $\mathbf{f}$-unsaturated (x, y)-path and, thus, an $\mathbf{f}$-incrementing path. The revised (x, y)-flow $\widetilde{\mathbf{f}}$ based on P is shown in Figure 4.8 (b), its value val $\widetilde{\mathbf{f}} = \text{val } \mathbf{f} + 2 = 6 + 2 = 8$. ∎

The $\mathbf{f}$-incrementing paths play an important role in describing the labelling method. The following theorem provides a basis.

Theorem 4.7 An (x, y)-flow $\mathbf{f}$ in $N = (G_{xy}, \mathbf{c})$ is maximum if and only if N contains no $\mathbf{f}$-incrementing path.

Proof If N contains an $\mathbf{f}$-incrementing path P, then $\mathbf{f}$ is not maximum since its revised flow $\widetilde{\mathbf{f}}$ based on P has a larger value than $\mathbf{f}$.

Conversely, suppose that N contains no $\mathbf{f}$-incrementing path. Our aim is to show that $\mathbf{f}$ is a maximum flow. To the end, let S denote the set of all vertices to which x is connected by an $\mathbf{f}$-unsaturated path in N. Clearly, $x \in S$. Also, since N has no $\mathbf{f}$-incrementing path, $y \notin S$. Thus $B = (S, \overline{S})$ is an (x, y)-cut in G.

We now show that $\mathbf{f}(a) = \mathbf{c}(a)$ for each $a \in (S, \overline{S})$. Suppose to the contrary that there exists an edge $a = (u, w)$, where $u \in S$ and $w \in \overline{S}$ such that $\mathbf{f}(a) < \mathbf{c}(a)$. Since $u \in S$, there exists an **f**-unsaturated xu-path Q and, thus, $Q + a$ is an **f**-unsaturated xw-path, which implies $w \in S$, a contradiction. Similarly, we can show that $\mathbf{f}(a) = 0$ for each $a \in (\overline{S}, S)$. It follows from the exercise 4.1.5 that **f** is a maximum (x, y)-flow and B is a minimum (x, y)-cut in N. ∎

We now describe the labelling method for finding a maximum (x, y)-flow and a minimum (x, y)-cut in a capacity network $N = (G_{xy}, \mathbf{c})$, the original proposed by Ford and Fulkerson [66] not efficient, slightly refined, as an efficient algorithm, by Edmonds and Karp [50], however, it is here stated in the form according to Chartrand and Lesniak [30].

Starting the procedure with a known (x, y)-flow, for instance, the zero flow, it recursively constructs a sequence of (x, y)-flows of increasing value, and terminates with a maximum (x, y)-flow. After the construction of each new (x, y)-flow **f**, a labelling procedure of the vertices of G is used to find an **f**-incrementing path, if one exists. A vertex z receives a label only if there is an **f**-unsaturated xz-path P. The label assigned to z is an ordered pair (α, β). If u is the vertex preceding z on P, then the first component α of the label is u^+ or u^-, depending on whether the edge preceding z is (u, z) or (z, u). The second component β of the label is $\sigma(z) = \sigma_P(z)$. If y is labelled, then P is an **f**-incrementing path, and the revised (x, y)-flow $\tilde{\mathbf{f}}$ bases on P is obtained, taken as the next (x, y)-flow in sequence, and the process is repeated. If y is not labelled, which implies that there is no **f**-incrementing path, then the algorithm terminates; by Theorem 4.7, **f** is a maximum (x, y)-flow, and the labels can be used to find a minimum (x, y)-cut.

Throughout the algorithm, a vertex is considered to be in one of three situations: unlabelled, labelled and unscanned, or labelled and scanned. Initially, all vertices are unlabelled. When a vertex receives a label, it is added to the bottom of the "labelled but unscanned" list L. These vertices are scanned on a "first-labelled first-scanned" basis, which ensures that a shortest **f**-incrementing path is selected.

Labelling Method

1. Assign values of an initial (x, y)-flow **f** to the edges of G. Label x with $(-, \infty)$ and add x to L.

2. Remove the first element u of L. If L is empty, then stop and, thus, $\mathbf{f}$ is a maximum (x, y)-flow. If L is nonempty, select an unlabelled vertex, say z, and an $\mathbf{f}$-unsaturated (x, z)-path P; add z to the end of L and count $\sigma(z) = \sigma_P(z)$.

(a) If $a = (u, z) \in E(G)$ and $\mathbf{f}(a) < \mathbf{c}(a)$, then label z with $(u^+, \sigma(z))$.

(b) If $a = (z, u) \in E(G)$ and $\mathbf{f}(a) > 0$, then label z with $(u^-, \sigma(z))$.

3. If y has been labelled, go to Step 4; otherwise, go to Step 2.

4. The labelled vertices describe an $\mathbf{f}$-incrementing path P:

$$x\, a_1\, x_1\, a_2\, x_2 \cdots x_{n-1}\, a_n\, y, \quad \text{where } x = x_0 \text{ and } y = x_n,$$

for each $i = 1, 2, \cdots, n$, vertex x_i be labelled with $(x^+_{i-1}, \sigma(x_i))$ if $a_i = (x_{i-1}, x_i)$ and with $(x^-_{i-1}, \sigma(x_i))$ if $a_i = (x_i, x_{i-1})$. A new (x, y)-flow $\widetilde{\mathbf{f}}$ is obtained by replacing $\mathbf{f}(a_i)$ by $\mathbf{f}(a_i) + \sigma(y)$ in the first case; by $\mathbf{f}(a_i) - \sigma(y)$ in the second case. Discard all labels, remove all vertices from L, replace $\mathbf{f}$ by $\widetilde{\mathbf{f}}$ and go to Step 1.

Example 4.4.2 As an application of the labelling method, we consider the capacity network $N = (G_{xy}, \mathbf{c})$ in Figure 4.9 (a). The two digits nearby each edge a are the flow $\mathbf{f}(a)$ on a and the capacity $\mathbf{c}(a)$ of a, respectively.

Choose an initial (x, y)-flow $\mathbf{f}$ as shown in Figure 4.9 (a), val $\mathbf{f} = 4$, label the vertex x with $(-, \infty)$, and set $L = \{x\}$.

As the algorithm proceeds through Step 2 the first time, the set of labelled vertices and the change of elements of L are shown in the following table

The labelling procedures	L
$x : (-, \infty)$	$\{x\}$
$\nearrow r : (x^+, 3)$ $x : (-, \infty) \leftarrow s : (x^-, 2)$	$\{r, s\}$
$\nearrow r : (x^+, 3) \rightarrow u : (r^+, 2)$ $x : (-, \infty) \quad \swarrow w : (s^-, 2)$ $\nwarrow s : (x^-, 2) \rightarrow t : (s^+, 2)$	$\{u, w, t\}$
$\nearrow r : (x^+, 3) \rightarrow u : (r^+, 2) \rightarrow y : (u^+, 2)$ $x : (-, \infty) \quad \swarrow w : (s^-, 2)$ $\nwarrow s : (x^-, 2) \rightarrow t : (s^+, 2)$	$\{w, t, y\}$

and Figure 4.9 (b), (c) and (d).

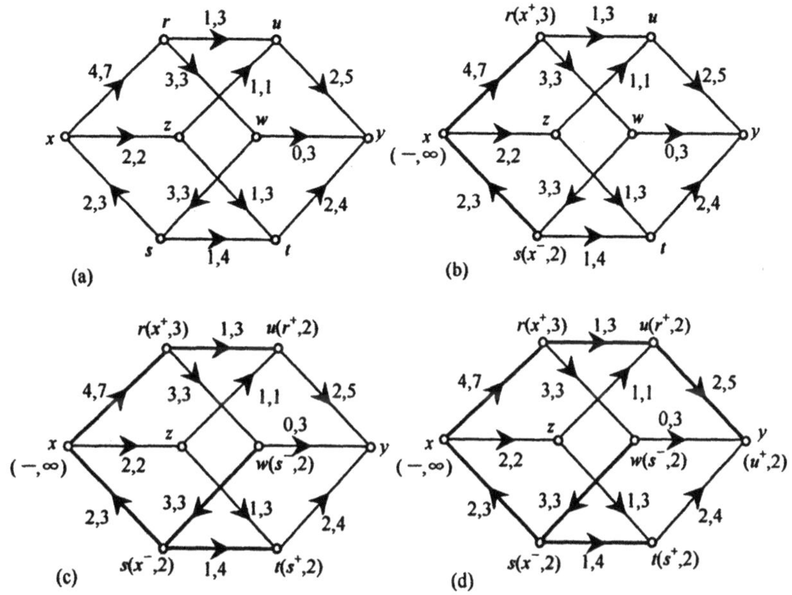

Figure 4.9: Applications of Labelling Method

Thus, we obtain an $\mathbf{f}$-incrementing path P:

$$P = x\ (x, r)\ r\ (r, u)\ u\ (u, y)\ y$$

with $\sigma_P(y) = 2$, the revised flow $\widetilde{\mathbf{f}}$ based on P is shown in Figure 4.10 (a).

Take $\widetilde{\mathbf{f}}$ as a new initial (x, y)-flow and repeat the process, see Figure 4.10 from (b) to (d) for the details. We obtain an $\widetilde{\mathbf{f}}$-incrementing path $\widetilde{P}$:

$$\widetilde{P} = x\ (s, x)\ s\ (w, s)\ w\ (w, y)\ y$$

with $\sigma_{\widetilde{P}}(y) = 2$, val $\widetilde{\mathbf{f}} = 6$. The revised flow $\mathbf{f}^*$ based on $\widetilde{P}$ is shown in Figure 4.10 (e).

Take $\widetilde{\mathbf{f}}^*$ as a new (x, y)-initial flow and repeat the process. From Figure 4.10 (f), we find L is empty when the vertex r is removed from L, and y has not been labelled. Thus, $\widetilde{\mathbf{f}}^*$ is a maximum (x, y)-flow, val $\widetilde{\mathbf{f}}^* = 8$, and the corresponding (x, y)-cut is $(S, \overline{S})$, where $S = \{x, r\}$. ∎

If the network in Figure 4.10 (e) is thought of as a transport system, then the (x, y)-flow $\mathbf{f}$ indicated in the figure is a transport scheme of commodities, for which the maximum freight volume is 8 units. The directed cut $(S, \overline{S})$ indicates "the neck of a bottle" in the system, which restricts the freight volume. To increase the freight volume one must improve the capacity of $(S, \overline{S})$. For example, if the capacity of $\mathbf{c}(x, z)$ is improved from 2 to 4, then the freight volume can be increased from 8 to 10.

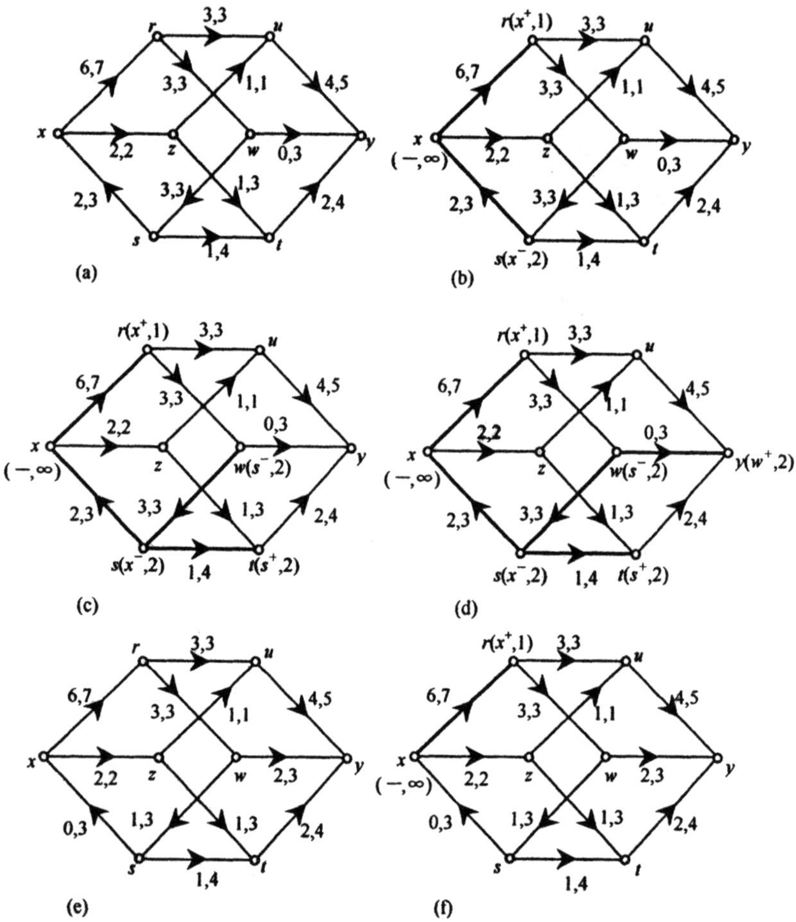

Figure 4.10: Applications of the labelling method

Theorem 4.8 If $N = (G_{xy}, \mathbf{c})$ is an integral capacity network, then the labelling method terminates with a maximum (x, y)-flow $\mathbf{f}$ in N. Furthermore, if S is the set of labelled vertices upon termination, then $(S, \overline{S})$ is a minimum (x, y)-cut.

Proof If, in Step 2, $L = \emptyset$, then there is no **f**-incrementing path in N, and so, by Theorem 4.6, **f** is a maximum (x, y)-flow and $(S, \overline{S})$ is a minimum (x, y)-cut. If $L \neq \emptyset$, Step 4 is completed each time, a new flow $\tilde{\mathbf{f}}$ in N with value larger than val **f** has been constructed. Note that N is an integral capacity network and val $\mathbf{f}' \leq$ cap $(E_G^+(x))$ for every (x, y)-flow $\mathbf{f}'$ in N. Thus, Step 4 can be repeated at most cap $(E_G^+(x))$ times. ∎

It should be pointed out that the labelling method may be not efficient if the principle of "first-labelled first-scanned" is not followed in the labelling procedure.

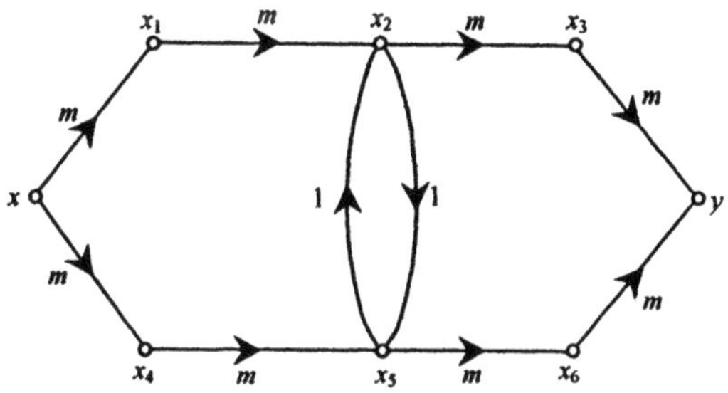

Figure 4.11: An illustration of Labelling Method

Example 4.4.3 Consider the network in Figure 4.11. Clearly, the value of a maximum (x, y)-flow is $2m$. The labelling methods will use the labelling procedure $2m + 1$ times if it starts with the zero flow and alternates between selecting $xx_1x_2x_5x_6y$ and $xx_4x_5x_2x_3y$ as an incrementing path; for, in each case, the flow value increases by exactly one. Since m is arbitrary, the number of computational steps required to implement the labelling method in this instance can be bounded by no function of v and ε. In other words, it is not an efficient algorithm. However, following the principle of "first-labelled first-scanned", the maximum (x, y)-flow in N would be found in just two iterations of the labelling procedure. ∎

It can be proved that the labelling method is an efficient algorithm and its complexity is $o(v\,\varepsilon^2)$, left to the reader as an exercise for the details (the exercise 4.4.5).

It is worth pointing out that the maximum (x, y)-flow **f** satisfies val **f**$=$ $\lambda_G(x, y)$ if $\mathbf{c}(a) \equiv 1$. In other words, the labelling method provides an efficient algorithm for finding the edge-connectivity of G.

Exercises

4.4.1 Prove that the $\widetilde{\mathbf{f}} \in \mathcal{E}(G)$ given in the expression (4.10) is an (x, y)-flow in N with val $\widetilde{\mathbf{f}} =$ val $\mathbf{f} + \sigma_P(y)$.

4.4.2 Find a maximum (x, y)-flow and a minimum (x, y)-cut in the following network.

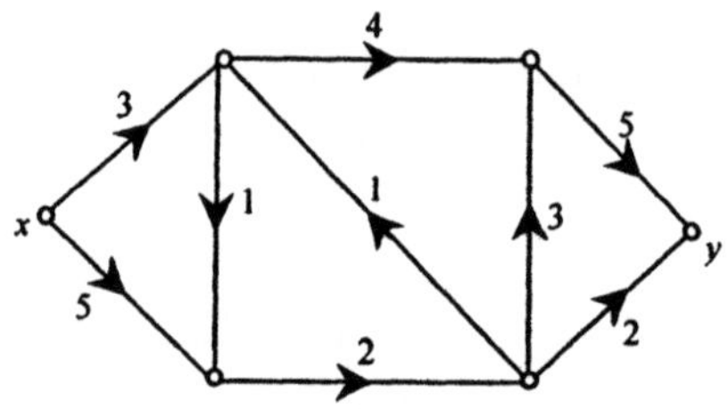

(the exercise 4.4.2)

4.4.3 Some commodities will be shipped from their producing areas x_1 and x_2 to their markets y_1, y_2, y_3 by the following transport system. Design transport scheme by which freight volume is as large as possible.

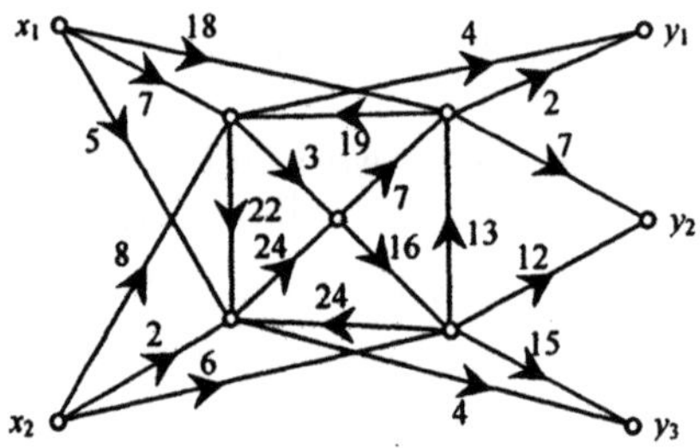

(the exercise 4.4.3)

4.4.4 Prove that the labelling method is avail for a nonnegative rational capacity.

4.4.5 Let N be an integral capacity network. Prove that

(a) if the revised flow $\widetilde{\mathbf{f}}$ is obtained from a shortest $\mathbf{f}$-incrementing path in the labelling method, then a maximum flow can be obtained by execution of the algorithm at most $\frac{1}{2}\,\upsilon\,\varepsilon$ times;

(b) the labelling method is an efficient algorithm and its complexity is $o(\upsilon\,\varepsilon^2)$.

4.5 Design of Optimal Transport Schemes

In preceding section, we have reduced the design of transport schemes to find a maximum flow in a given capacity network by the labelling method. In which, the value of a flow is only considered, however, the cost is not considered or only subject to the same cost of each unit of flow per section. In practical problems, the cost is a very important factor to be considered. Generally, the cost varies as the means of communication at different section. Thus, it is desired to provide a transport scheme with freight volume as large as possible and cost as small as possible. Such a transport scheme is called an *optimal transport scheme.*

We denote the transport system by $N = (G_{xy}, \mathbf{b}, \mathbf{c})$, called a *cost-capacity network*, where $\mathbf{b}, \mathbf{c} \in \mathcal{E}(G)$ are the cost function and the capacity function, which are positive. Figure 4.12(a) shows such a network, where the ordered pair (b, c) of digits nearby an edge a are the values of $\mathbf{b}$ and $\mathbf{c}$ on the edge a, respectively, that is, $b = \mathbf{b}\,(a)$, the cost of sending a unit of flow along a, and $c = \mathbf{c}\,(a)$. For an (x, y)-flow $\mathbf{f}$ in $N = (G_{xy}, \mathbf{b}, \mathbf{c})$, we define

$$\mathbf{b}(\mathbf{f}) = \sum_{a \in E(G)} \mathbf{f}(a)\mathbf{b}(a)$$

as the cost of the flow $\mathbf{f}$. An (x, y)-flow $\mathbf{f}$ is said to be a *minimum-cost flow* if $\mathbf{b}(\mathbf{f}) \leq \mathbf{b}(\mathbf{f}')$ for any (x, y)-flow $\mathbf{f}'$ with val $\mathbf{f}'$ = val $\mathbf{f}$.

Thus, the design of optimal transport schemes can be reduced to finding the minimum-cost maximum-flow in $N = (G_{xy}, \mathbf{b}, \mathbf{c})$. We, in this section, introduce an algorithm, Klein's algorithm, for finding such a kind of flow.

Without loss of generality, suppose that G contains a cycle since our problem is trivial for a tree. Let C be a cycle in G with a specified orientation. We write C^+ and C^- for the sets of forward and reverse edges of C, respectively. Define the *cost* of C as follows

$$\mathbf{b}(C) = \sum_{a \in C^+} \mathbf{b}(a) - \sum_{a \in C^-} \mathbf{b}(a).$$

For an (x, y)-flow $\mathbf{f}$ in N, define

$$\sigma_{\mathbf{f}}(a) = \begin{cases} \mathbf{c}(a) - \mathbf{f}(a), & \text{for } a \in C^+; \\ \mathbf{f}(a), & \text{for } a \in C^-, \end{cases}$$

$$\sigma_{\mathbf{f}}(C) = \min\{\sigma_{\mathbf{f}}(a) : a \in E(C)\}.$$

A cycle C is said to be $\mathbf{f}$-*incrementing* if it can be specified an orientation such that $\sigma_{\mathbf{f}}(C) > 0$. For any σ with $0 < |\sigma| \leq \sigma_{\mathbf{f}}$, define $\widetilde{\mathbf{f}}_\sigma \in \mathcal{E}(G)$ as follows:

$$\widetilde{\mathbf{f}}_\sigma(a) = \begin{cases} \mathbf{f}(a) + \sigma, & \text{for } a \in C^+; \\ \mathbf{f}(a) - \sigma, & \text{for } a \in C^-; \\ \mathbf{f}(a), & \text{otherwise.} \end{cases} \tag{4.11}$$

It can be easily verified that $\widetilde{\mathbf{f}}_\sigma$ is an (x, y)-flow and val $\widetilde{\mathbf{f}}_\sigma$ = val $\mathbf{f}$, left to the reader as an exercise for the details (the exercise 4.5.1). We will refer to $\widetilde{\mathbf{f}}_\sigma$ as the *revised flow of* $\mathbf{f}$ *based on* C.

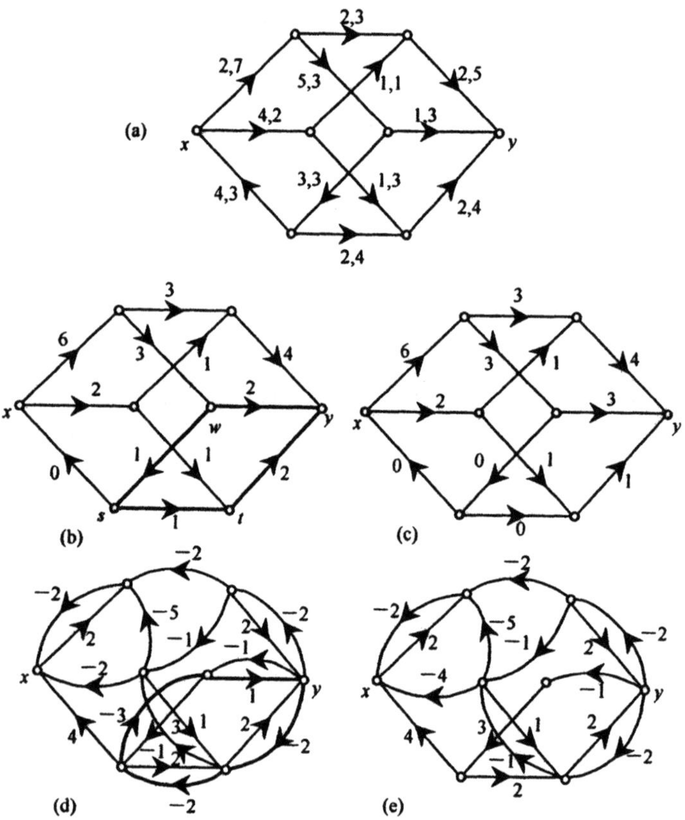

Figure 4.12: Applications of Klein's algorithm

Example 4.5.1 Consider the network in Figure 4.12 (a). We have found a maximum (x, y)-flow $\mathbf{f}$ in Section 4.4 shown in Figure 4.12 (b). $C =$

$s\,(w,s)\,w\,(w,y)\,y\,(t,y)\,t\,(s,t)\,s$ is a cycle of G. If we specify its orientation as the clockwise, then

$C^+ = \{(w,y)\}$ and $C^-\{(w,s),(t,y),(s,t)\}$,
$\sigma_{\mathbf{f}}(w,y) = \mathbf{c}(w,y) - \mathbf{f}(w,y) = 3 - 2 = 1$,
$\sigma_{\mathbf{f}}(w,s) = \mathbf{f}(w,s) = 1,\ \sigma_{\mathbf{f}}(t,y) = \mathbf{f}(t,y) = 2,\ \sigma_{\mathbf{f}}(s,t) = \mathbf{f}(s,t) = 1$,
$\sigma_{\mathbf{f}}(C) = \min\{\sigma_{\mathbf{f}}(a) :\ a \in E(C)\} = \min\{1,2\} = 1 > 0$.

Thus, C is an f-incrementing cycle, a revised flow $\widetilde{\mathbf{f}}_1$ based on C shown in Figure 4.12 (c), val $\widetilde{\mathbf{f}}_1$ = val $\mathbf{f} = 8$, and $\mathbf{b}(\mathbf{f}) = 62$, and $\mathbf{b}(\widetilde{\mathbf{f}}_1) = 56$. ∎

In general, for any f-incrementing cycle C and σ with $0 < |\sigma| \leq \sigma_{\mathbf{f}}$, we have

$$\begin{aligned} \mathbf{b}(\widetilde{\mathbf{f}}_\sigma) = &\ \mathbf{b}(\mathbf{f}) + \sigma\left(\sum_{a\in C^+} \mathbf{b}(a) - \sum_{a\in C^-} \mathbf{b}(a)\right) \\ = &\ \mathbf{b}(\mathbf{f}) + \sigma\,\mathbf{b}(C). \end{aligned} \tag{4.12}$$

Noting that $\sigma_{\mathbf{f}}(C) > 0$ and choosing $\sigma = \sigma_{\mathbf{f}}(C)$, we, from (4.12), immediately obtain a necessary condition for $\mathbf{f}$ to be a minimum-cost flow: $\mathbf{b}(C) \geq 0$ for any f-incrementing cycle C. We can show that the condition is also sufficient, and state this result as the following theorem, which is the basis of Klein's algorithm.

Theorem 4.9 An (x,y)-flow $\mathbf{f}$ in N is a minimum-cost flow if and only if $\mathbf{b}(C) \geq 0$ for any f-incrementing cycle C.

Proof By the above statement, we need to only prove the sufficiency. Let $\mathbf{f}$ be an (x,y)-flow in N and $\mathbf{b}(C) \geq 0$ for any f-incrementing cycle C. To show that $\mathbf{f}$ has minimum cost, we need to only prove $\mathbf{b}(\mathbf{f}^*) \geq \mathbf{b}(\mathbf{f})$ for any minimum-cost (x,y)-flow $\mathbf{f}^*$ with val $\mathbf{f}$. To the end, we construct a new digraph G' obtained by adding a new edge $a' = (y,x)$ to G. Adding

$$\mathbf{c}(a') = +\infty,\ \mathbf{b}(a') = 0,\ \mathbf{f}(a') = \mathbf{f}^*(a') = \mathrm{val}\,\mathbf{f},$$

we obtain a new network $N' = (G'_{xy}, \mathbf{b}, \mathbf{c})$, and $\mathbf{f}, \mathbf{f}^* \in \mathcal{C}(G')$.

Let T be a spanning tree of G' with $E(\overline{T}) = \{a_1, a_2, \cdots, a_n, a_{n+1}\}$, and $\{\mathbf{f}_1, \mathbf{f}_2, \cdots, \mathbf{f}_{n+1}\}$ be the basic vectors of the cycle-space $\mathcal{C}(G')$ corresponding to T, where $n = \varepsilon(G) - v(G) + 1$, $a_{n+1} = a'$, and $\mathbf{f}_i$ is a cycle vector corresponding to the cycle C_i obtained by adding the edge a_i to T , whose orientation is so specified that $\mathbf{f}_i(a_i) = 1$ for each $i = 1, 2, \cdots, n, n+1$. Then

$$\mathbf{f} = \sum_{i=1}^{n+1} \mathbf{f}(a_i)\ \mathbf{f}_i, \qquad \mathbf{f}^* = \sum_{i=1}^{n+1} \mathbf{f}^*(a_i)\ \mathbf{f}_i.$$

For each $i = 1, 2, \cdots, n, n+1$, define

$$\alpha_i = \mathbf{f}^*(a_i) - \mathbf{f}(a_i), \quad \text{and} \quad \sigma_i = \sigma_{\mathbf{f}}(C_i) - \sigma_{\mathbf{f}^*}(C_i).$$

It follows that $\alpha_i \geq \sigma_i$ for each $i = 1, 2, \cdots, n, n+1$ and

$$\mathbf{f}^* = \sum_{i=1}^{n+1} (\mathbf{f}(a_i) + \alpha_i)\, \mathbf{f}_i = \mathbf{f} + \sum_{i=1}^{n+1} \alpha_i\, \mathbf{f}_i. \tag{4.13}$$

Note that C_i is an $\mathbf{f}$-incrementing cycle if $\sigma_i > 0$ for some i, and so $b(C_i) \geq 0$ by the hypothesis. We now claim $b(C_i) = 0$ if $\sigma_i < 0$ for some i. In fact, if $\sigma_i < 0$, then C_i is an $\mathbf{f}^*$-incrementing cycle and so $b(C_i) \geq 0$ by the necessity. If $b(C_i) > 0$, then the revised flow $\widetilde{\mathbf{f}}^*_{\sigma_i}$ based on C_i satisfies $b(\widetilde{\mathbf{f}}^*_{\sigma_i}) - b(\mathbf{f}^*) = \sigma_i b(C_i) < 0$ by (4.12), which contradicts our hypothesis that $\mathbf{f}^*$ is a minimum-cost flow. It follows from (4.13) that

$$\begin{aligned} \mathbf{b}(\mathbf{f}^*) &= \mathbf{b}(\mathbf{f}) + \sum_{i=1}^{n+1} \alpha_i\, \mathbf{b}(\mathbf{f}_i) \geq \mathbf{b}(\mathbf{f}) + \sum_{i=1}^{n+1} \sigma_i\, \mathbf{b}(\mathbf{f}_i) \\ &= \mathbf{b}(\mathbf{f}) + \sum_{i=1}^{n+1} \sigma_i\, \mathbf{b}(C_i) \geq \mathbf{b}(\mathbf{f}) \end{aligned}$$

as desired. ∎

Theorem 4.9 provides an algorithm for finding minimum-cost maximum-flow in a network. This algorithm was first proposed by Klein [108] in 1967. Start with any maximum (x, y)-flow $\mathbf{f}$, check every $\mathbf{f}$-incrementing cycle. If the costs of all $\mathbf{f}$-incrementing cycles are non-negative, then $\mathbf{f}$ is a minimum-cost maximum-flow by Theorem 4.9. If there exists an $\mathbf{f}$-incrementing cycle C with negative cost, then replace $\mathbf{f}$ by the revised flow $\widetilde{\mathbf{f}}_{\sigma_{\mathbf{f}}}$ based on C and repeat the preceding process.

The key to Klein's algorithm is to find an $\mathbf{f}$-incrementing cycle with negative cost. We can do it by constructing a new weighted digraph $(G'(\mathbf{f}), \mathbf{w})$ with the same vertex-set as G. The edge-set E' of $G'(\mathbf{f})$ and the weighted function $\mathbf{w}$ are defined according to the following rules. For $a \in E(G)$,

if $\mathbf{f}(a) = 0$, then $a \in E'$ and $\mathbf{w}(a) = \mathbf{b}(a)$;

if $0 < \mathbf{f}(a) < \mathbf{c}(a)$, then $a, \overleftarrow{a} \in E'$, $\mathbf{w}(a) = \mathbf{b}(a)$ and $\mathbf{w}(\overleftarrow{a}) = -\mathbf{b}(a)$;

if $\mathbf{f}(a) = \mathbf{c}(a)$, then $\overleftarrow{a} \in E'$ and $\mathbf{w}(\overleftarrow{a}) = -\mathbf{b}(a)$,

which $\overleftarrow{a}$ denotes a new directed edge with the same end-vertices as a and the opposite direction to a.

For example, for the network and the (x, y)-flow $\mathbf{f}$ in Figure 4.12 (b), the corresponding $(G'(\mathbf{f}), \mathbf{w})$ is shown in Figure 4.12 (d), where the curves denote the edges of the form $\overleftarrow{a}$.

Let C be $\mathbf{f}$-incrementing cycle in $(G_{xy}, \mathbf{b}, \mathbf{c})$, $\overleftarrow{C}$ be a corresponding directed cycle in $G'(\mathbf{f})$ such that $C^+ \subseteq E(\overleftarrow{C})$. Then, by simply counting, $\mathbf{b}(C) = \mathbf{w}(\overleftarrow{C})$. Conversely, let $\overleftarrow{C}$ be a directed cycle in $G'(\mathbf{f})$, then there is an $\mathbf{f}$-incrementing cycle C such that $\mathbf{w}(\overleftarrow{C}) = \mathbf{b}(C)$. The details are left to the reader as an exercise (the exercise 4.5.2).

A directed cycle in $(G, \mathbf{w})$ is called to be *negative* if the sum of its weights is negative. Thus, making use of the above observation, Theorem 4.8 can be equivalently stated as the following theorem.

Theorem 4.10 An (x, y)-flow $\mathbf{f}$ in $N = (G_{xy}, \mathbf{b}, \mathbf{c})$ is a minimum-cost flow if and only if $(G(\mathbf{f}), \mathbf{w})$ contains no negative cycle. ∎

Yamada and Kinoshita [183] proposed an efficient algorithm for finding all the negative cycles a digraph. We now describe Klein's algorithm as follows.

Klein's Algorithm

1. Find a maximum (x, y)-flow $\mathbf{f}$ in N.

2. Construct $(G(\mathbf{f}), \mathbf{w})$.

3. If $(G(\mathbf{f}), \mathbf{w})$ contains no negative cycle, then, stop, by Theorem 4.10, $\mathbf{f}$ is a minimum-cost maximum-flow. If $(G(\mathbf{f}), \mathbf{w})$ contain a negative cycle $\overleftarrow{C}$, then, C corresponding to $\overleftarrow{C}$ is an $\mathbf{f}$-incrementing cycle, whose direction identifies with $\overleftarrow{C}$'s, replace $\mathbf{f}$ by the revised flow $\widetilde{\mathbf{f}}_{\sigma_\mathbf{f}}$ based on C, and go to Step 1.

Example 4.5.2 Consider the cost-capacity network in Figure 4.12 (a), a maximum (x, y)-flow $\mathbf{f}$ is shown in (b). By Klein's algorithm, $(G(\mathbf{f}), \mathbf{w})$ is shown in (d), in which the heavy edges denote a negative cycle. The corresponding $\mathbf{f}$-incrementing cycle C is shown in (b), and the revised flow $\widetilde{\mathbf{f}}_1$ based on C is shown in (c). Repeat Klein's algorithm for $\widetilde{\mathbf{f}}_1$, the corresponding $(G(\widetilde{\mathbf{f}}_1), \mathbf{w})$ is shown in (e), it contains no negative cycle. By Theorem 4.10, $\widetilde{\mathbf{f}}_1$ is a minimum-cost maximum-flow, for which $\mathbf{b}(\widetilde{\mathbf{f}}_1) = 56 < 62 = \mathbf{b}(\mathbf{f})$. ∎

It is not difficult to show that Klein's algorithm is efficient. More discussions on the minimum-cost flows are referred to Ford and Fulkerson [67].

Exercises

4.5.1 Prove that the function $\widetilde{\mathbf{f}} \in \mathcal{E}(G)$ defined by (4.11) is an (x, y)-flow in N, and val $\widetilde{\mathbf{f}}$ = val $\mathbf{f}$.

4.5.2 Prove Theorem 4.10 and prove that Klein's algorithm is efficient.

4.5.3 Use Klein's algorithm to find minimum-cost maximum (x, y)-flows in the following networks, respectively, where the ordered pair (b, c) of digits nearby the edge a denotes the values of the cost function $\mathbf{b}$ and the capacity function $\mathbf{c}$ on a, respectively, that is, $b = \mathbf{b}(a)$ and $c = \mathbf{c}(a)$.

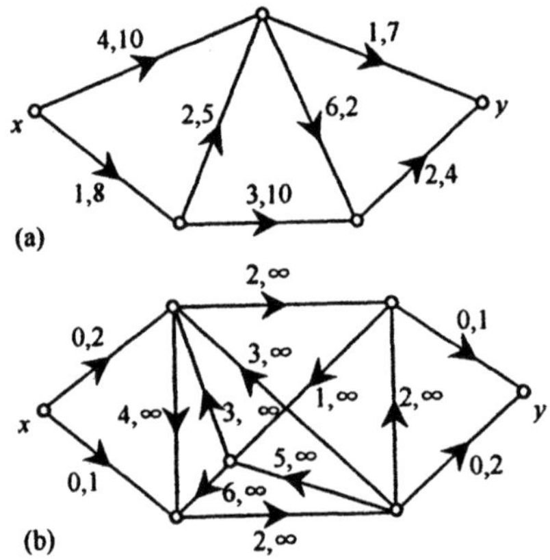

(the exercise 4.5.3)

4.5.4 Suppose that $\mathbf{f}$ is an (x, y)-flow in a network $N = (G_{xy}, \mathbf{b}, \mathbf{c})$, P an $\mathbf{f}$-incrementing path with cost as little as possible, $\widetilde{\mathbf{f}}$ the revised flow based on P.

(a) Prove that, if $\mathbf{f}$ is a minimum-cost maximum-flow, then so is $\widetilde{\mathbf{f}}$.

(b) Prove that finding an $\mathbf{f}$-incrementing path with minimum cost in N is equivalent to finding a shortest xy-path in $G(\mathbf{f})$.

(c) Using (a) and (b), design an algorithm for finding a minimum-cost maximum-flow in N, and by which shows that the flow obtained by the exercise 4.5.3 is a minimum-cost maximum-flow.

(R.G.Busacker and P.J.Gowen, 1961)

4.6 The Chinese Postman Problem

In his job, a postman picks up mail at the post office, delivers it, and then returns to the post office. He must, of course, cover each street in his area at least once. Subject to this condition, he wishes to choose his route in such a way that walks as little as possible. This problem is known as the *Chinese postman problem*, since it was first considered by a Chinese mathematician, Guan [80] in 1960.

We refer to the street system as a weighted graph $(G, \mathbf{w})$ whose vertices represent the intersections of the streets, whose edges represent the streets (one-way or two-way), and the weight represents the distance between two intersections, of course, a positive real number. A closed walk that covers each edge at least once in G is called a *postman tour*. Clearly, the Chinese postman problem is just that of finding a minimum-weight postman tour. We will refer to such a postman tour as an *optimal tour*.

There are many real-world situations that can be reduced as the Chinese postman problem. For example, a driver of a watering car or a garbage truck, he wishes to choose his route in such a way that traverses as little as possible. In this section, we introduce an efficient algorithm for solving the Chinese postman problem, due to Edmonds and Johnson [49].

First consider simple case that G is eulerian. Then any an Euler circuit is an optimal tour since it traverses each edge exactly once. The Chinese postman problem is easily solved in this case, since there exists an efficient algorithm determining an Euler circuit in an eulerian graph, no matter that it is directed or undirected.

We here consider G a directed eulerian graph. Then G is strongly connected, and so there must be a spanning out-tree rooted at x_0 in G for any $x_0 \in V(G)$ (by the exercise 2.1.10). Making use of Dijsktra's algorithm, we can find a spanning out-tree T rooted at x_0. Based on this tree T, Edmonds and Johnson's algorithm can find an Euler directed circuit in G.

Edmonds and Johnson's Algorithm

1. Choose arbitrarily a vertex x_0 in G, find a spanning out-tree T rooted at x_0 in G, and set $P_0 = x_0$.

2. Suppose that a directed trail $P_i = x_i a_i x_{i-1} a_{i-1} \cdots a_2 x_1 a_1 x_0$ has been chosen. Then choose an edge a_{i+1} from $E(G) \setminus \{a_1, a_2, \cdots, a_i\}$ in such

a way that

(i) $\psi_G(a_{i+1}) = (x_{i+1}, x_i)$;

(ii) $a_{i+1} \notin E(T)$ unless there is no alternative.

3. Stop when Step 2 can no longer be implemented.

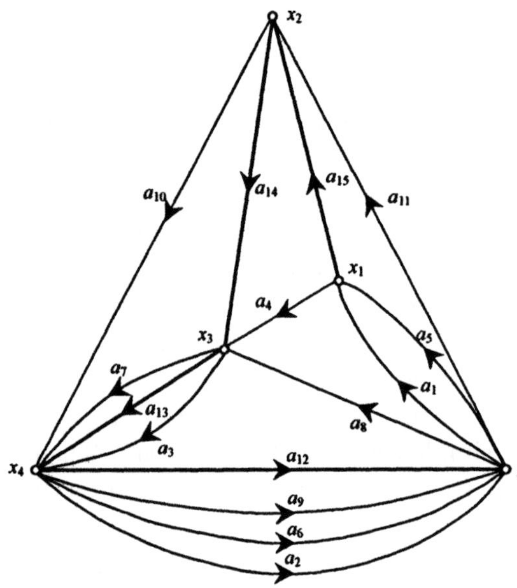

Figure 4.13: A spanning out-tree T rooted at x_1 in G

Theorem 4.11 If G is an eulerian digraph, then any directed trail in G constructed by the above algorithm is an Euler directed circuit in G.

Proof Let G be an eulerian digraph, and let $P_n = x_n a_n x_{n-1} a_{n-1} \cdots a_2 x_1 a_1 x_0$ be a directed trail in G constructed by the above algorithm. Since G is eulerian, G is balanced by Theorem 1.7, and so $x_n = x_0$.

Suppose, now, that P_n is not an Euler circuit of G. Then there is $b_1 \in E(G)$, but $b_1 \notin E(P_n)$. Let $\psi_G(b_1) = (x_i, x_j)$. Then, by Step 2 (ii), without loss of generality suppose $b_1 \in E(T)$. Since x_i is balanced in G and P_n, there is $b_2 \in E(G)$, but $b_2 \notin E(P_n)$ with $\psi_G(b_2) = (x_k, x_i) \in E(T)$. Similarly, there is $b_3 \in E(G)$, but $b_3 \notin E(P_n)$ with $\psi_G(b_3) = (x_l, x_k) \in E(T)$ and so on. Thus, there is a sequence of edges $b_1, b_2, b_3, \cdots$, which can be traced back to $x_0 = x_n$ along an (x_0, x_j)-path of T in reverse direction. Since x_n is balanced in G and P_n, there is $a \in E(G)$ with head x_n, but $a \notin E(P_n)$. This contradicts Step 3. Therefore, P_n is an Euler circuit of G. ∎

Example 4.6.1 Consider the digraph G in Figure 4.13. Since G is connected and balanced, by Theorem 1.7, G is eulerian. A spanning out-tree T rooted at x_1 in G is denoted by heavy edges. An Euler circuit constructed by Edmonds and Johnson's algorithm is as follows:

$$\begin{aligned} P = \; & x_1\ a_{15}\ x_2\ a_{14}\ x_3\ a_{13}\ x_4\ a_{12}\ x_5\ a_{11}\ x_2\ a_{10}\ x_4\ a_9 \\ & x_5\ a_8\ x_3\ a_7\ x_4\ a_6\ x_5\ a_5\ x_1\ a_4\ x_3\ a_3\ x_4\ a_2\ x_5\ a_1\ x_1. \end{aligned}$$ ∎

We, now, suppose that G is not eulerian. The digraph G in Figure 4.14 has no postman tour since it contains no directed path from $\{y_1, y_2, y_3\}$ to $\{x_1, x_2, x_3\}$. Thus, we first discuss the existence of post-tours.

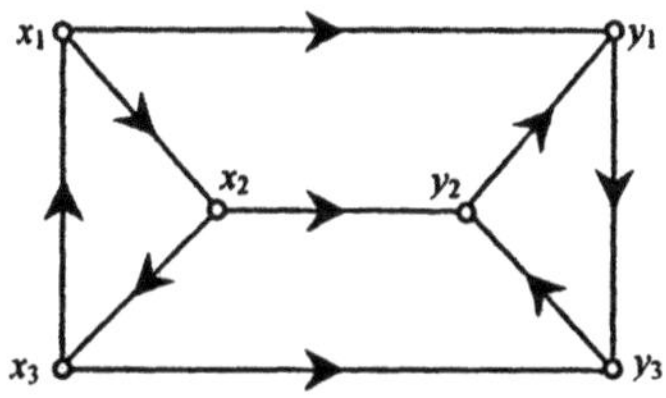

Figure 4.14: A digraph that contains no postman tour

Theorem 4.12 A digraph G contains a postman tour if and only if G is strongly connected.

Proof The necessity holds clearly. We suppose that G is strongly connected. Then G must contains a directed cycle. Choose a closed directed walk C that contains edges of G as many as possible. If C were not a postman tour, there should be $a \in E(G) \setminus E(C)$. Let $\psi_G(a) = (x, y)$ and choose arbitrarily a vertex u in C. There are one (u, x)-path P and one (y, u)-path Q. Then $C' = C \oplus (P + a) \oplus Q$ is a closed directed walk in G that contains edges of G more than C does, a contradiction. ∎

Suppose, now, that $(G, \mathbf{w})$ is a non-balanced and strongly connected weighted digraph, P is a postman tour in G. Thus, P must traverse each edge of G once or more. We denote by $p(a)$ the number of times that the edge a repeatedly occurs in P. Let G^* denote the supergraph of G obtained by adding extra $p(a)$ copies of a for each edge $a \in E(G)$ to G. Clearly, G^* is balanced and the postman tour P in G corresponds to an Euler circuit in G^*. Thus, a basic outline of solving the Chinese postman problem can be described as follows.

For a given strongly connected weighted digraph $(G, \mathbf{w})$,
(i) construct a balanced supergraph G^* of G such that the added edges have the sum of weight as little as possible;
(ii) finding an Euler directed circuit in G^*.

Edmonds and Johnson's algorithm described above solves (ii). We will introduce an algorithm for solving (i), also due to Edmonds and Johnson [49].
For $x \in V(G)$, let $\rho(x) = d^-_G(x) - d^+_G(x)$, and let

$$X = \{x \in V(G) : \ \rho(x) > 0\}, \quad \text{and} \quad Y = \{y \in V(G) : \ \rho(y) < 0\}.$$

Since G is non-balanced, by Theorem 1.1, we have $X \neq \emptyset$, $Y \neq \emptyset$ and

$$\sum_{x \in X} \rho(x) = -\sum_{y \in Y} \rho(y).$$

Denote by $\rho(G)$ the above value.

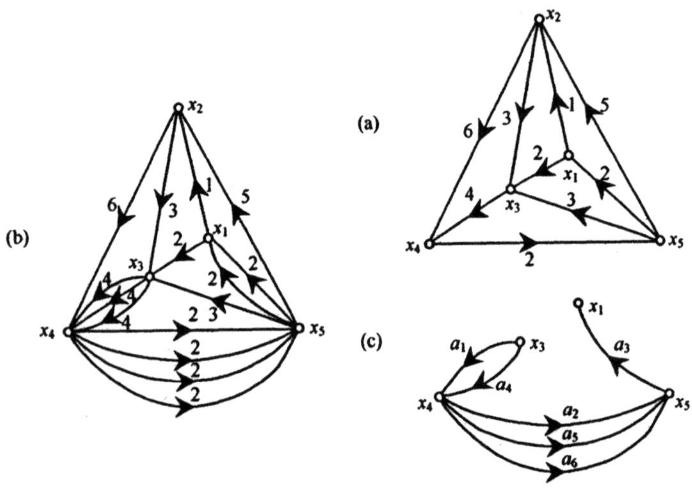

Figure 4.15: (a) A weighted digraph $(G, \mathbf{w})$; (b) $(G^*, \mathbf{w}^*)$; (c) $H = G^*[E^*]$

For example, for the digraph G in Figure 4.15 (a), we have

$$\rho(x_1) = -1,\ \rho(x_2) = 0,\ \rho(x_3) = 2,\ \rho(x_4) = 1,\ \rho(x_5) = -2,$$

and so

$$X = \{x_3, x_4\}, \quad Y = \{x_1, x_5\}, \quad \text{and} \quad \rho(G) = 3.$$

Suppose that, subject to (i), a supergraph G^* and the collection E^* of added edges have been chosen. For example, the digraph G^* in Figure 4.15 (b) is a supergraph G^* of G in (a), where E^* consists of the curved lines. Let $H = G^*[E^*]$. Then H consists of $\rho(G)$ edge-disjoint (X, Y)-paths. See, for example, Figure 4.15 (c), H consists of three edge-disjoint directed paths:

$$P_1 = x_3\, a_1\, x_4\, a_2\, x_5\, a_3\, x_1, \quad P_2 = x_3\, a_4\, x_4\, a_5\, x_5, \quad P_3 = x_4\, a_6\, x_5.$$

Conversely, the set of edges E^* of any $\rho(G)$ edge-disjoint (X, Y)-paths with minimum weight (if some edge a is used m times, then the weight of a must be computed m times) is a solution of (i) (since, in the present case, G^* is balanced). Thus, the solution of (i) is reduced to choosing ρ edge-disjoint (X, Y)-paths $P_1, P_2, \cdots, P_\rho$ in G such that the sum of their weights $\mathbf{w}(P_1) + \mathbf{w}(P_2) + \cdots + \mathbf{w}(P_\rho)$ is as little as possible, where $\rho = \rho(G)$.

To the end, we construct a cost-capacity network $N = (G'_{x_0 y_0}, \mathbf{b}, \mathbf{c})$, where G' is obtained from G by adding two new vertices x_0 and y_0, then joining x_0 to each vertex x in X by a directed edge with cost 0 and capacity $\rho(x)$; joining each vertex y in Y to y_0 by a directed edge with cost 0 and capacity $-\rho(y)$; $\mathbf{b}(a) = \mathbf{w}(a)$ and $\mathbf{c}(a) = \infty$ for each $a \in E(G)$.

Figure 4.16 (a) illustrates G' and $N = (G'_{x_0 y_0}, \mathbf{b}, \mathbf{c})$, where G is shown in Figure 4.15 (a).

Thus, each unit of (x_0, y_0)-flow $\mathbf{f}_0$ in N denotes an (x, y)-path P_0 in G, where $x \in X$ and $y \in Y$, $\mathbf{w}(P_0) = \mathbf{b}(\mathbf{f}_0)$. Since both $E^+_{G'}(x_0)$ and $E^-_{G'}(x_0)$ are (x_0, y_0)-cuts in G' and admit capacity $\rho(G)$, but all of other (x_0, y_0)-cuts admit capacity ∞, it follows that $E^+_{G'}(x_0)$ and $E^-_{G'}(x_0)$ are minimum (x_0, y_0)-cuts in G'. By Theorem 4.1, there exists a maximum (x_0, y_0)-flow $\mathbf{f}$ with value val $\mathbf{f} = \rho(G)$. Thus, finding a solution of (i) is reduced to finding a minimum-cost maximum-flow in N, the latter has been solved by Klein's algorithm described in the preceding section.

Summing up the above statement, for a given weighted digraph $(G, \mathbf{w})$, we can describe Edmonds-Johnson algorithm for solving the Chinese postman problem as follows.

Edmonds-Johnson's Algorithm

1. Construct G' and $N = (G'_{x_0 y_0}, \mathbf{b}, \mathbf{c})$;
2. Find a minimum-cost maximum-flow in N;
3. Construct a supergraph G^* of G;
4. Find an Euler directed circuit in G^*, which is an optimal postman tour in $(G, \mathbf{w})$.

(a)

(b)

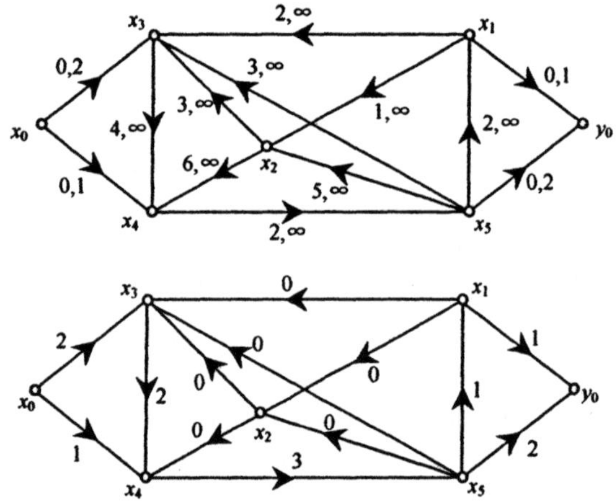

Figure 4.16: (a) $N = (G'_{x_0 y_0}, \mathbf{b}, \mathbf{c})$; (b) a minimum-cost maximum-flow in N

Example 4.6.2 Consider the weighted digraph $(G, \mathbf{w})$ in Figure 4.15 (a). G' and $N = (G'_{x_0 y_0}, \mathbf{b}, \mathbf{c})$ are shown in Figure 4.16 (a). A minimum-cost maximum-flow $\mathbf{f}$ is shown in Figure 4.16 (b), where $\mathbf{f}(a)$ denotes the times that the edge a appears in E^*. A supergraph G^* of G is shown in Figure 4.13 or Figure 4.15 (b); an optimal tour is

$$P = (x_1, x_2, x_3, x_4, x_5, x_2, x_4, x_5, x_3, x_4, x_5, x_1, x_3, x_4, x_5, x_1)$$

with weight $\mathbf{w}(P) = 44$. ∎

It is not difficult to see the algorithm is efficient since execution of its each step can be completed in a polynomial time, the detail is left to the reader as an exercise (the exercise 4.6.3).

Exercises

4.6.1 Find an Euler directed circuit in the following digraph, applying the first Edmonds-Johnson's algorithm.

4.6.2 Find an optimal tour of the following weighted digraph and its weight-sum.

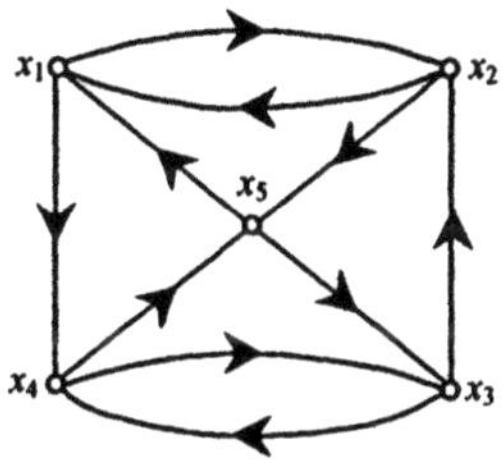

(the exercise 4.6.1)

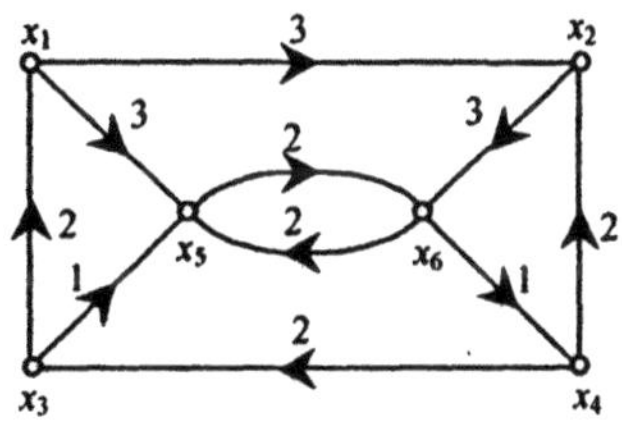

(the exercise 4.6.2)

4.6.3 Prove that Edmonds-Johnson's algorithm for finding an optimal postman tour in a non-eulerian weighted digraph is efficient.

4.6.4 Prove that a postman tour P of a weighted undirected graph $(G, \mathbf{w})$ is optimal if and only if it satisfies the following conditions:

(i) no edge in G appears in P than two times;

(ii) for any cycle C of G, the weight-sum of the edges in C that belong to multiple-edges is at most $\frac{1}{2}\mathbf{w}(C)$. (M.G.Guan, 1960)

4.6.5 Find an optimal tour of the following weighted undirected graph and its weight-sum.

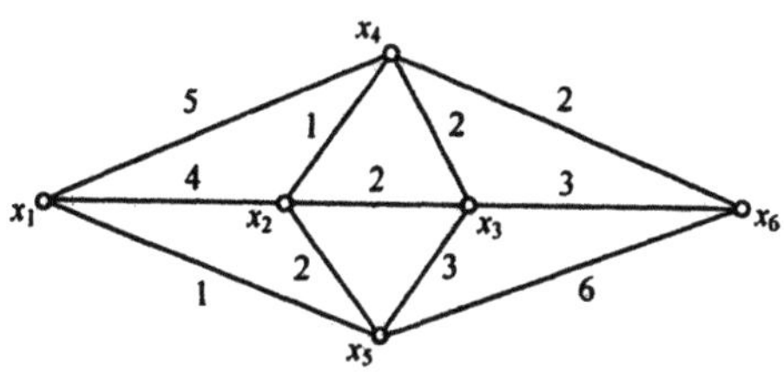

(the exercise 4.6.5)

4.7 Construction of Squared Rectangles

In this section, we will introduce a famous combinatorial problem, squared rectangles (squares), arising from recreational mathematics whose solution is related to connectivity of planar graphs and theory of network flows.

A *squared rectangle* (*square*) is a rectangle (square) dissected into a finitely many (but at least two) squares. A squared rectangle is called to be *perfect* if no two of the squares in the dissection have the same size. The *order* of a squared rectangle is the number of squares into which it is dissected. A squared rectangle is *simple* if it does not contains a rectangle which is self squared. Clearly, every squared rectangle is composed of ones that are simple.

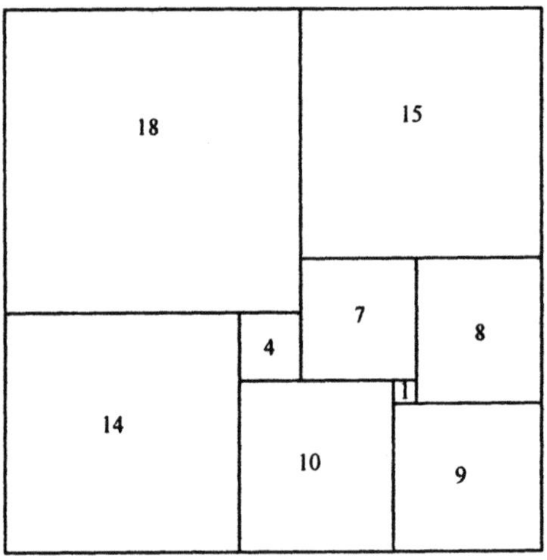

Figure 4.17: The perfect squaring of the 33 × 32 rectangle of order 9

Figure 4.17 shows a perfect rectangle of order 9, due to by Moroń [136], in which the digit associated with a square is the length of its side.

Since the beginning of the 20th century, the problem of squared rectangles (squares) has received much attention. In 1940, with the aid of theory of graphs, Brooks *et al.* [22] developed a systematic method for constructing of a squared rectangle, proved the lowest order of a perfect squared rectangle is 9 and gave a list of perfect squared rectangles of order 9 through 11. In 1964, Bouwkamp et al. gave a list of all perfect squared rectangles of order 9 order is as follows.

Order	9	10	11	12	13	14	15	16	17	18
Number	2	6	22	67	213	744	2609	9016	31427	110384

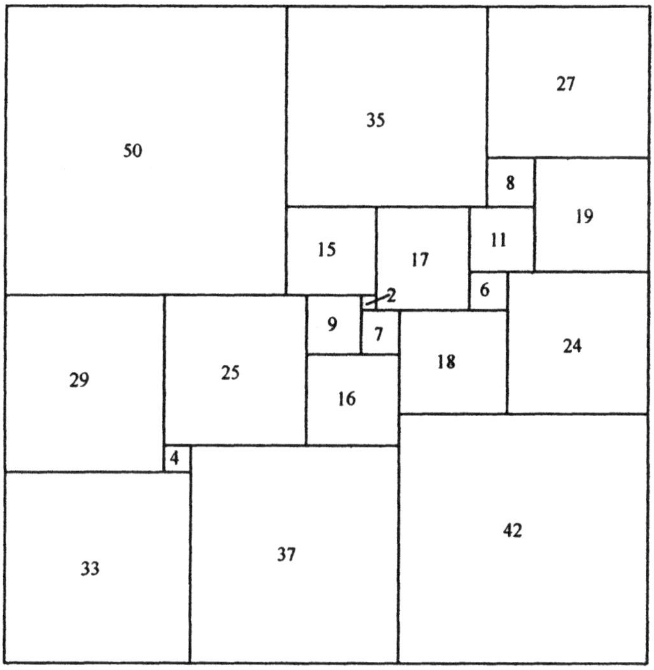

Figure 4.18: The perfect 112-square of order 21

For a long time no perfect square was known, and it was conjectured that such squares might not exist. In 1939, Sprague [157] was the first to publish an example of a perfect 4205-square of order 55 obtained by a make-up of several known perfect squared rectangles. In 1940, in this manner Brooks *et al.* [22] constructed a perfect 608-square of order 26. Wilson constructed a perfect 112-square of order 25 by a computer. In 1978, Duijvestijn [44] found a perfect 112-square of order 21 also with the help of a computer, shown in Figure 4.18. There are none below order 21 and only one of order 21. The number of known perfect squares of lower order is as follows.

Order	21	22	23	24	25	26	27	28	29	30	31
Number	1	?	?	?	8	28	6	?	?	?	4

We now introduce the method of Brooks *et al.* for constructing a perfect squared rectangle. The treatment stated here follows Bondy and Murty [18].

We first show how a simple digraph D can be associated with a given squared rectangle R of order n. A horizontal line segment of the dissection of R is called a *horizontal dissector* of R. For example, In Figure 4.19 (a), the horizontal dissectors are indicated by solid line segments.

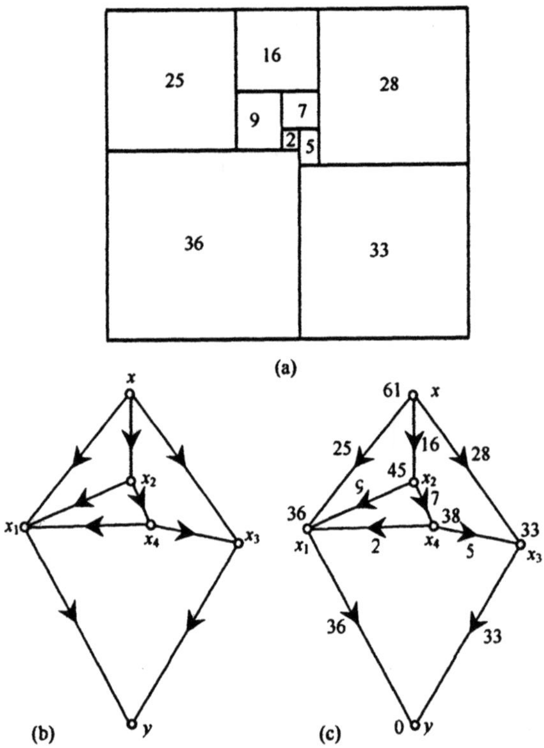

Figure 4.19: A squared rectangle and an associated digraph

Let $H_1, H_2, \cdots, H_m$ be all horizontal dissectors of R. Define a simple digraph $D = (V, E)$ as follows. $V = \{x_1, x_2, \cdots, x_m\}$ and $(x_i, x_j) \in E$ if and only if H_i and H_j are the upper and lower sides of some square of in R. The vertices corresponding to the upper and lower sides of R are denoted by x

and y, respectively. Clearly, $\varepsilon(D) = n$. Figure 4.19 (b) shows the digraph D associated with the squared rectangle R in Figure 4.19 (a).

Define a vector $\mathbf{p} \in \mathcal{V}(D)$ with $\mathbf{p}(x_i)$ equal to the height (above the lower side of R) of the corresponding horizontal dissector, see Figure 4.19 (c). We can regard D as a network with enough large capacity and the source x and the sink y. It is easily verified (the exercise 4.7.2) that the bond-vector $\mathbf{g} \in \mathcal{E}(D)$ defined by

$$\mathbf{g}(a) = \mathbf{p}(x_i) - \mathbf{p}(x_j), \quad \forall\, a = (x_i, x_j) \in E(D) \tag{4.14}$$

is an (x, y)-flow. See Figure 4.19 (c), for example, the digits nearby edges determine an (x, y)-flow $\mathbf{g}$ with value 69.

Let D be the digraph corresponding to a squared rectangle R, and let G be the underlying graph of D. The graph $G + xy$ is called the *horizontal graph* of R. Brooks *et al.* [22] showed that the horizontal graph of any simple squared rectangle is a 3-connected planar graph. They also showed that, conversely, if H is a 3-connected planar graph and $xy \in E(H)$, then an (x, y)-flow defined by any bond-vector in $\mathcal{E}(H - xy)$ determines a squared rectangle. Thus, a possible way of searching for perfect rectangles of order n is to

1. list all 3-connected planar graphs with $n + 1$ edges, and

2. for each such graph H and each edge xy of H, determine an (x, y)-flow defined by a bond-vector in $\mathcal{E}(H - xy)$.

We now introduce how to compute such an (x, y)-flow in D. Suppose $\mathbf{g} \in \mathcal{B}(D)$ is an (x, y)-flow with value σ. Then

$$\sum_{a \in E(D)} m_x(a)\mathbf{g}(a) = \sigma, \tag{4.15}$$

and for any $x_i \in V(D) \setminus \{x, y\}$,

$$\sum_{a \in E(D)} m_{x_i}(a)\mathbf{g}(a) = 0. \tag{4.16}$$

By Theorem 2.7, $\mathbf{g}$ is orthogonal to every cycle-vector of $\mathcal{E}(D)$, that is,

$$\mathbf{C}\,\mathbf{g}^T = 0, \tag{4.17}$$

where $\mathbf{C}$ is a basis matrix of the cycle-space $\mathcal{C}(D)$ corresponding to a spanning tree T of D and $\mathbf{g}^T$ is the transpose of the vector $\mathbf{g}$. Equations (4.15) – (4.17) together give the matrix equation

$$\begin{pmatrix} \mathbf{K} \\ \mathbf{C} \end{pmatrix} \mathbf{g}^T = \begin{pmatrix} \sigma \\ \mathbf{O} \end{pmatrix}, \tag{4.18}$$

where $\mathbf{K}$ is the matrix obtained from the incidence matrix $\mathbf{M}$ of D by deleting the row m_y. The equation can be solved using Cramér's rule. Note that, since (the exercise 2.3.1)

$$\det \begin{pmatrix} \mathbf{K} \\ \mathbf{C} \end{pmatrix} = \pm\varsigma(D),$$

we can obtain an integral solution if $\sigma = \varsigma(D)$. Thus, in computing the (x, y)-flow $\mathbf{g}$, it is convenient to take $\operatorname{val}\mathbf{g} = \varsigma(D)$.

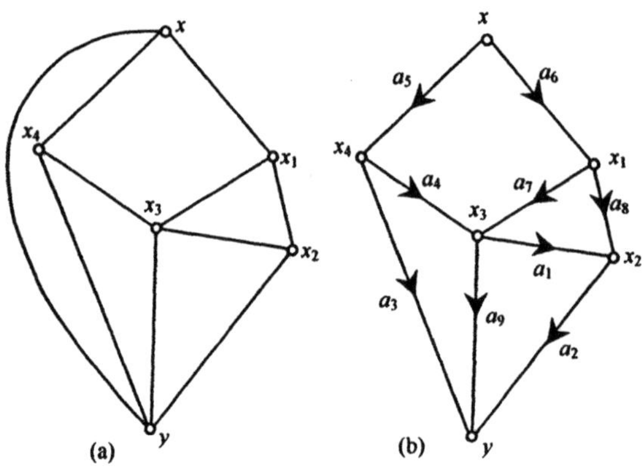

Figure 4.20: An illustration in Example 4.7.1

We illustrate the above procedure.

Example 4.7.1 Consider the 3-connected planar graph in Figure 4.20 (a). On deleting the edge xy and orienting each edge we obtain the digraph D of Figure 4.20 (b). The matrix

$$\mathbf{K} = \begin{pmatrix} 0 & 0 & 0 & 0 & 0 & -1 & 1 & 1 & 0 \\ -1 & 1 & 0 & 0 & 0 & 0 & 0 & -1 & 0 \\ 1 & 0 & 0 & -1 & 0 & 0 & -1 & 0 & 1 \\ 0 & 0 & 1 & 1 & -1 & 0 & 0 & 0 & 0 \\ 0 & 0 & 0 & 0 & 1 & 1 & 0 & 0 & 0 \end{pmatrix}.$$

It has been computed in Example 2.3.3 that $\varsigma(D) = \det \mathbf{K}\mathbf{K}^T = 66$.

Choose the spanning tree T of D induced by the set of edges $\{a_5, a_6, a_7, a_8, a_9\}$. Then the basis matrix $\mathbf{C}$ of the cycle-space $\mathcal{C}(D)$ corresponding to a spanning tree T of D is

$$\mathbf{C} = \begin{pmatrix} 1 & 0 & 0 & 0 & 0 & 0 & 1 & -1 & 0 \\ 0 & 1 & 0 & 0 & 0 & 0 & -1 & 1 & -1 \\ 0 & 0 & 1 & 0 & 1 & -1 & -1 & 0 & 0 \\ 0 & 0 & 0 & 1 & 1 & -1 & -1 & 0 & 0 \end{pmatrix}.$$

Suppose that the required bond-vector $\mathbf{g} \in \mathcal{E}(D)$ is

$$\mathbf{g} = (g_1, g_2, g_3, g_4, g_5, g_6, g_7, g_8, g_9),$$

where

$$g_i = \mathbf{g}(a_i), \quad \text{for } i = 1, 2, \cdots, 9.$$

We obtain the following nine equations, as in (4.18):

$$\begin{pmatrix} 0 & 0 & 0 & 0 & 0 & -1 & 1 & 1 & 0 \\ -1 & 1 & 0 & 0 & 0 & 0 & 0 & -1 & 0 \\ 1 & 0 & 0 & -1 & 0 & 0 & -1 & 0 & 1 \\ 0 & 0 & 1 & 1 & -1 & 0 & 0 & 0 & 0 \\ 0 & 0 & 0 & 0 & 1 & 1 & 0 & 0 & 0 \\ 1 & 0 & 0 & 0 & 0 & 0 & 1 & -1 & 0 \\ 0 & 1 & 0 & 0 & 0 & 0 & -1 & 1 & -1 \\ 0 & 0 & 1 & 0 & 1 & -1 & -1 & 0 & 0 \\ 0 & 0 & 0 & 1 & 1 & -1 & -1 & 0 & 0 \end{pmatrix} \begin{pmatrix} g_1 \\ g_2 \\ g_3 \\ g_4 \\ g_5 \\ g_6 \\ g_7 \\ g_8 \\ g_9 \end{pmatrix} = \begin{pmatrix} 66 \\ 0 \\ 0 \\ 0 \\ 0 \\ 0 \\ 0 \\ 0 \\ 0 \end{pmatrix}$$

The solution to this system of equations is given by

$$\mathbf{g} = (g_1, g_2, g_3, g_4, g_5, g_6, g_7, g_8, g_9) = (36, 30, 14, 16, 20, 2, 18, 28, 8).$$

The squared rectangle based on this (x, y)-flow is just the one in Figure 4.17 with all dimensions doubled.

For further results and a historical overview of the squared rectangles and squared squares problems, see the survey article by Federico [61] and the recent paper by Duijvestijn [45].

Exercises

4.7.1 Prove that the ratio of two neighboring sides of any squared rectangle is rational. (This result is due to M. Dehn (1930), and generalized by Sprague (1940) who showed that a rectangle has a perfect squaring if and only if the ratio of two neighboring sides is rational.)

4.7.2 Prove that $\mathbf{g} \in \mathcal{B}(D)$ defined by (4.14) is an (x, y)-flow.

4.7.3 Using the way described this section, find an (x, y)-flow $\mathbf{g} \in \mathcal{B}(D)$, where the digraph D is shown in Figure 4.19 (b), from which construct a squared rectangle.

4.7.4 It has been shown that there are two essentially different squared rectangles of order 9; the one shown in Figure 4.17, the other indicated as follows, which is a 69×61 rectangle. Find the side length of each square.

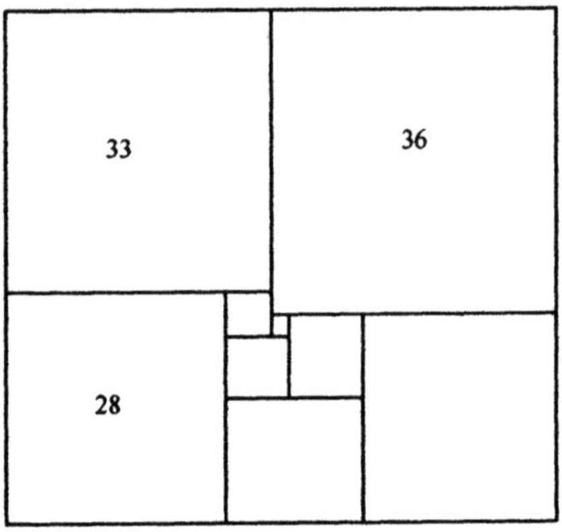

(the exercise 4.7.4)

4.7.5 Prove that there is no perfect squared rectangle of order less than 9.

4.7.6 A *perfect equilateral triangle* is an equilateral triangle dissected into a finitely many (but at least two) smaller equilateral triangles, no two of the same size. Prove that there exists no perfect equilateral triangle.

4.7.7 A *perfect cube* is a cube dissected into a finitely many (but at least two) smaller cubes, no two of the same size. Prove that there exists no perfect cube.

Chapter 5

Matchings and Independent Sets

Central to the content of this chapter is to consider two special classes of subsets in a graph, that is, matchings and independent sets.

A matching of a graph is a subset of its edges such that no two elements of the subset are adjacent. We will present two basic results on matching theory: Hall's theorem and Tutte's theorem, which characterize the existence of a perfect matching in a bipartite graph and a general graph, respectively, and a related result, Köning's theorem on vertex-covering. These theorems are equivalent to Manger's theorem. Thus, in essence, the matching theory provided in this chapter is applications and extensions of Menger's theorem.

The concept of independent sets is analogous to that of matchings, replacing edges by vertices. There exists, however, no theory of independent sets comparable to that of matchings. We discuss the relationships between independence number and connectivity as well as hamiltonicity of a graph.

A number of real-world problems can be described as a matching in a direct and natural way by choosing a appropriate graph. As applications of matching theory, we will present three classical efficient algorithms: Hungarian method for find a maximum matching of a bipartite graph to solve the personnel assignment problem, Kuhn-Munkres' algorithm for finding a maximum-weight perfect matching of a weighted bipartite graph to solve the optimal assignment problem, and Christofides' algorithm for giving an approximate solution of the travelling salesman problem.

5.1 Matchings

Let G be a nonempty and loopless graph. A nonempty subset M of $E(G)$ is called a *matching* in G if no two of elements of M are adjacent in G; the two end-vertices of an edge in M are said to be *matched under* M. A matching M *saturates* a vertex x, or x is M*-saturated*, if some edge of M is incident with x; otherwise, x is M*-unsaturated.* A matching M is *prefect* if it saturates every vertex of G; and *maximum* if $|M| \geq |M'|$ for any matching M' in G. It is clear that a prefect matching is maximum, the converse, however, is not true in general.

The concept of matching has no relation to orientations of edges. Therefore, in discussing properties of matching, we will restrict ourselves to undirected graphs. A maximum matching and a prefect matching in the graphs are indicated in Figure 5.1 by heavy lines.

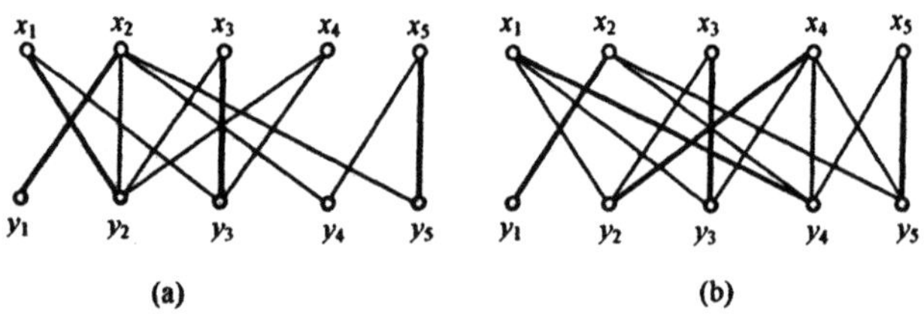

Figure 5.1: (a) A maximum matching; (b) a prefect matching

Necessary and sufficient conditions for the existence of a prefect matching in G is what we are mainly concerned with in this section. Such a condition for a bipartite graph was first established by Hall [82] in 1935. We deduce it from Menger's theorem.

Theorem 5.1 (Hall's theorem) Let G be a bipartite graph with bipartition $\{X, Y\}$. Then G contains a matching that saturates every vertex in X if and only if

$$|S| \leq |N_G(S)| \quad \text{for all } S \subseteq X. \tag{5.1}$$

Proof Suppose that G contains a matching M which saturates every vertex in X and let S be a subset of X. Since the vertices in S are matched under M with distinct vertices in $N_G(S)$, we have $|S| \leq |N_G(S)|$ clearly.

Conversely, suppose that M is a maximum matching in G. We need to only prove $|M| \geq |X|$. To the end, we construct a digraph D obtained from

G by adding two new vertices x and y, joining x to each vertex x' in X by a directed edge (x, x'); joining each vertex y' in Y to y by a directed edge (y', y); then assigning a direction to each edge $x'y'$ of G from x' to y', where $x' \in X$ and $y' \in Y$. Such a construction is illustrated in Figure 5.2.

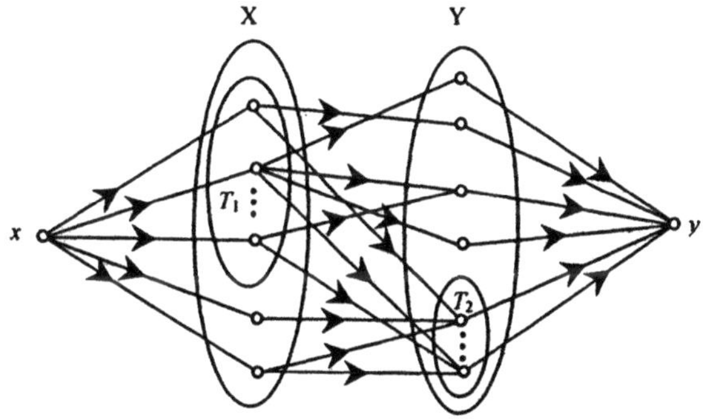

Figure 5.2: An illustration of the proof of Theorem 5.1

Clearly, $\zeta_D(x, y) = |M|$. Let T be a minimum (x, y)-separating set of D. Then by Menger's theorem (Theorem 4.3), we have

$$|M| = \zeta_D(x, y) = \kappa_D(x, y) = |T|.$$

Let $T_1 = T \cap X$ and $T_2 = T \cap Y$ (see Figure 5.2). Then $E_D(X \setminus T_1, Y \setminus T_2) = \emptyset$, which implies $N_D^+(X \setminus T_1) \subseteq T_2$. It follows from this fact and (5.1) that

$$\begin{aligned} |M| = & \ |T| = |T_1| + |T_2| \geq |T_1| + |N_D^+(X \setminus T_1)| \\ = & \ |T_1| + |N_G(X \setminus T_1)| \geq |T_1| + |X \setminus T_1| = |X| \end{aligned}$$

as desired. ∎

Corollary 5.1.1 (Marriage theorem, Forbenius [64]) A bipartite graph G with bipartition $\{X, Y\}$ has a prefect matching if and only if $|X| = |Y|$ and $|S| \leq |N_G(S)|$ for any subset $S \subseteq X$ or Y. ∎

Corollary 5.1.2 (König [111]) If G is a k-regular bipartite graph with $k > 0$, then G has a prefect matching.

Proof Let G be a k-regular bipartite graph with bipartition $\{X, Y\}$. Then $k|X| = k|Y|$ and so $|X| = |Y|$ since $k > 0$. Now let S be any subset of X and denote by E_1 and E_2 the sets of edges incident with vertices in S and $N_G(S)$, respectively. By definition of $N_G(S)$, $E_1 \subseteq E_2$ and therefore

$$k|S| = |E_1| \leq |E_2| = k|N_G(S)|.$$

It follows that $|S| \leq |N_G(S)|$ and hence, by Hall's theorem (5.1), that G has a matching M saturating every vertex in X. Since $|X| = |Y|$, M is a perfect matching in G. ∎

Corollary 5.1.3 Let G be an equally bipartite simple graph of order $2n$. If $\delta(G) \geq \frac{n}{2}$, then G has a prefect matching.

Proof Let G be an equally bipartite simple graph with bipartition $\{X, Y\}$. Suppose that there exists a subset S of X such that $|S| > |N_G(S)|$. Then $Y \setminus N_G(S) \neq \emptyset$ since $|X| = |Y|$, and $|S| > |N_G(S)| \geq \delta(G) \geq \frac{n}{2}$ since G is simple. Let $u \in Y \setminus N_G(S)$. Then $N_G(u) \subseteq X \setminus S$. We can deduce a contradiction as follows.

$$\delta(G) \leq d_G(u) = |N_G(u)| \leq |X| - |S| < \frac{n}{2}.$$

Therefore, $|S| \leq |N_G(S)|$ for all $S \subseteq X$. By Hall's theorem (5.1) and $|X| = |Y|$, G has a prefect matching. ∎

A necessary and sufficient condition for general graph to have a prefect matching was established by Tutte [163] in 1947. The proof given here is due to Mader [124].

A component of a graph G is *odd* or *even* according as it has an odd or even number of vertices. We denote by $o(G)$ the number of odd components of G.

Theorem 5.2 (Tutte's theorem) A graph G has a perfect matching if and only if

$$o(G - S) \leq |S| \quad \text{for all } S \subset V(G). \tag{5.2}$$

Proof Clearly, we need to only prove the theorem for a simple graph.

Suppose that G has a prefect matching M. Let S be a proper subset of $V(G)$, and let $G_1, G_2, \cdots, G_n$ be all odd components of $G - S$, where $n = o(G - S)$. Then, since $v(G_i)$ is odd, there must exist some $u_i \in V(G_i)$ and $w_i \in S$ such that $\{u_i w_i : i = 1, 2, \cdots, n\} \subseteq M$. Thus

$$o(G - S) = n = |\{w_1, w_2, \cdots, w_n\}| \leq |S|.$$

Conversely, suppose that G satisfies (5.2). In particular, setting $S = \emptyset$, we see that $o(G - S) = 0$, and so $v = v(G)$ is even. We prove that G has a perfect matching by induction $v \geq 2$. If $v = 2$, then the conclusion holds clearly. Suppose that every graph of order fewer than v has a perfect matching provided that it satisfies (5.2) and suppose that G has an even order v (≥ 4) and satisfies (5.2).

Let U be a maximum subset of $V(G)$ for which the equality in (5.2) holds, and let $|U| = m$. By the induction hypothesis, we can suppose $m \neq 0$. Let $G_1, G_2, \cdots, G_m$ be all odd components of $G - U$. We first show the following three propositions.

(i) $G - U$ has no even component. In fact, if H is an even component of $G - U$, then, for any $x \in V(H)$,

$$m + 1 \leq o(G - (U \cup \{x\})) \leq |U \cup \{x\}| = m + 1,$$

which implies that $o(G - (U \cup \{x\})) = |U \cup \{x\}|$. This contradicts the choice of U.

(ii) $G_i - x$ has a perfect matching for every $x \in V(G_i)$. Otherwise, by the induction hypothesis, there exists $S \subset V(G_i - x)$ such that $o((G_i - x) - S) > |S|$. Since $o((G_i - x) - S)$ and $|S|$ are of the same parity, it follows $o((G_i - x) - S) \geq |S| + 2$, and so

$$\begin{aligned} |U| + 1 + |S| = & \ |U \cup S \cup \{x\}| \geq o(G - (U \cup S \cup \{x\})) \\ = & \ o(G - U) - 1 + o((G_i - x) - S) \\ \geq & \ |U| + 1 + |S|, \end{aligned}$$

which contradicts the choice of U.

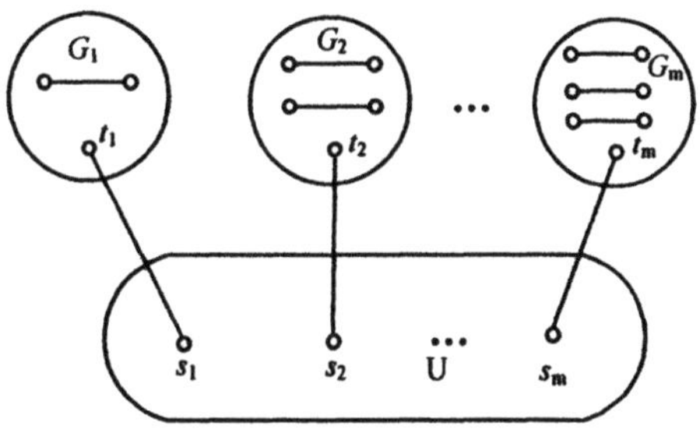

Figure 5.3: An illustration of the proof of Theorem 5.2

(iii) G has a matching $M = \{t_i u_i : \ t_i \in V(G_i), \ u_i \in U, \ i = 1, 2, \cdots, m\}$. Construct a bipartite graph H with bipartition $\{X, U\}$, where $X = \{G_1, G_2,$

$\cdots, G_m\}$; G_i and some $u \in U$ are linked by an edge in H if and only if G has an edge joining u and some vertex of G_i. Arbitrarily choose $A \subseteq X$, and let $B = N_H(A) \subseteq U$. Then, since every element in A is an odd component of $G - B$ and by (5.2), we have

$$|A| \leq o(G - B) \leq |B| = |N_H(A)|,$$

that is, H satisfies Hall's theorem (5.1), and so (iii) follows.

By (i), (ii), (iii) (see Figure 5.3) and the induction principle, the theorem follows. ∎

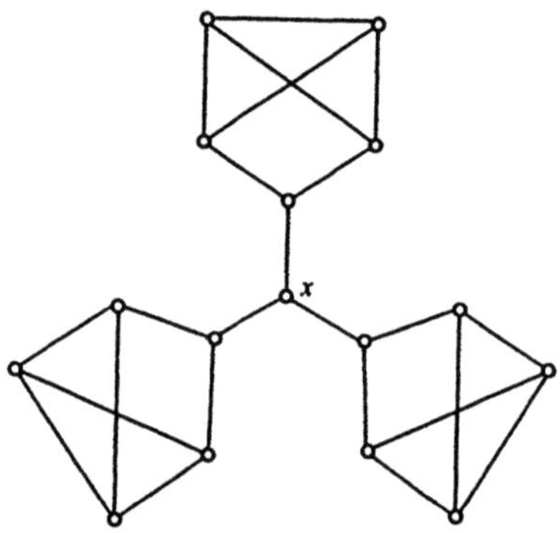

Figure 5.4: A 3-regular graph having no perfect matching

Corollary 5.2.1 Every k-regular $(k-1)$-edge connected graph of even order has a perfect matching for any integer $k \geq 1$.

Proof Suppose that G is, k-regular and $(k-1)$-edge connected graph of even order. If $k = 1$, then the result holds clearly. Suppose $k \geq 2$ below, and let S be any proper subset of $V(G)$. By Tutte's theorem (Theorem 5.2), it is sufficient to prove that S satisfies the condition (5.2).

If $S = \emptyset$, then there is nothing to do. Suppose $S \neq \emptyset$ below. Let $G_1, G_2, \cdots, G_n$ be all odd components of $G - S$ and let $m_i = |E_G(V(G_i), S)|$, $v_i = V(G_i)$ for each $i = 1, 2, \cdots, n$. Then, since $\lambda(G) \geq k - 1$, we have $m_i \geq k - 1$ for each $i = 1, 2, \cdots, n$. If there exists some $i\,(1 \leq i \leq n)$ such that $m_i = k - 1$, then

$$\varepsilon(G_i) = \frac{1}{2}(kv_i - k + 1) = \frac{1}{2}k(v_i - 1) + \frac{1}{2}$$

is not an integer, a contradiction. Therefore, $m_i \geq k$ for each $i = 1, 2, \cdots, n$. It follows that

$$o(G - S) = n \leq \frac{1}{k} \sum_{i=1}^{n} m_i \leq \frac{1}{k} \sum_{u \in S} d_G(u) = |S|$$

as desired. ∎

Corollary 5.2.2 (Petersen [143]) Every 2-edge connected and 3-regular graph has a perfect matching. ∎

A 3-regular graph with cut edges does not always have a perfect matching. For example, it follows from Tutte's theorem that the graph G of Figure 5.4 has no perfect matching, since $o(G - x) = 3$.

We have deduced Hall's theorem from Menger's theorem, and Tutte's theorem from Hall's theorem. In fact, these theorems are equivalent (the exercise 5.1.12). In combinatorics, there exist another important result, known as König's theorem which is equivalent to theses theorems. To state König's theorem, we need the following notion.

Let G be a loopless graph. A nonempty set K of $V(G)$ is a *(vertex-) covering* of G if every edge of G has at least one end-vertex in K. A covering K is *minimum* if $|K| \leq |K'|$ for any covering K' of G; and *minimal* if $K \setminus \{x\}$ is not a covering of G for any $x \in K$.

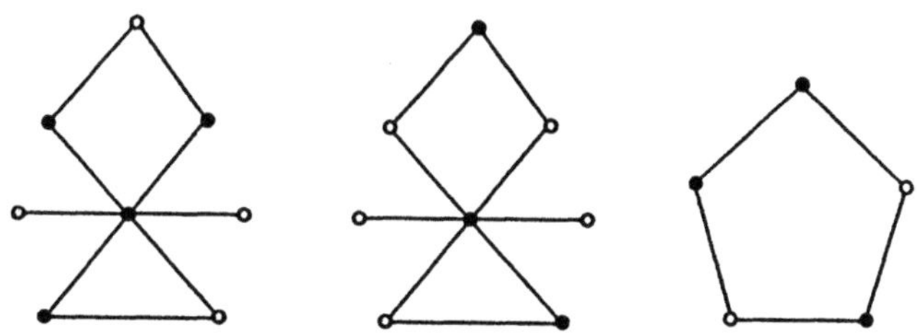

(a) (b) (c)

Figure 5.5: Minimal and minimum coverings

It is clear that every minimum covering is minimal, but the converse is not true in general. Figure 5.5 illustrates tree minimal coverings, where (b) and (c) are minimum coverings.

The number of vertices in a minimum covering of G is called the *covering number* of G and denoted by $\beta(G)$; the number of edges in a maximum matching of G is called the *matching number* of G and denoted by $\alpha'(G)$. For example,

$$\beta(K_n) = n-1,\ \alpha'(K_n) = \lfloor \tfrac{n}{2} \rfloor;$$
$$\beta(C_n) = \lceil \tfrac{n}{2} \rceil,\ \alpha'(C_n) = \lfloor \tfrac{n}{2} \rfloor,$$
$$\beta(K_{m,n}) = \min\{m,n\} = \alpha'(K_{m,n}).$$

If K is a covering of G, and M is a matching of G, then K contains at least one end-vertex of each of edges in M. Thus, we obtain the following inequality for any loopless graph G immediately:

$$\alpha'(G) \le \beta(G). \tag{5.3}$$

In general, the equality in (5.3) does not always holds. For example, $\beta(K_n) = n-1 > \lfloor \frac{n}{2} \rfloor = \alpha'(K_n)$ for $n \ge 3$. However, for a bipartite graph, König [112] proved that equality in (5.3) holds always. There are many direct and indirect proofs of König's theorem. We will here deduce this result from Hall's theorem.

Theorem 5.3 (König's Theorem) for any bipartite graph G,

$$\alpha'(G) = \beta(G).$$

Proof By (5.3), we need to only prove $\alpha'(G) \ge \beta(G)$ for any bipartite graph G. Let $\{X, Y\}$ be a bipartition of G, K a minimum covering of G, and let

$$S = K \cap X, \quad T = K \cap Y, \quad S' = X \setminus S, \quad T' = Y \setminus T.$$

It follows from definition of the covering that there is no edge between S' and T' in G. Consider the subgraph $H = G[S \cup T']$. Since K is a minimum covering of G, $|R| \le |N_H(R)|$ for any subset R of S. By Hall's theorem, H has a matching M_1 saturating S. Similarly, $G[S' \cup T]$ has a matching M_2 saturating T. It is clear that $M_1 \cup M_2$ is a matching of G and $M_1 \cap M_2 = \emptyset$. It follows that

$$\beta(G) = |K| = |S| + |T| = |M_1| + |M_2| = |M_1 \cup M_2| \le \alpha'(G)$$

as desired. ∎

Corollary 5.3 Let G be an equally bipartite simple graph of order $2n$. If $\varepsilon > (k-1)n$ for $k \ge 1$, then $\alpha'(G) = \beta(G) \ge k$.

Proof Since G is a bipartite graph, it is sufficient to prove $\beta(G) \ge k$ by König's theorem. Also since G is simple and equally bipartite, every vertex of G covers at most n edges of G. Suppose that $\beta(G) \le k-1$, then from the hypothesis, we can deduce a contradiction as follows.

$$(k-1)\,n < \varepsilon(G) \le \beta(G)\,n \le (k-1)\,n.$$

Therefore, $\beta(G) \ge k$ and so the result follows. ∎

There is a great volume of literature associated with matching theory and its applications. See, for example, the book of Lovász and Plummer [122]. Several classic applications of matching theory will be presented in later sections this chapter. As simple applications of matching theory, we here give two examples.

Example 5.1.1 It is impossible, using 1×2 rectangles, to exactly cover an 4×4 square from which two opposite 1×1 corner squares have been removed.

Proof Label 14 squares with $1, 2, \cdots, 14$. Construct a simple graph G with vertex-set $\{1, 2, \cdots, 14\}$, and $ij \in E(G)$ if and only if two 1×1 squares i and j have a common line (see Figure 5.6). An edge of G denotes a 1×2 rectangle. Thus, the problem can be reduced to proving that G has no perfect matching. It is clear that G is a bipartite graph with bipartition $\{X, Y\}$, where $X = \{1, 3, 4, 6, 8, 11, 12, 14\}$ and $Y = \{2, 5, 7, 8, 10, 13\}$. Because $|X| = 8 > 6 = |Y|$, G has no perfect matching by Corollary 5.1.1. ∎

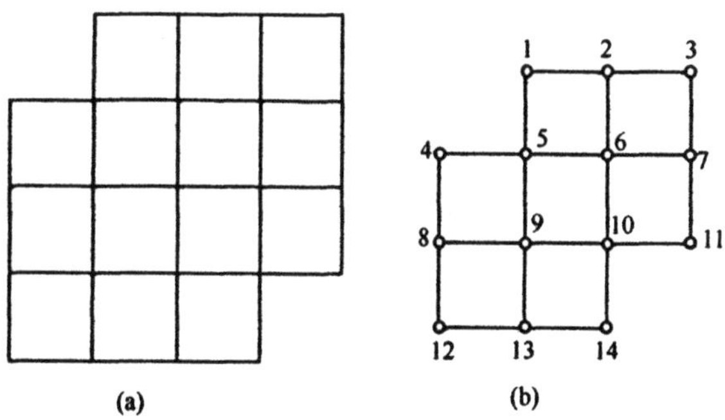

Figure 5.6: The illustration of Example 1.5.1

Example 5.1.2 Let $\mathbf{Q}$ be a nonnegative real matrix. If its row-sum and column-sum both are 1 (such a matrix is called to be *double stochastic*), then

(a) $\mathbf{Q}$ is necessarily square;

(b) each row and each column of $\mathbf{Q}_n$ have a common positive entry;

(c) $\mathbf{Q}$ can be expressed as a convex linear combination of permutation matrices; that is,

$$\mathbf{Q} = c_1\mathbf{P}_1 + c_2\mathbf{P}_2 + \cdots + c_k\mathbf{P}_k,$$

where each $\mathbf{P}_i$ is a permutation matrix, each c_i is a nonnegative real number, and $c_1 + c_2 + \cdots + c_k = 1$. (G. Birkhoff, 1946; and J. von Neuman, 1953)

Proof (a) Let $\mathbf{Q} = (q_{ij})_{m\times n}$. Then, by definition,

$$n = \sum_{j=1}^{n}\left(\sum_{i=1}^{m} q_{ij}\right) = \sum_{i=1}^{m}\left(\sum_{j=1}^{n} q_{ij}\right) = m.$$

(b) Construct an equally bipartite simple graph $G = (X \cup Y, E)$, where X and Y are the sets of row-labels and column-labels of a double stochastic matrix $\mathbf{Q}_n$, respectively, $i \in X$ and $j \in Y$ are linked by edge $ij \in E(G)$ if and only if $q_{ij} > 0$. Arbitrarily choose $S \subseteq X$. Without loss of generality, suppose that $S = \{1, 2, \cdots, k\}$ and $Y \setminus N_G(S) = \{1, 2, \cdots, l\}$. Then $q_{ij} = 0$ for each $i = 1, 2, \cdots, k$ and each $j = 1, 2, \cdots, l$. It follows that

$$l = \sum_{j=1}^{l}\left(\sum_{i=1}^{n} q_{ij}\right) = \sum_{i=k+1}^{n}\left(\sum_{j=1}^{l} q_{ij}\right) \le \sum_{i=k+1}^{n}\left(\sum_{j=1}^{n} q_{ij}\right) = n - k,$$

that is

$$|S| = k \le n - l = |N_G(S)|, \quad \forall\ S \subseteq X.$$

By Hall's theorem, G has a matching M saturating X. Since $|X| = |Y|$ by (a), thus, M is a perfect matching of G, which implies that each row and each column of $\mathbf{Q}_n$ have a common positive entry.

(c) Let $t = t(\mathbf{Q_n})$ denote the number of positive entries in $\mathbf{Q_n}$. It is clear that $t \ge n$ from definition of $\mathbf{Q}_n$. We will prove (c) by induction on $t \ge n$. For $t = n$, $\mathbf{Q}_n$ is a permutation matrix, and the conclusion holds clearly. Below suppose $t \ge n + 1$.

By (b) let $q_{1j_1}, q_{2j_2}, \cdots, q_{nj_n}$ be n common positive entries of each row and each column of $\mathbf{Q}_n$, and let

$$c_1 = \min\{q_{kj_k} :\ 1 \le k \le n\}.$$

Since $\mathbf{Q}_n$ has $t\,(\ge n+1)$ positive entries, and their sum is n, thus, $0 < c_1 < 1$. Let P_1 be the permutation matrix corresponding to $\binom{1\ 2\ \cdots\ n}{j_1\ j_2\ \cdots, j_n}$ and let $\mathbf{Q}_1 =$

$\mathbf{Q} - c_1\mathbf{P}_1$. Then $\mathbf{Q}_1$ is nonnegative, and $\frac{1}{1-c_1}\,\mathbf{Q}_1$ is double stochastic. Since $t(\mathbf{Q}_1) < t(\mathbf{Q})$, by the induction hypothesis, there exist permutation matrices $\mathbf{P}_2, \mathbf{P}_3, \cdots, \mathbf{P}_k$ and $c_2', c_3', \cdots, c_k'$ with $c_2' + c_3' + \cdots + c_k' = 1$ such that

$$\frac{1}{1-c_1}\mathbf{Q}_1 = c_2'\mathbf{P}_2 + c_3'\mathbf{P}_3 + \cdots + c_k'\mathbf{P}_k.$$

It follows that

$$\begin{aligned}
\mathbf{Q} = {} & c_1\mathbf{P}_1 + \mathbf{Q}_1 = c_1\mathbf{P}_1 + (1-c_1)\left(\tfrac{1}{1-c_1}\mathbf{Q}_1\right) \\
= {} & c_1\mathbf{P}_1 + (1-c_1)(c_2'\mathbf{P}_2 + c_3'\mathbf{P}_3 + \cdots + c_k'\mathbf{P}_k) \\
= {} & c_1\mathbf{P}_1 + (1-c_1)c_2'\mathbf{P}_2 + (1-c_1)c_3'\mathbf{P}_3 + \cdots + (1-c_1)c_k'\mathbf{P}_k \\
= {} & c_1\mathbf{P}_1 + c_2\mathbf{P}_2 + c_3\mathbf{P}_3 + \cdots + c_k\mathbf{P}_k,
\end{aligned}$$

where

$$c_i = (1-c_1)c_i' \quad \text{for each } i = 2, 3, \cdots, k$$

and

$$c_1 + c_2 + \cdots + c_k = c_1 + (1-c_1)(c_2' + c_3' + \cdots + c_k') = 1.$$

By the induction principle, the result follows. ∎

We conclude this section with presenting some self-contained proofs of Hall's theorem (Theorem 5.1), Tuttle's theorem (Theorem 5.2) and Köning's theorem (Theorem 5.3), at the first reading some readers may wish to skip them.

There are many directed proofs of Hall's theorem, two of them are presented below, the first proof is due to Halmos and Vaughan [83] and the second due to Rado [148]. It suffices to prove the sufficiency of the condition (5.1). By definition of matchings, in the following proofs, we need only consider a simple undirected bipartite graph G with bipartition $\{X, Y\}$.

The first direct proof of Hall's theorem By induction on $|X| \geq 1$. For $|X| = 1$ the condition (5.1) is sufficient clearly, so we proceed to the induction step.

Assume that $|N_G(S)| \geq |S| + 1$ for any nonempty subset S of X. Choose $x \in S$ and $y \in N_G(x)$. Then $G' = G - \{x, y\}$ is bipartite and satisfies the condition (5.1). By the induction hypothesis, G' has a matching M' saturating $(X \setminus \{x\})$. Clearly, $M = M' \cup \{xy\}$ is a matching saturating X in G.

Assume there is some nonempty subset S of X such that $|N_G(S)| = |S|$. Let $G_1 = G[S \cup N_G(S)]$ and $G_2 = G - (S \cup N_G(S))$. Then both G_1 and G_2 are bipartite. G_1 satisfied the condition (5.1) clearly. Arbitrarily choose a nonempty subset $S' \subseteq X \setminus S$. Then

$$|N_{G_2}(S')| \geq |N_G(S \cup S')| - |N_{G_1}(S)| \geq |S \cup S'| - |S| = |S'|,$$

that is, G_2 satisfies the condition (5.1). By the induction hypothesis, let M_1 be a matching saturating S in G_1 and M_2 be a matching saturating $X \setminus S$ in G_2. Then, $M_1 \cap M_2 = \emptyset$ clearly, and so $M_1 \cup M_2$ is a matching saturating X in G. ∎

The second direct proof of Hall's theorem Suppose that G satisfies the condition (5.1) and has as few edges as possible. We will show $E(G)$ is exactly a matching of G consisting of $|X|$ edges.

By contradiction. Then there exist two edges x_1y, x_2y of G such that $x_1, x_2 \in X$ and $y \in Y$. By the minimality of G, any graph obtained from G by deleting either of the edges x_1y and x_2y does not satisfy the condition (5.1). Thus there are two subset X_1, X_2 of X such that $|N_G(X_i)| = |X_i|$ and x_i is only vertex in X_i that is adjacent to y for each $i = 1, 2$. It follows that

$$\begin{aligned} |N_G(X_1) \cap N_G(X_2)| &= |N_G(X_1 \setminus \{x_1\}) \cap N_G(X_2 \setminus \{x_2\})| + 1 \\ &\geq |N_G(X_1 \cap X_2)| + 1 \geq |X_1 \cap X_2| + 1. \end{aligned}$$

This gives

$$\begin{aligned} |N_G(X_1 \cup X_2)| &= |N_G(X_1) \cup N_G(X_2)| \\ &= |N_G(X_1)| + |N_G(x_2)| - |N_G(X_1 \cap X_2)| \\ &\leq |X_1| + |X_2| - |X_1 \cap X_2| - 1 = |X_1 \cup X_2| - 1. \end{aligned}$$

This contradicts to the condition (5.1). ∎

We now present a direct proof of Tutte's theorem, due to Lovśz [121]. It suffices to prove the sufficiency of the condition (5.2). By definition of matchings, we need only consider that G is a simple undirected graphs.

Direct proof of Tutte's theorem By contradiction. Suppose that G satisfies the condition (5.2) but has no perfect matching. Then G is a spanning subgraph of a maximal graph G^* having no perfect matching. Since $G - S$ is a spanning subgraph of $G^* - S$ we have $o(G^* - S) \leq o(G - S)$ and so, by the condition (5.2), we have

$$o(G^* - S) \le |S| \quad \text{for all} \quad S \subset V(G^*). \tag{5.2}'$$

In particular, setting $S = \emptyset$, we have that $o(G^*) = 0$, and so $v(G^*)$ is even. Let

$$U = \{u \in V(G^*) : \ d_{G^*}(u) = v - 1\}.$$

Then $U \neq V(G^*)$ since G^* has no perfect matching. Since $v(G^*)$ is even it is necessary that $|U|$ and $o(G^* - U)$ be of the same parity. We will show that each of all connected components of $G^* - U$ is a complete graph. Suppose to the contrary that some component G_i is not complete. Then $v(G_i) \ge 3$ and so there are vertices x, y and z such that $xy, yz \in E(G^*)$ and $xz \notin E(G^*)$ (see Example 1.6.1). Moreover, since $y \notin U$, there is a vertex w in $G^* - U$ such that $yw \notin E(G^*)$.

Since G^* is a maximal graph containing no perfect matching, $G^* + e$ has a perfect matching for any $e \notin E(G^*)$. Let M_1 and M_2 be perfect matchings in $G^* + xz$ and $G^* + yw$, respectively, and denote by H the subgraph of $G^* \cup \{xz, yw\}$ induced by $M_1 \Delta M_2$. Since each vertex of H has degree two, H is a disjoint union of cycles. Furthermore, all of these cycles are even. We distinguish two cases:

Case 1 xz and yw are in different components of H. Then, if yw in the cycle C of H, the edges of M_1 in C, together with the edges of M_2 not in C, constitute a perfect matching in G^*, contradicting the choice of G^*.

Case 2 xz and yw are in the same component C of H. By symmetry of x and z, we may assume that the vertices x, y, w and z occur in that order on C. Then the edges of M_1 in the section $yw \cdots z$ of C, together with the edge yz and the edges of M_2 not in the section $yw \cdots z$ of C, constitute a perfect matching in G^*, again contradicting the choice of G^*.

Since both case 1 and case 2 lead to contradiction, it follows that every components of $G^* - U$ is indeed a complete graph.

Let $G_1, G_2, \cdots, G_\omega$ are all components of $G^* - U$, where the first o components are odd. For $x_i \in V(G_i)$, set $G_i' = G_i - x_i$ for $i = 1, 2, \cdots, o$. Thus, each of $G_i', G_2', \cdots, G_o'$ and $G_{o+1}, \cdots, G_\omega$ is a complete graph of even order and so has a perfect matching M_i $(1 \le i \le \omega)$. Now, by $(5.2)'$, $o(G^* - U) \le |U|$. Arbitrarily choose o vertices $y_1, y_2, \cdots, y_o$ in U. Then $x_i y_i \in E(G)$ as $d_G(y_i) = v - 1$. Thus $M_{\omega+1} = \{x_i y_i : \ 0 \le i \le o\}$ is a matching of G^*. Since the subgraph of G^* induced by $U \setminus \{y_1, y_2, \cdots, y_o\}$ is a complete graph of even order, there is a perfect matching $M_{\omega+1}$ in the subgraph. It follows that

$$M = M_1 \cup M_2 \cup \cdots \cup M_{\omega+1} \cup M_{\omega+2}$$

is a perfect matching of G^*. Since G^* is assume to have no perfect matching we have obtained the desired contradiction. Thus G does indeed have a perfect matching. ∎

We now present two directed proofs of König's theorem, the first proof is due to Lovśz [121] and the second due to Rizzi [151]. By the inequality (5.3) it suffices to prove $\beta(G) \leq \alpha'(G)$ for any undirected bipartite graph G with bipartition $\{X, Y\}$.

The first direct proof of König's theorem By induction on $\varepsilon \geq 1$. For $\varepsilon = 1$ the inequality holds clearly, so we proceed to the induction step. Without loss of generality suppose that there is $x \in X$ with $d_G(x) \geq 2$. Choose $e_1, e_2 \in E(G)$ such that $\psi_G(e_1) = xy$ and $\psi_G(e_2) = xz$. Assume that there are a covering S_1 of $G - e_1$ and a covering S_2 of $G - e_2$ such that $|S_1| = |S_2| = \beta(G) - 1$. Clearly $x \notin S_1 \cup S_2$, $y \in S_2 \setminus S_1$ and $z \in S_1 \setminus S_2$. It follows that

$$|((S_1 \cap S_2) \cap X) \cup ((S_1 \cup S_2) \cap Y)| \geq \beta(G),$$
$$|((S_1 \cup S_2 \cup \{x\}) \cap X) \cup ((S_1 \cap S_2) \cap Y)| \geq \beta(G).$$

Summing up the two inequality, we obtain

$$|S_1 \cup S_2| + |S_1 \cap S_2| + 1 \geq 2\beta(G).$$

It follows that $2\beta(G) - 2 = |S_1| + |S_2| \geq 2\beta(G) - 1$, which is impossible. Thus, $\beta(G - e_1) = \beta(G)$ or $\beta(G - e_2) = \beta(G)$. Without loss of generality, assume the former holds. By the induction hypothesis, we have

$$\beta(G) = \beta(G - e_1) \leq \alpha'(G - e_1) \leq \alpha'(G)$$

as desired. ∎

The second direct proof of König's theorem Let G be a minimal counterexample, that is, $\beta(G) > \alpha'(G)$ but $\beta(G - e) \leq \alpha'(G - e)$ for any edge $e \in E(G)$. Then G is connected, is not a cycle, nor a path. So, G has a vertex of degree at least 3. Let x be such a vertex and y one of its neighbors. If $\alpha'(G - y) < \alpha(G)$, then, by minimality, $G - y$ has a covering S' with $|S'| = \beta(G - y) \leq \alpha'(G - y) < \alpha(G)$. Hence, $S = S' \cup \{y\}$ is a covering of G with cardinality $\alpha(G)$ at most, that is, $\beta(G) \leq |S| \leq \alpha'(G)$, contradicting the choice of G. Assume, therefore, there exists a maximum matching M of G having no edge incident at y. Let e be an edge of $G-M$ incident with x but not with y. Let S' be a covering of $G-e$, determined by M, with $|S'| = \alpha'(G)$. Since no edge of M is incident with y, it follows that S' does not contain y. So S' contains x and is a covering of G, and so $\beta(G) \leq |S| = \alpha'(G)$, contradicting the choice of G. ∎

Exercises

5.1.1 Prove that $\left\lceil \frac{v}{1+\Delta} \right\rceil \leq \alpha'(G) \leq \left\lfloor \frac{v}{2} \right\rfloor$ for any graph G without isolated vertices. (Weinstein, 1963)

5.1.2 Prove that

(a) n-cube Q_n has n edge-disjoint perfect matchings;

(b) complete graph K_{2n} has $(2n-1)$ edge-disjoint perfect matchings;

(c) complete graph K_{2n} has $(2n-1)!!$ distinct perfect matchings;

(d) $K_{n,n}$ has $n\,!$ different perfect matchings;

(e) for any $k\ (\geq 2)$, there exists a k-regular simple graph that contains no perfect matching.

5.1.3 Prove that every plane triangulation of order $v\ (\geq 4)$ contains a bipartite subgraph with $\frac{2}{3}v$ edges.

5.1.4 A *line* of a matrix is a row or a column of the matrix. Prove that the minimum number of lines containing all the 1's of a $(0,1)$-matrix is equal to the maximum number of 1's, no two of which are in the same line.

5.1.5 Let $A_1, A_2, \cdots, A_m$ be subsets of a set S. A *system of distinct representatives* for the family $\{A_1, A_2, \cdots, A_m\}$ is a subset $\{a_1, a_2, \cdots, a_m\}$ of S such that $a_i \in A_i$ for each $i = 1, 2, \cdots, m$, and $a_i \neq a_j$ for $i \neq j$. Prove that $\{A_1, A_2, \cdots, A_m\}$ has a system of distinct representatives if and only if $|\bigcup_{i\in J} A_i| \geq |J|$ for all subsets J of $\{1, 2, \cdots, m\}$. (P.Hall, 1935)

5.1.6 Prove that a tree T has a perfect matching if and only if $o(T-x) = 1$ for any $x \in V(T)$.

5.1.7 Prove that, if G is a bipartite graph with bipartition $\{X, Y\}$, then

$$\alpha'(G) = |X| - \max\{|S| - |N_G(S)| : \forall\ S \subseteq X\}. \qquad \text{(O.Ore, 1955)}$$

5.1.8 Prove that, if G is a graph and $r = \max\{o(G-S) - |S| : \forall\ S \subset V(G)\}$, then $\alpha'(G) = \frac{1}{2}(v - r)$. (C.Berge, 1958)

5.1.9 Let Γ be a finite group and H be a subgroup of Γ. Prove that there exist elements $a_1, a_2, \cdots, a_n \in \Gamma$ such that $a_1H, a_2H, \cdots, a_nH$ are the left cosets of H and $Ha_1, Ha_2, \cdots, Ha_n$ are the right cosets of H.

(P.Hall, 1935)

5.1.10 Let $\mathbf{A}$ be an $m \times n\,(m \le n)$ matrix. The *permanent* of $\mathbf{A}$, denoted by $\text{Per}\,(\mathbf{A})$, is defined as the sum of products of m entries from different rows and columns of $\mathbf{A}$. Prove that, if $\mathbf{A}$ is the adjacency matrix of a bipartite graph G with bipartition $\{X, Y\}$ and $|X| \le |Y|$, then

(a) the number of matchings of G saturating X is equal to $\text{Per}\,(\mathbf{A})$;

(b) $K_{n,n}$ has $n!$ different perfect matchings.

5.1.11 A k-regular spanning subgraph of G is called a *k-factor* of G. A graph G is called to be *k-factorable* if it can be expressed as the union of edge-disjoint k-factors. Prove that

(a) G has 1-factor if and only if G has a perfect matching;

(b) Petersen graph is not 1-factorable, but the union of one 1-factor and one 2-factor;

(c) K_{2n} and $K_{n,n}$ is 1-factorable;

(d) K_{2n+1} is 2-factorable;

(e) a simple graph G is 2-factorable if and only if G is $2k$-regular;

(f) every $2k$-regular graph is 2-factorable for $k \ge 1$.

5.1.12 Deduce the following theorems

(a) Hall's theorem (5.1) from Tutte's theorem (5.2);

(b) Hall's theorem (5.1) from König's theorem (5.3);

(c) König's theorem (5.3) from Menger's theorem (4.3);

(d) König's theorem (5.3) from the max-flow min-cut theorem (4.1);

(e) Menger's theorem (4.3) from Hall's theorem (5.1).

5.2 Independent Sets

Let G be a loopless graph. A nonempty subset I of $V(G)$ is an *independent set* of G if no two elements of I are adjacent in G. An independent set I of G is called to be *maximum* if $|I| \geq |I'|$ for any independent set I' of G; and *maximal* if $I \cup \{x\}$ is not an independent set of G for any $x \in V(G) \setminus I$.

From definition, it is clear that the concept of independent sets has no relation to orientations of edges and parallel edges. Therefore, in discussing properties of independent sets, we will restrict ourselves to simple undirected graphs. Maximal and maximum independent sets are shown in Figure 5.7 by solid points.

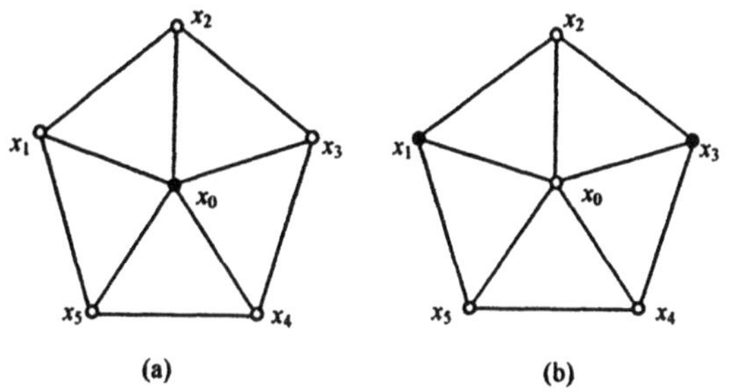

Figure 5.7: (a) A maximal independent set; (b) a maximum independent set

The number of vertices in a maximum independent set of G is called the *independence number* of G, denoted by $\alpha(G)$. For example,

$$\alpha(K_n) = 1, \quad \alpha(C_{2n}) = n = \alpha(C_{2n+1}), \quad \alpha(K_{m,n}) = \max\{m, n\}.$$

It is clear that $\alpha(G) = 1$ if and only if G contains a complete graph as its spanning subgraph. The problem determining $\alpha(G)$ for a given graph G is extremely difficult in general.

The following result, due to Gallai [70], is simple, however, explores closed relationship between independent sets and coverings.

Theorem 5.4 A set $I \subset V(G)$ is an independent set of a loopless graph G if and only if $V(G) \setminus I$ is a covering of G.

Proof By definition, I is an independent set of G if and only if no edge of G has both end-vertices in I or, equivalently, if and only if each edge has at least one end-vertex in $V(G) \setminus I$ since G is loopless, equivalently, if and only if $V(G) \setminus I$ is a covering of G. ∎

Corollary 5.4.1 A subset I of $V(G)$ is a maximal (reps. maximum) independent set of G if and only if $V(G) \setminus I$ is a minimal (reps. minimum) covering of G. ∎

Corollary 5.4.2 $\alpha(G) + \beta(G) = v(G)$ for any loopless graph G. ∎

The edge analogue of an independent set is a set of edges no two of which are adjacent, that is, a matching defined in the preceding section. The edge analogue of a covering is called an edge-covering. An *edge-covering* of G is a subset L of $E(G)$ such that each vertex of G is an end-vertex of some edge in L. Note that edge-coverings do not always exist; however, it is clear that G has an edge-covering if and only if G contains no isolated vertex. The minimum number of edges in an edge-covering of G is called the *edge-covering number* of G, denoted by $\beta'(G)$. For example,

$$\beta'(K_n) = \beta'(C_n) = \left\lceil \tfrac{n}{2} \right\rceil, \quad \beta'(K_{m,n}) = \max\{m, n\}.$$

Matchings and edge-coverings are not related to one another as simply as are independent sets and coverings; the complement of a matching is not always an edge-covering, nor is the complement of an edge-covering necessarily a matching. However, it so happens that the parameters α' and β' are related in precisely the same manner as are α and β, found by Gallai [70].

Theorem 5.5 $\alpha'(G) + \beta'(G) = v(G)$ for any graph G with $\delta(G) > 0$.

Proof Let M be a maximum matching of G and U be the set of M-unsaturated vertices of G. Since G contains no isolated vertex, there exists a set E' of $|U|$ edges, one incident with each vertex in U. Clearly, $M \cup E'$ is an edge-covering of G, and so

$$\alpha' + \beta' \leq |M| + |M \cup E'| = \alpha' + [\alpha' + (v - 2\alpha')] = v.$$

On the other hand, let L be a minimum edge-covering of G, $H = G[L]$ and let M be a maximum matching of H. Denote by U the set of M-unsaturated vertices in H. Then $H[U]$ has no edge and so

$$|L| - |M| = |L \setminus M| \geq |U| = v - 2|M|.$$

Since H is a subgraph of G, M is a matching of G and so

$$\alpha' + \beta' \geq |M| + |L| \geq v.$$

The theorem follows. ∎

Corollary 5.5 $\alpha(G) = \beta'(G)$ for any bipartite graph G with $\delta(G) > 0$.

Even though the concept of an independent set is analogous to that of a matching, there is no theory of independent sets comparable to the theory of matchings presented the preceding section. However, two parameters α and α' can be closely related by line graphs, that is, if G is nonempty,

$$\alpha'(G) = \alpha(L(G)).$$

The proof is not difficult, left to the reader as an exercise (the exercise 5.2.4).

The following theorem establishes the relationship between two parameters α and κ, due to Bondy [16].

Theorem 5.6 Let G be a simple undirected graph. If $d_G(x)+d_G(y) \geq v$ for any two nonadjacent vertices x and y of G, then $\alpha(G) \leq \kappa(G)$.

Proof It is not difficult to see that G is connected subject to the given condition (see the exercise 1.5.8). If G is a complete graph, the theorem holds clearly. Suppose below that G is not complete. We prove $\alpha(G) \leq \kappa(G)$.

Set $\alpha = \alpha(G)$ and $\kappa = \kappa(G)$. Suppose to the contrary that $\alpha \geq \kappa + 1$, and let I and S be a maximum independent set and a κ-separating set of G, respectively. Then $|I| = \alpha \geq 2$ and

$$|N_G(x) \cup N_G(y)| \leq v - \alpha, \quad \forall\, x, y \in I. \tag{5.4}$$

It follows from the hypothesis and (5.4) that for any two vertices $x, y \in I$

$$\begin{aligned} |N_G(x) \cap N_G(y)| = &\ |N_G(x)| + |N_G(y)| - |N_G(x) \cup N_G(y)| \\ \geq &\ v - (v - \alpha) = \alpha \geq \kappa + 1 > |S|. \end{aligned}$$

This implies that only one of all components of $G - S$, say G_1, may contain vertices in I, that is, $I \subseteq V(G_1) \cup S$. Since $\alpha \geq \kappa + 1$, there exists $x \in I \cap V(G_1)$. Choose a vertex z in other component, say G_2, of $G - S$. Then

$$|N_G(x) \cup N_G(z)| \leq v - 2 - |I \cap V(G_1)| + 1 = v - \alpha + |I \cap S| - 1. \tag{5.5}$$

Since $N_G(x) \cap N_G(z) \subseteq S \setminus I$, thus,

$$|N_G(x) \cap N_G(z)| \leq \kappa - |I \cap S|. \tag{5.6}$$

Combining (5.5) with (5.6), we can deduce a contradiction as follows.

$$\begin{aligned} v \leq &\ d_G(x) + d_G(z) = |N_G(x) \cup N_G(z)| + |N_G(x) \cap N_G(z)| \\ \leq &\ (v - \alpha + |I \cap S| - 1) + (\kappa - |I \cap S|) \\ = &\ v - \alpha + \kappa - 1 \leq v - 2. \end{aligned}$$

Therefore, $\alpha(G) \leq \kappa(G)$ and so the theorem follows. ∎

Corollary If G is a simple graph with $\delta(G) \geq \frac{v}{2}$, then $\alpha(G) \leq \kappa(G)$. ∎

The following theorem establishes the relations between parameters α, κ and hamiltoncity, due to Chvàtal and Erdös [33].

Theorem 5.7 Let G be a simple undirected graph of order $v \geq 3$. If $\alpha(G) \leq \kappa(G)$, then G is hamiltonian.

Proof Since if $\alpha(G) = 1$ then G is complete, thus, suppose $\alpha(G) \geq 2$. Then G contains a cycle since $\delta(G) \geq \kappa(G) \geq \alpha(G) \geq 2$. Let C be a longest cycle in G. We will prove that C is a Hamilton cycle in G. Suppose to the contrary that C is not a Hamilton cycle. Then $V(G) \setminus V(C) \neq \emptyset$. Let H be a component of $G - V(C)$, and let $N_G(V(H)) \cap V(C) = \{x_1, x_2, \cdots, x_s\}$, which is a separating set of G. Then

$$s \geq \kappa(G) \geq 2. \tag{5.7}$$

Furthermore, any two of $\{x_1, x_2, \cdots, x_s\}$ are not adjacent in C, otherwise we can construct a longer cycle than C.

Specify C a direction to obtained a directed cycle $\overrightarrow{C}$. Let $Y = \{y_i : (x_i, y_i) \in E(\overrightarrow{C}), i = 1, 2, \cdots, s\}$. Then $|Y| = s \geq 2$. We will show that Y is an independent set of G. In fact, if there exists $y_iy_j \in E(G)$, choosing an x_ix_j-path P_{ij} going through H, then $C - x_iy_i - x_jy_j + y_iy_j + P_{ij}$ is a cycle in G whose length is larger than C's, contradicting to the choice of C.

Since y_i and x_i are adjacent in G, y_i is not adjacent with any vertex in H for each $i = 1, 2, \cdots, s$. Thus, for any $y_0 \in V(H)$, $I = \{y_0, y_1, y_2, \cdots, y_s\}$ is an independent set of G, and so, by (5.7), $\alpha(G) \geq |I| = s + 1 \geq \kappa(G) + 1$, contradicting to the hypothesis. Therefore, C is a Hamilton cycle in G, and the theorem follows. ∎

Note that a combination of Theorem 5.6 and Theorem 5.7 implies Theorem 1.9 immediately.

Before concluding this section, we mention another important concept of graph theory. A nonempty subset $S \subset V(G)$ is called a *dominating set* of G if every vertex not in S is adjacent to at least one vertex in S. A dominating set is called to be *minimal* if none of its proper subsets is a dominating set. The *domination number* of a graph G, denoted by $\gamma(G)$, is the minimum cardinality of a dominating set of G.

It is clear from definition that if S is a minimal dominating set of G then $V(G) \setminus S$ is also a dominating set. Hence $\gamma(G) \leq \frac{1}{2}v(G)$. It is also clear that an independent set of G is a dominating set of G if and only if it is a maximal independent set. Hence $\gamma(G) \leq \alpha(G)$.

For further results and applications of dominating sets in graphs, see Haynes, Hedetniemi and Slater [89, 90].

Exercises

5.2.1 Prove that G is a bipartite graph if and only if

(a) $\alpha(H) \geq \frac{1}{2}v(H)$ for any subgraph H of G;

(b) $\alpha(H) = \beta'(H)$ for any subgraph H of G without isolated vertices.

5.2.2 Let $\{V_1, V_2, \cdots, V_n\}$ be a partition of $V(G)$ and V_i is a maximal independent set of G for each $i = 1, 2, \cdots, n$. Let H be a simple graph with vertex-set $\{u_1, u_2, \cdots, u_n\}$ and $u_iu_j \in E(H) \Longleftrightarrow E_G(V_i, V_j) \neq \emptyset$. Prove that H is a complete graph.

5.2.3 Prove that $\alpha'(G) = \alpha(L(G))$, where $L(G)$ is the line graph of a nonempty graph G.

5.2.4 Construct graphs to show that

(a) the condition "$d_G(x) + d_G(y) \geq v$" in Theorem 5.6 can not be improved as "$d_G(x) + d_G(y) \geq v - 1$";

(b) the condition "$\delta(G) \geq \frac{1}{2}v$" in Corollary 5.6 can not be improved as "$\delta(G) \geq \left\lfloor \frac{1}{2}v \right\rfloor$".

5.2.5 Prove that a simple graph G is hamiltonian if it satisfies one of the following conditions.

(a) G is k-regular k-connected and $v = 2k + 1$;

(b) $T = \{x \in V(G):\ d_G(x) = v - 1\}$ and $|T| \geq \alpha(G)$.

5.2.6 Let G be a simple graph. Prove that, if $\delta(G) \geq \frac{1}{3}[v(G) + \kappa(G)]$, then $\alpha(G) \leq \delta(G)$.

5.2.7 Let G be a loopless digraph. Prove that G contains an independent set I and $x \in I$ such that $d_G(x, y) \leq 2$ for any $y \in V(G) \setminus I$.

(V.Chvàtal, and L.Lovász, 1974)

Applications

5.3 The Personnel Assignment Problem

In a certain company, n workers $x_1, x_2, \cdots, x_n$ are available for n jobs $y_1, y_2, \cdots, y_n$, each worker being qualified for one or more of these jobs. Can all the men be assigned, one worker per job, to jobs for which they are qualified? This is the *personnel assignment problem.*

We construct a simple bipartite graph G with bipartition $\{X, Y\}$, where $X = \{x_1, x_2, \cdots, x_n\}$, $Y = \{y_1, y_2, \cdots, y_n\}$, and $x_i y_j \in E(G)$ if and only if worker x_i is qualified for job y_j. The personnel assignment problem becomes one of determining whether or not G has a matching saturating every vertex in X, equivalently, determining whether or not G has a perfect matching.

Hall's theorem can not lead to an efficient algorithm to determine whether or not G has a matching saturating every vertex in X as the number of subsets $S \subset X$ is 2^n, which is exponentially large on order of G. Actually, our proof of Hall's theorem has provided a way to find a maximum matching in a general bipartite graph using flow techniques. For a given bipartite graph G, construct a capacity network $N = (D_{xy}, \mathbf{c})$ with a unit of capacity for each edge, such a construction is illustrated in Figure 5.2. Finding a maximum matching in G can become finding a maximum (x, y)-flow in N; while the latter has been solved by the labelling method described in Section 4.4. If $\mathbf{f}$ is a maximum (x, y)-flow, then the set of the edges $\{e \in E(G) : \mathbf{f}(e) = 1\}$ is a maximum matching in G.

We will here describe an efficient algorithm for testing whether there exists a perfect matching in a given bipartite graph or, equivalently, for solving the personnel assignment problem, known as the *Hungarian method.* The algorithm is based on the following concepts and results on alternating paths.

Let G be a graph, and let M and M' be two disjoint nonempty and proper subsets of $E(G)$. An *(M, M')-alternating path* in G is a path whose edges are alternately in M and M'. An *M-alternating path* is an $(M, \overline{M})$-alternating path, where $\overline{M} = E(G) \setminus M$. If M is a matching in G, then an M-alternating path connecting two M-unsaturated vertices is said an *M-augmenting path.*

Lemma 5.8 Let M and M' be two distinct matchings in G, $H = G[M \Delta M']$, where $M \Delta M'$ denotes the symmetric difference of M and M'.

Then every component of H must be one of the following three types:

(i) an isolated vertex;

(ii) an (M, M')-alternating cycle of even order;

(iii) an (M, M')-alternating path.

Proof Since each vertex of H can be incident with at most one edge of M and one edge of M', thus $0 \leq \Delta(H) \leq 2$, furthermore, $d_H(x) = 2$ if and only if x is incident with exactly one edge of M and one edge of M' (see Figure 5.8).

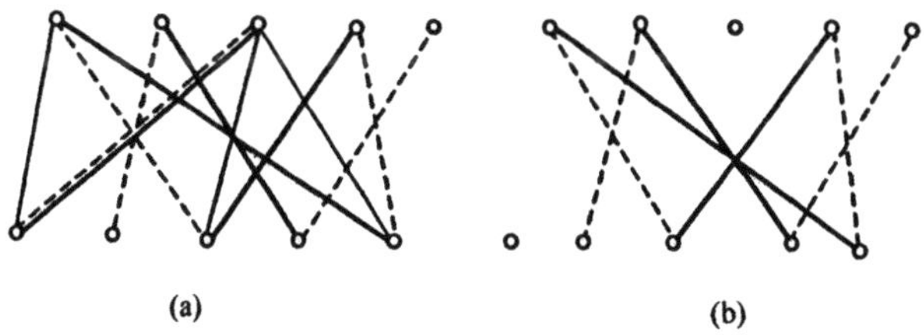

Figure 5.8: (a) Two matchings M (heavy) and M' (broken); (b) $H = G[M \Delta M']$

Suppose that P is a component of H. Then $d_P(x) = d_H(x)$ for any vertex x in P. If $\Delta(P) = 0$, then P consists of a single vertex. Suppose that $1 \leq \Delta(P) \leq 2$ below. If all of vertices in P are of degree two, then each of them is incident with one edge of M and one edge of M', and so P is an (M, M')-alternating cycle of even order.

Choose $x \in V(P)$ such that $d_H(x) = 1$. Then there exists at least one vertex, say y, of degree one other than x by Corollary 1.1. As $\Delta(P) \leq 2$, P is a path connecting x and y, internal vertices (if exist) are of degree two. Thus, P is an (M, M')-alternating path. ∎

Theorem 5.8 (Berge [9]) A matching M in G is maximum if and only if G contains no M-augmenting path.

Proof Let M be a maximum matching in G, and suppose to the contrary that G contains an M-augmenting path $P = x_0e_1x_2e_2 \cdots e_mx_m$. Then m is odd, $e_2, e_4, \cdots, e_{m-1} \in M$ and $e_1, e_3, \cdots e_m \notin M$. Define $M' \subset E(G)$ by

$$M' = M \Delta E(P) = (M \setminus \{e_2, e_4, \cdots, e_{m-1}\}) \cup \{e_1, e_3, \cdots e_m\}.$$

Then M' is a matching in G, and $|M'| = |M| + 1$. Thus M is not a maximum matching of G, contradicting to the hypothesis.

Conversely, suppose that M is not maximum, but M' be a maximum matching of G. Then $|M'| > |M|$. Let $H = G[M\Delta M']$. Then, by Lemma 5.8, every component of H must be an isolated vertex, or an (M, M')-alternating even-cycle, or an (M, M')-alternating path. Moreover, $|E(H) \cap M'| > |E(H) \cap M|$ since $|M'| > |M|$. Therefore, there exists a component P of H which is an (M, M')-alternating path starting and ending with edges of M'. Two end-vertices of P are M'-saturated in H, but M-unsaturated in G. Thus P is an M-augmenting path in G, a contradiction. ∎

Theorem 5.8 naturally suggests an algorithm for finding a maximum matching in a general graph. Starting from an arbitrary matching we repeatedly carry out augmentations along M-augmenting paths until no such path exists. This process is bound to terminate because a maximum matching has finite edges and each augmentation increases the cardinality of the current matching by one. The only practical problem is to specify a systematic search for M-augmenting paths. Before we do this it is interesting to emphasis relationship between this algorithm applied to bipartite graph and the labelling method for finding maximum flows descried Section 4.4.

As mentioned above, to find a maximum matching in a bipartite graph G, we may consider an (x, y)-flow in the capacity network $N = (D_{xy}, \mathbf{c})$ with $\mathbf{c} \equiv 1$, illustrated in Figure 5.2. Suppose that $\mathbf{f}$ is an (x, y)-flow obtained at some intermediate stage of the labelling method. The flow $\mathbf{f}$ corresponds to a matching M in G which consists of those edges with a unit of flow. Any $\mathbf{f}$-incrementing path coincides with some M-augmenting path in G. Moreover, if we separately carry out an $\mathbf{f}$-increment in the labelling method and an M-augmentation in other algorithm, then there is still a one-to-one coincidence between $\mathbf{f}$-incrementing paths in N and M-augmenting paths in G. Moreover, an $\mathbf{f}$-incrementing path in N exists if and only if an M-augmenting path in G does. Clearly, both algorithms are different aspects of essentially the same process.

The flow point of view as described here can not be used for non-bipartite graphs although the idea of M-augmentations has quite general applicability. Edmonds [47] proposed an efficient method for searching for M-augmenting paths in a general graph, which requires some lengthy explanations. We here, therefore, describe it for the spacial case of bipartite graphs. This algorithm is based on the following theorem.

Theorem 5.9 Let M be a matching in a bipartite graph G with bipartition $\{X, Y\}$, $x \in X$ be an M-unsaturated vertex and Z be the set of all vertices connected to x by M-alternating paths, $S = Z \cap X$, $T = Z \cap Y$. Then

(a) $T \subseteq N_G(S)$, and

(b) the following three statements are equivalent:

(i) G contains no M-augmenting path with origin x;

(ii) x is the unique M-unsaturated vertex in Z;

(iii) $T = N_G(S)$ and $|T| = |S| - 1$.

Proof (a) For any $y \in T$, G contains an M-alternating xy-path P. Let $z \in N_P(y)$. Since G is bipartite and $y \in T \subseteq Y$, thus $z \in Z \cap X = S$, that is, $y \in N_G(S)$. Therefore $T \subseteq N_G(S)$.

(b) (i) $\Rightarrow$ (ii) Suppose to the contrary that $y \in Z$ is an M-unsaturated vertex other than x, and P an M-alternating xy-path in G. Then P is an M-augmenting with origin x, which contradicts the hypothesis given in (i).

(ii) $\Rightarrow$ (iii) For $y \in N_G(S) \subseteq Y$, there exist $u \in S = Z \cap X$ and $e \in E(G)$ such that $\psi_G(e) = uy$. If $u = x$, then $y \in T$ clearly. Suppose $u \neq x$ below. Let P be an M-alternating xu-path in G. Then u is M-saturated since x is the unique M-unsaturated in Z. If P contains y, then $y \in T$ clearly. If P does not contain y, then $e \notin M$. By the definition of Z, $y \in Z \cap Y = T$, and so $N_G(S) \subseteq T$. By (a), we have $T = N_G(S)$.

By the hypothesis, x is the unique M-unsaturated vertex in Z, thus, all vertices in T are M-saturated. Since all vertices in X matched under M are in S and $T = N_G(S)$, the vertices in $S \setminus \{x\}$ are matched under M. Therefore $|T| = |S| - 1$.

(iii) $\Rightarrow$ (i) For any $z \in S \setminus \{x\}$, let P be an M-alternating xz-path in G. Then $x, z \in X$ and P is of even length since G is bipartite. Moreover, z is M-saturated since x is M-unsaturated. By the arbitrary choice of z, all vertices in $S \setminus \{x\}$ are M-saturated. By the hypothesis, $T = N_G(S)$ and $|T| = |S| - 1$, thus all vertices in T are M-saturated, which implies that x is the unique M-unsaturated vertex in Z. Thus, G contains no M-augmenting path with origin x since any M-augmenting path contains exactly two M-unsaturated vertices. ∎

Corollary 5.9 Every bipartite graph contains a matching saturating all maximum degree vertices.

Proof Let G be a bipartite graph with bipartition $\{X, Y\}$ and M be a maximum matching saturating maximum degree vertices of G as many as possible.

Suppose to the contrary that x is a maximum degree vertex of G that is M-unsaturated. Let Z be the set of all vertices connected to x by M-alternating paths in G. Suppose, without loss of generality, that $x \in X$, and let $S = Z \cap X$ and $T = Z \cap Y$. Since M is a maximum matching in G, by Theorem 5.8, G contains no M-augmenting path with origin x. Also by Theorem 5.9, $T = N_G(S)$ and $|T| = |S| - 1$.

If all of vertices in S are maximum degree vertices of G, then we have

$$\Delta|S| = \sum_{u \in S} d_G(u) = |[S,T]| = \sum_{u \in T} d_G(u) \leq \Delta|T|,$$

that is, $|S| \leq |T| = |S|-1$, a contradiction. Thus, S contains a non-maximum degree vertex, say $z \neq x$.

Let $P = x\, e_1\, x_1\, e_2 \cdots e_{m-1}\, x_{m-1}\, e_m\, z$ be an M-alternating xz-path in G. Then m is even because $x, z \in X$. Also since x is M-unsaturated, $e_1, e_3, \cdots, e_{m-1} \notin M$ and $e_2, e_4, \cdots, e_m \in M$. So z is an M-saturated vertex. Thus $M' = M \Delta E(P)$ is a matching in G with $|M| = |M'|$, and so M' is a maximum matching in G. However, M' saturates maximum degree vertices more than M does, which contradicts the choice of M. ∎

Corollary 5.9.1 (König [111]) The edge-set of every bipartite graph can be partitioned into Δ edge-disjoint matchings.

We can now describe the Hungarian method for determining whether or not there exists a matching in a bipartite graph G with bipartition $\{X, Y\}$ that saturates all vertices of X. The basic idea behind the algorithm is very simple. We start with an arbitrary matching M. If M saturates every vertex in X, then there is nothing to do. If not, we choose an M-unsaturated vertex $x \in X$. The method will systematically search for an M-augmenting path with origin x. If such a path P exists, then, by Theorem 5.8, M is not a maximum matching, but $M' = M \Delta E(P)$ is a larger matching than M, and hence saturates more vertices in X. We then repeat the procedure with M' instead of M. If such a path does not exist, the set Z of all vertices which are connected to x by an M-alternating path is found. Then $S = Z \cap X$ satisfies $|N_G(S)| < |S|$. By Hall's theorem, G contains no matching that saturates all vertices of X.

An outline of the Hungarian method can be simply described as follows.

Hungarian Method

1. Arbitrarily choose a matching M in G. If M saturates X, stop. Otherwise, let x be an M-unsaturated vertex in X. Set $S = \{x\}$ and $T = \emptyset$.

2. If $N_G(S) = T$, stop, in this case, G contains no matching that saturates all vertices of X. Otherwise, choose $y \in N_G(S) \setminus T$.

3. If y is M-saturated, then there exists $z \in X \setminus S$ such that $yz \in M$. Replace S by $S \cup \{z\}$ and T by $T \cup \{y\}$ and go to Step 2. Otherwise, let P be an M-augmenting xy-path. Replace M by $M' = M \Delta E(P)$ and go to Step 1.

Next, we give two simple examples to show applications of the Hungarian method.

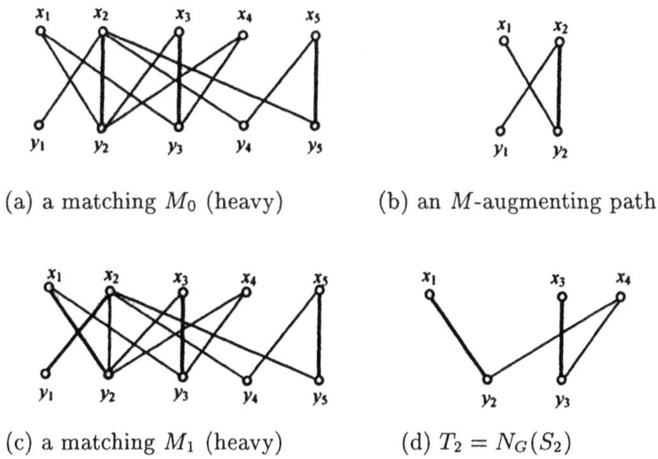

(a) a matching M_0 (heavy) (b) an M-augmenting path

(c) a matching M_1 (heavy) (d) $T_2 = N_G(S_2)$

Figure 5.9: An application of Hungarian method

Example 5.3.1 Consider the bipartite graph G with bipartition $\{X, Y\}$ in Figure 5.9 (a), where $X = \{x_1, x_2, \cdots, x_5\}$ and $Y = \{y_1, y_2, \cdots, y_5\}$.

Choose a matching M_0, for example, $M_0 = \{x_2y_2, x_3y_3, x_5y_5\}$. The vertex $x_1 \in X$ is M_0-unsaturated. Set $S_0 = \{x_0\}$ and $T_0 = \emptyset$. Since $T_0 \subset \{y_2, y_3\} = N_G(S_0)$, choose $y_2 \in N_G(S_0) \setminus T_0$. The vertex y_2 is M_0-saturated and the edge $x_2y_2 \in M_0$.

Set $S_1 = \{x_1, x_2\}$ and $T_1 = \{y_2\}$. Then $T_1 \subset \{y_1, y_2, y_3, y_4, y_5\} = N_G(S_1)$. Choose $y_1 \in N_G(S_1) \setminus T_1$. The vertex y_1 is an M_0-unsaturated vertex, $P_0 = (x_1, y_2, x_2, y_1)$ is an M_0-augmenting path in G (see Figure 5.9 (b)). Set (see Figure 5.9 (c))

$$M_1 = M_0 \Delta E(P_0) = \{x_1y_2, x_2y_1, x_3y_3, x_5y_5\}.$$

Replace M_0 by M_1 and go to Step 1. The vertex $x_4 \in X$ is M_1-unsaturated. Set $S_0 = \{x_4\}$ and $T_0 = \emptyset$. Then $T_0 \subset \{y_2, y_3\} = N_G(S_0)$. Choose $y_2 \in N_G(S_0) \setminus T_0$. The vertex y_2 is M_1-saturated and the edge $x_2y_2 \in M_0$. Set $S_1 = \{x_1, x_4\}$ and $T_1 = \{y_2\}$. Then $T_1 \subset \{y_2, y_3\} = N_G(S_1)$. Choose $y_3 \in N_G(S_1) \setminus T_1$. The vertex y_3 is M_1-saturated and the edge $x_3y_3 \in M_1$. Set $S_2 = \{x_1, x_3, x_4\}$ and $T_2 = \{y_2, y_3\}$. Then $T_2 = N_G(S_2)$ (see Figure 5.9 (d)). Stop. G contains no matching that saturates all vertices of X since $|N_G(S_2)| < |S_2|$ by Hall's theorem. ∎

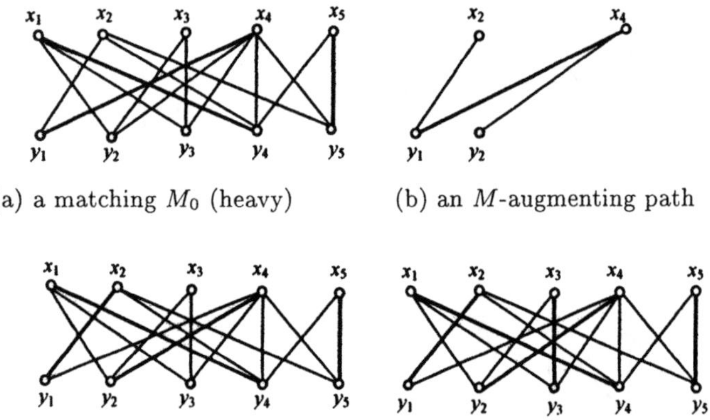

(a) a matching M_0 (heavy) (b) an M-augmenting path

(c) a matching M_1 (heavy) (d) a perfect matching M_2

Figure 5.10: An application of Hungarian method

Example 5.3.2 Consider the bipartite graph G in Figure 5.10 (a). Choose a matching $M_0 = \{x_1y_4, x_4y_1, x_5y_5\}$. The vertex $x_2 \in X$ is M_0-unsaturated. Set $S_0 = \{x_2\}$ and $T_0 = \emptyset$. Then $T_0 \subset \{y_1, y_4, y_5\} = N_G(S_0)$. Choose $y_1 \in N_G(S_0) \setminus T_0$. The vertex y_1 is M_0-saturated and the edge $x_4y_1 \in M_0$.

Set $S_1 = \{x_2, x_4\}$ and $T_1 = \{y_1\}$. Then $T_1 \subset \{y_1, y_2, y_3, y_4, y_5\} = N_G(S_1)$. Choose $y_2 \in N_G(S_1) \setminus T_1$. The vertex y_2 is an M_0-unsaturated vertex, $P_0 =$

(x_2, y_1, x_4, y_2) is an M_0-augmenting path in G (see Figure 5.10 (b)). Set (see Figure 5.10 (c))

$$M_1 = M_0 \Delta E(P_0) = \{x_1y_4, x_2y_1, x_4y_2, x_5y_5\}.$$

Replace M_0 by M_1 and go to Step 1. The vertex $x_3 \in X$ is M_1-unsaturated. Set $S_0 = \{x_3\}$ and $T_0 = \emptyset$. Then $T_0 \subset \{y_2, y_3\} = N_G(S_0)$. Choose $y_3 \in N_G(S_0) \setminus T_0$. The vertex y_3 is M_1-unsaturated, $P_1 = (x_3, y_3)$ is an M_1-augmenting path. Let (see Figure 5.10 (d))

$$M_2 = M_1 \Delta E(P_1) = \{x_1y_4, x_2y_1, x_3y_3, x_5y_5\}.$$

Then M_2 is a matching that saturates X, and so M_2 is a perfect matching in G. ∎

It is not difficult to show that the Hungarian method is an efficient algorithm with the complexity of $o(v\varepsilon^2)$ (the exercise 5.3.3 (a)). We can find a maximum matching in a bipartite graph by slightly modifying the above procedure (the exercise 5.3.3 (b)).

It is interesting to note that the approach used in the algorithm presented above for finding a perfect matching in a bipartite graph has come to be called the Hungarian method since it sees to have first appeared in the works of Egerváry [51], a Hungarian mathematician, for finding a maximum matching in a bipartite graph. Kuhn [115] presented the first formal procedures for finding a perfect matching in a bipartite graph. It seems to have been Kuhn who at time first used the phrase "Hungarian Method" to distinguish algorithm of this type.

Matching in non-bipartite graphs turned out to be substantially more difficult. The necessary and sufficient condition provided by Tutte's theorem (5.2) does not lead to an efficient algorithm to determine whether or not a graph contains a perfect matching. This is because that the number of subsets $S \subset V$ is exponentially large on order v of G. Edmonds [47] found the first efficient algorithm to find a maximum matching in such a graph. This algorithm also motivated him to propose polynomial time as a measure of "goodness" of algorithms, a point of view which has proved extremely fruitful in theoretical computer science.

Exercises

5.3.1 (a) Let M and N be two disjoint matchings of G, and $|M| > |N|$. Prove that G has two disjoint matchings M' and N' such that $M' \cup N' = M \cup N$, $|M'| = |M| - 1$ and $|N'| = |N| + 1$.

(b) Let G be a bipartite graph. Prove that if $p \geq \Delta$, then there exist p disjoint matchings $M_1, M_2, \cdots, M_p$ such that $E(G) = M_1 \cup M_2 \cup \cdots \cup M_p$ and $\lfloor \frac{\varepsilon}{p} \rfloor \leq |M_i| \leq \lceil \frac{\varepsilon}{p} \rceil$ for any $1 \leq i \leq p$.

5.3.2 Give another proof of Hall's theorem (Theorem 5.1) by making use of Theorem 5.8.

5.3.3 (a) Prove that the Hungarian method is an $o(v\varepsilon^2)$ algorithm.

(b) Describe an efficient algorithm for finding a maximum matching in a bipartite graph.

5.3.4 Test whether or not the following two graphs have perfect matchings. If no perfect matching exists, then finding a maximum matching such that it contains all maximum degree vertices.

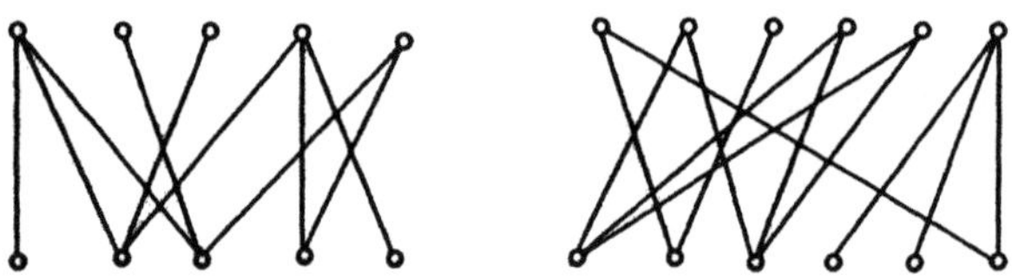

(a) (b)

(the exercise 5.3.4)

5.3.5 A *Latin rectangle* is an $m \times n$ matrix in which the entries are integers in the range from 1 to n. No entry appears more than once in any row or any column. A Latin rectangle is called a *Latin square* if $m = n$.

(a) Add two other two rows to the matrix such which is a Latin square.

$$\mathbf{A} = \begin{pmatrix} 1 & 2 & 3 & 4 & 5 \\ 3 & 4 & 2 & 5 & 1 \\ 5 & 1 & 4 & 2 & 3 \end{pmatrix}$$

(b) Prove that for any $m \times n \, (m < n)$ Latin rectangle $\mathbf{A}$, other $n - m$ rows can be added to $\mathbf{A}$ such which is a Latin square.

5.4 The Optimal Assignment Problem

The algorithms described in preceding section, in particular, the Hungarian method, provide efficient ways of determining a feasible assignment of workers to jobs, if one exists. However if such assignments are more than one, one may, in addition, wish to take into account the effectiveness of the workers in their various jobs (measured, perhaps, by the profit to the company). In this case, one is interested in an assignment that maximizes the total effectiveness of the workers. The problem of finding such an assignment is known as the *optimal assignment problem.*

Consider a weighted bipartite graph $(G, \mathbf{w})$ of bipartition $\{X, Y\}$, where $X = \{x_1, x_2, \cdots, x_n\}$, $Y = \{y_1, y_2, \cdots, y_n\}$ and an edge $x_i y_j$ has the weight $w_{ij} = \mathbf{w}(x_i y_j)$, the effectiveness of worker x_i in job y_j. The optimal assignment problem is clearly equivalent to that of finding a maximum-weight perfect matching in G if G has a perfect matching. We will refer to such a matching as an *optimal matching* shortly.

We suppose that G is a complete bipartite graph $K_{n,n}$, that is, each worker is qualified for very jobs.

To solve the optimal assignment problem it is, of course, possible to enumerate all $n!$ perfect matchings in $K_{n,n}$ (see the exercise 5.1.10 (b)) and find an optimal one among them. However, for large n, such a procedure would clearly be most inefficient. In this section we will present an efficient algorithm, due to Kuhn [115] and, independently, Munkres [137], for finding an optimal matching in a weighted complete bipartite graph. For $\mathbf{l} \in \mathcal{V}(G)$, if

$$\mathbf{l}(x) + \mathbf{l}(y) \geq \mathbf{w}(xy), \quad \forall\, x \in X,\, y \in Y,$$

then $\mathbf{l}$ is said a *feasible labelling* of $(G, \mathbf{w})$. There exists a feasible labelling of $(G, \mathbf{w})$, for example, a *trivial labelling*, $\mathbf{l} \in \mathcal{V}(G)$ defined by

$$\begin{cases} \mathbf{l}(x) = \max\limits_{y \in Y} \mathbf{w}(xy), & \text{if } x \in X; \\ \mathbf{l}(y) = 0, & \text{if } y \in Y. \end{cases}$$

For a feasible labelling $\mathbf{l} \in \mathcal{V}(G)$, set

$$E_l = \{xy \in E(G) : \ \mathbf{l}(x) + \mathbf{l}(y) = \mathbf{w}(xy)\}.$$

The spanning subgraph of G with edge-set E_l is referred to as the *equality subgraph* corresponding to the feasible labelling $\mathbf{l}$, and denoted by G_l.

Example 5.4.1 Consider a weighted complete bipartite graph $(G, \mathbf{w})$ with partition $\{X, Y\}$, where $X = \{x_1, x_2, \cdots, x_5\}$, and $Y = \{y_1, y_2, \cdots, y_5\}$, the weight function is the matrix $\mathbf{W}$ shown in Figure 5.11 (a). Its trivial labelling $\mathbf{l}$ is shown in Figure 5.11 (b) by placing the label $\mathbf{l}(x_i)$ to the right of row i of the matrix and the label $\mathbf{l}(y_j)$ below column j. The entries corresponding to edges of the associated equality subgraph are indicated by the bold types; the equality subgraph G_l itself is depicted in Figure 5.11 (c).

$$\text{(a)}\quad \mathbf{W} = \begin{pmatrix} 3 & 5 & 5 & 4 & 1 \\ 2 & 2 & 0 & 2 & 2 \\ 2 & 4 & 4 & 1 & 0 \\ 0 & 1 & 1 & 0 & 0 \\ 1 & 2 & 1 & 3 & 3 \end{pmatrix}$$

$$\text{(b)}\quad \mathbf{W} = \begin{array}{c} \begin{pmatrix} 3 & \mathbf{5} & \mathbf{5} & 4 & 1 \\ \mathbf{2} & \mathbf{2} & 0 & \mathbf{2} & \mathbf{2} \\ 2 & \mathbf{4} & \mathbf{4} & 1 & 0 \\ 0 & \mathbf{1} & \mathbf{1} & 0 & 0 \\ 1 & 2 & 1 & \mathbf{3} & \mathbf{3} \end{pmatrix} \\ \begin{matrix} 0 & 0 & 0 & 0 & 0 \end{matrix} \end{array} \begin{array}{c} 5 \\ 2 \\ 4 \\ 1 \\ 3 \\ \\ \end{array}$$

(c) 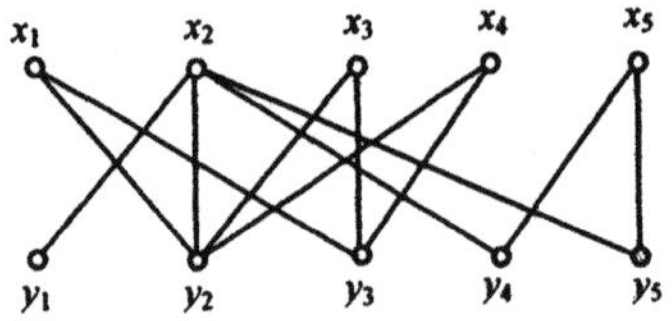

the equality subgraph G_l of $\mathbf{l}$

$$\text{(d)}\quad \mathbf{W} = \begin{array}{c} \begin{pmatrix} 3 & \mathbf{5} & \mathbf{5} & \mathbf{4} & 1 \\ \mathbf{2} & 2 & 0 & \mathbf{2} & \mathbf{2} \\ 2 & \mathbf{4} & \mathbf{4} & 1 & 0 \\ \mathbf{0} & \mathbf{1} & \mathbf{1} & \mathbf{0} & \mathbf{0} \\ 1 & 2 & 1 & \mathbf{3} & \mathbf{3} \end{pmatrix} \\ \begin{matrix} 0 & 1 & 1 & 0 & 0 \end{matrix} \end{array} \begin{array}{c} 4 \\ 2 \\ 3 \\ 0 \\ 3 \\ \\ \end{array}$$

(e) 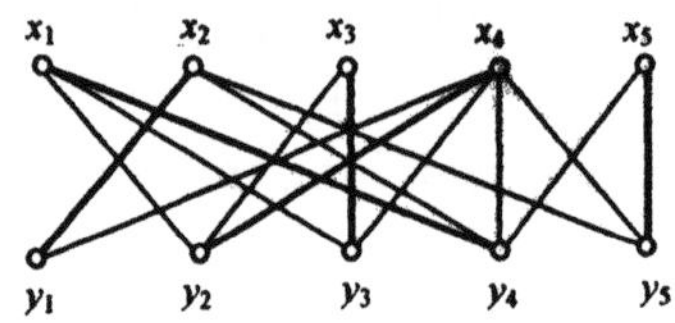

the equality subgraph $G_{\hat{l}}$ of $\hat{\mathbf{l}}$

Figure 5.11: An application of Kuhn-Munkres' algorithm

The connection between equality subgraphs and optimal matchings is provided by the following theorem.

Theorem 5.10 Let $\mathbf{l}$ be a feasible labelling of G. If the equality subgraph G_l contains a perfect matching M^*, then M^* is an optimal matching of G.

Proof Suppose that M^* is a perfect matching of the equality subgraph G_l, then M^* is also a perfect matching of G since G_l is a spanning subgraph of G. Now

$$\mathbf{w}(M^*) = \sum_{e \in M^*} \mathbf{w}(e) = \sum_{x \in V} \mathbf{l}(x) \tag{5.8}$$

since each $e \in M^*$ belongs to G_l and the end-vertices of edges of M^* cover each vertex of G exactly once. On the other hand, if M is any perfect matching of G, then

$$\mathbf{w}(M) = \sum_{e \in M} \mathbf{w}(e) \leq \sum_{x \in V} \mathbf{l}(x) \tag{5.9}$$

It follows from (5.8) and (5.9) that $\mathbf{w}(M^*) \geq \mathbf{w}(M)$. Thus M^* is an optimal matching of G. ∎

The basic idea of Kuhn-Munkres' algorithm for finding an optimal matching is as follows.

Start with an arbitrary feasible labelling $\mathbf{l}$ (for example, the trivial labelling), then determine the equality subgraph G_l and apply the Hungarian method to G_l. If a perfect matching of G_l can be found, then it is optimal by Theorem 5.10. Otherwise, the Hungarian method terminates in $S \subset X$, $T \subset Y$ with $N_{G_l}(S) = T$. Set

$$\alpha_l = \min\{\mathbf{l}(x) + \mathbf{l}(y) - \mathbf{w}(xy) : \ x \in S, \, y \in Y \setminus T\}.$$

Then $\alpha_l > 0$, and the function $\hat{\mathbf{l}}$ defined by

$$\hat{\mathbf{l}} = \begin{cases} \mathbf{l}(u) - \alpha_l, & u \in S; \\ \mathbf{l}(u) + \alpha_l, & u \in T; \\ \mathbf{l}(u), & \text{otherwise} \end{cases}$$

is a new feasible labelling of G and $T \subset N_{G_{\hat{l}}}(S)$ (the exercise 5.4.1). The labelling $\hat{\mathbf{l}}$ is called a *revised feasible labelling* of $\mathbf{l}$. Replace $\mathbf{l}$ by $\hat{\mathbf{l}}$. Such modifications in the feasible labelling are made whenever necessary, until a perfect matching is found in some equality subgraph.

Example 5.4.2 Consider the equality subgraph G_l shown in Figure 5.11 (c). We have known from Example 5.3.1 that it contains no perfect matching and Hungarian method terminates in $S = \{x_1, x_3, x_4\}$ and $T = \{y_2, y_3\} = N(S)$. By counting, we have

$$\alpha_l = \min\{\mathbf{l}(x) + \mathbf{l}(y) - \mathbf{w}(xy) : \ x \in S, \, y \in Y \setminus T\} = 1.$$

The revised feasible labelling $\hat{\mathbf{l}}$ of $\mathbf{l}$ is shown on the right of the matrix and below in Figure 5.11 (d). The equality subgraph $G_{\hat{l}}$ induced by the edges corresponding to the bold entries in the matrix is depicted in Figure 5.11 (e), which is what is considered in Example 5.3.2. We have found a perfect matching by Hungarian method

$$M = \{x_1y_4, x_2y_1, x_3y_3, x_4y_2, x_5y_3\},$$

which is an optimal matching in $(G, \mathbf{w})$ by Theorem 5.10, and the weight $\mathbf{w}(M) = 14$. ∎

We now describe a general outline of Kuhn-Munkres' algorithm as follows.

Kuhn-Munkres' Algorithm

1. Start with an arbitrary feasible labelling $\mathbf{l}$ (for example, the trivial labelling), then determine the equality subgraph G_l and choose a matching M of G_l. If M saturates X, then stop since M is an optimal matching in $(G, \mathbf{w})$ by Theorem 5.10. Otherwise go to Step 2.

2. Hungarian method terminates in $S \subset X$, $T \subset Y$ with $N_{G_l}(S) = T$. Count α_l and obtain a revised feasible labelling $\hat{\mathbf{l}}$, then replace $\mathbf{l}$ by $\hat{\mathbf{l}}$, and go to Step 1.

Kuhn-Munkres' algorithm is efficient (the exercise 5.4.1). It should be noted that optimal matchings are not unique. For example, $M' = \{x_1y_3, x_2y_5, x_3y_2, x_4y_1, x_5y_4\}$ is another matching of the equality subgraph graph shown in Figure 5.11 (e). Kuhn-Munkres' algorithm can be also used in finding a minimum-weight perfect matching of a weighted complete bipartite graph. The connection between minimum-weight perfect matchings and maximum-weight matchings is provided by the following theorem, the proof is simple and left to the reader as an exercise (the exercise 5.4.1).

Theorem 5.11 Let $(K_{n,n}, \mathbf{w})$ be a weighted complete bipartite graph with weighted matrix $\mathbf{W} = (w_{ij})$, a a maximum entry in $\mathbf{W}$, $\mathbf{J}_n$ an n-square matrix whose entries all are one, and let $\mathbf{W}^* = a\,\mathbf{J}_n - \mathbf{W}$ be the weighted matrix of $(K_{n,n}, \mathbf{w}^*)$. Then M is a minimum-weight perfect matching of $(K_{n,n}, \mathbf{w})$ if and only if M is a maximum-weight matching of $(K_{n,n}, \mathbf{w}^*)$. Moreover, $\mathbf{w}(M) = n\,a - \mathbf{w}^*(M)$. ∎

Example 5.4.3 Consider the weighted complete bipartite graph $(K_{5,5}, \mathbf{w})$, where the weighted matrix $\mathbf{W}$ is shown in Figure 5.11 (a). In this case, $a = 5$, and set

$$\mathbf{W}^* = 5\mathbf{J}_5 - \mathbf{W} = \begin{pmatrix} 2 & 0 & 0 & 1 & 4 \\ 3 & 3 & 5 & 3 & 3 \\ 3 & 1 & 1 & 4 & 5 \\ 5 & 4 & 4 & 5 & 5 \\ 4 & 3 & 4 & 2 & 2 \end{pmatrix}.$$

$$\mathbf{W}^* = \begin{pmatrix} 2 & 0 & 0 & 1 & \mathbf{4} \\ 3 & 3 & \mathbf{5} & 3 & 3 \\ 3 & 1 & 1 & \mathbf{4} & \mathbf{5} \\ \mathbf{5} & \mathbf{4} & 4 & \mathbf{5} & 5 \\ \mathbf{4} & \mathbf{3} & \mathbf{4} & 2 & 2 \end{pmatrix} \begin{matrix} 2 \\ 4 \\ 3 \\ 4 \\ 3 \end{matrix}$$
$$\begin{matrix} 1 & 0 & 1 & 1 & 2 \end{matrix}$$

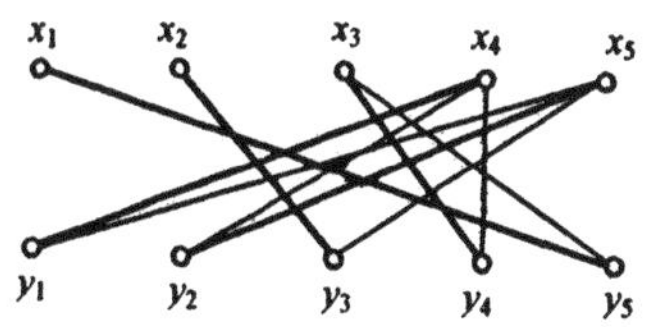

Figure 5.12: An illustration of Example 5.4.3

Applying the Kuhn-Munkres' algorithm to $(G, \mathbf{w}^*)$, we find a maximum-weight matching $M = \{x_1y_5, x_2y_3, x_3y_4, x_4y_1, x_5y_2\}$, indicated by the heavy edges in the graph shown in Figure 5.12, with the weight $\mathbf{w}^*(M) = 21$. By Theorem 5.11, M is a minimum-weighted perfect matching of $(K_{5,5}, \mathbf{w})$ with the weight $\mathbf{w}(M) = 25 - \mathbf{w}^*(M) = 25 - 21 = 4$.

Comparing this matching with one in Example 5.4.2, for any perfect matching M of $(K_{5,5}, \mathbf{w})$, we have $4 \leq \mathbf{w}(M) \leq 14$. This fact shows that to choose an optimal personnel assignment is very necessary to maximize the profit to the company. ∎

Example 5.4.4 As an application of the above method, we consider a sequencing problem of jobs.

A number of different jobs $J_1, J_2, \cdots, J_n$ have to be processed on one machine. After each job, the machine must be adjusted to fit the requirements of the next job. If the time of adaptation from job J_i to job J_j is t_{ij}, find a sequencing of the jobs such that the total machine adjustment time is as small as possible.

For example, there are six jobs to be processed, the time t_{ij} of adaptation from job J_i to job J_j is as follows:

$$\mathbf{T} = (t_{ij}) = \begin{pmatrix} 0 & 5 & 3 & 4 & 2 & 1 \\ 1 & 0 & 1 & 2 & 3 & 2 \\ 2 & 5 & 0 & 1 & 2 & 3 \\ 1 & 4 & 4 & 0 & 1 & 2 \\ 1 & 3 & 4 & 5 & 0 & 5 \\ 4 & 4 & 2 & 3 & 1 & 0 \end{pmatrix}.$$

These jobs processed in the order

$$J_1 \xrightarrow{5} J_2 \xrightarrow{1} J_3 \xrightarrow{1} J_4 \xrightarrow{1} J_5 \xrightarrow{5} J_6$$

requires 13 units of adjustment time; and in the order

$$J_1 \xrightarrow{4} J_4 \xrightarrow{4} J_3 \xrightarrow{5} J_2 \xrightarrow{3} J_5 \xrightarrow{5} J_6$$

requires 21 units of adjustment time. Thus, to maximizes the total effectiveness of the machine, find a sequencing of the jobs that the total machine adjustment time is as small as possible is very necessary.

To the end, construct a weighted simple digraph $(D, \mathbf{w})$ with vertex-set $V(D) = \{J_1, J_2, \cdots, J_6\}$, and $(J_i, J_j) \in E(D)$ if and only if $t_{ij} \leq t_{ji}$, and $\mathbf{w}(J_i, J_j) = t_{ij}$, see Figure 5.13.

$$\mathbf{W} = \begin{pmatrix} 0 & 0 & 0 & 0 & 0 & 1 \\ 1 & 0 & 1 & 2 & 3 & 2 \\ 2 & 0 & 0 & 1 & 2 & 0 \\ 1 & 0 & 0 & 0 & 1 & 2 \\ 1 & 3 & 0 & 0 & 0 & 0 \\ 0 & 0 & 2 & 0 & 1 & 0 \end{pmatrix}$$

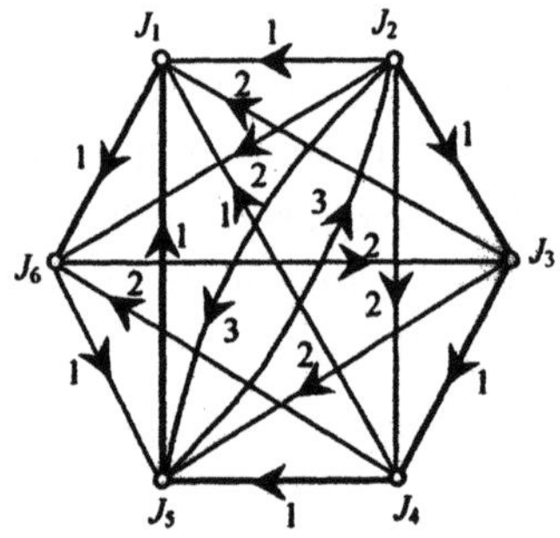

Figure 5.13: An illustration of Example 5.4.4

The digraph D contains a tournament as its spanning subgraph. Thus D contains a Hamilton directed path by Theorem 1.2. It is clear that any Hamilton directed path provides a sequencing of the jobs, and vice versa. For example, $(J_2, J_3, J_4, J_5, J_1, J_6)$ is a Hamilton directed path in D (indicated by heavy edges in Figure 5.13). The sequence corresponding to this path requires only five units of adjustment time. Thus, the sequencing problem of jobs can be clearly referred to one of finding a Hamilton directed paths in D with minimum weight, no efficient algorithm for its solution is known (see the exercise 1.9.10). It is therefore desirable to have a method for obtaining a reasonably good (but not necessarily optimal) solution. We can describe an approximation method by making use of Kuhn-Munkres' algorithm.

Suppose that P^* is a Hamilton directed path with minimum weight in $(D, \mathbf{w})$, corresponding to an optimal sequencing of the jobs. Consider the associated bipartite graph G with D (defined in Section 1.2) with the weighted function $\mathbf{w}$, see Figure 5.14 (a), $\mathbf{w}(J_i'J_j'') = t_{ij}$. Note that D is loopless and contains no edge of the form $J_i'J_i''$ for each $i = 1, 2, \cdots, 6$.

(a)

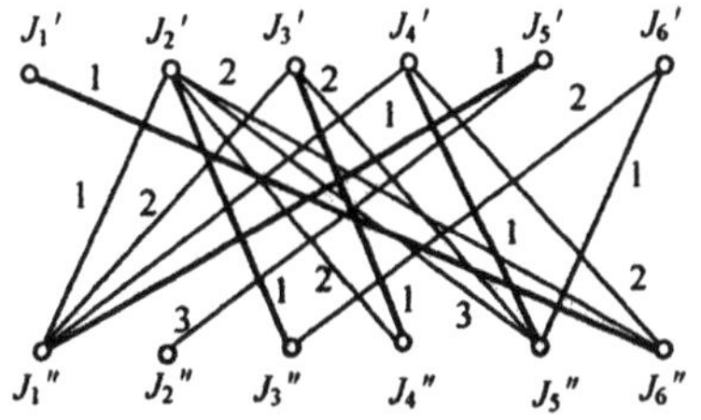

$$\mathbf{W}' = \begin{pmatrix} 0 & 0 & 2 & 1 & 3 & \mathbf{4} \\ \mathbf{4} & 0 & \mathbf{4} & 3 & 2 & 3 \\ 3 & 0 & 0 & \mathbf{4} & 3 & 2 \\ \mathbf{4} & 1 & 1 & 0 & \mathbf{4} & 3 \\ \mathbf{4} & 2 & 1 & 0 & 0 & 0 \\ 1 & 1 & 3 & 2 & \mathbf{4} & 0 \end{pmatrix} \begin{matrix} 4 \\ 4 \\ 4 \\ 4 \\ 4 \\ 4 \end{matrix}$$

$$\begin{matrix} 0 & 0 & 0 & 0 & 0 & 0 \end{matrix}$$

(b)

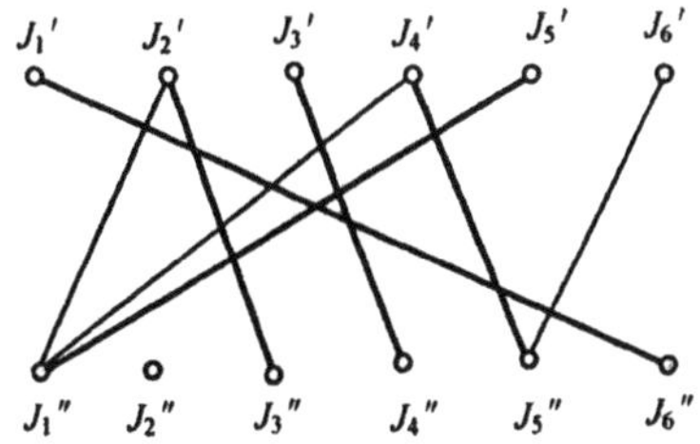

Figure 5.14: An illustration of Example 5.4.4

It is clear that a Hamilton (J_i, J_j)-path P in D corresponds a perfect matching M in $G - \{J_j', J_i''\}$ and $\mathbf{w}(P) = \mathbf{w}(M)$. For example, $P = (J_2, J_3, J_4, J_5, J_1, J_6)$ is a Hamilton (J_2, J_6)-path in D (heavy edges in Figure 5.13), corresponding to a perfect matching $M = \{J_1'J_6'', J_2'J_3'', J_3'J_4'', J_4J_5'', J_5'J_1''\}$ in $G - \{J_6', J_2''\}$ (heavy edges in Figure 5.14) (a)) and $\mathbf{w}(P) = 5 = \mathbf{w}(M)$. The converse, however, is not true. For example, $M' = \{J_1'J_6'', J_2'J_5'', J_3'J_4'', J_5'J_2'', J_6'J_3''\}$ is a perfect matching in $G - \{J_4', J_1''\}$, but no Hamilton directed path in D corresponds it. This fact brings us difficulty in solving the problem. If we can find two vertices J_j'' and J_i' $(i \neq j)$ of G and a minimum-weight perfect matching M in $G - \{J_j', J_i''\}$ corresponding to a Hamilton (J_i, J_j)-path P in D, then P may think of as an approximately optimal sequencing, which requires $\mathbf{w}(P) = \mathbf{w}(M)$ units of adjustment time.

We intend to find such a perfect matching by Kuhn-Munkres' algorithm. To the end, we consider the weighted complete bipartite graph $(K_{6,6}, \mathbf{w})$ with weighted matrix

$$\mathbf{W} = \mathbf{T} + 5\mathbf{I} = \begin{pmatrix} 5 & 5 & 3 & 4 & 2 & 1 \\ 1 & 5 & 1 & 2 & 3 & 2 \\ 2 & 5 & 5 & 1 & 2 & 3 \\ 1 & 4 & 4 & 5 & 1 & 2 \\ 1 & 3 & 4 & 5 & 5 & 5 \\ 4 & 4 & 2 & 3 & 1 & 5 \end{pmatrix}.$$

To find a minimum-weight perfect matching in $(K_{6,6}, \mathbf{w})$, by Theorem 5.11 we need to only find a maximum-weight perfect matching in $(K_{6,6}, \mathbf{w}')$, where $\mathbf{w}' = 5\mathbf{J} - \mathbf{W}$ (see Figure 5.14). By Kuhn-Munkres' algorithm, for a trivial labelling $\mathbf{l}$, the equality subgraph G_l (see Figure 5.14 (b)) contains a perfect matching (the heavy edges)

$$M = \{J_1'J_6'', J_2'J_3'', J_3'J_4'', J_4'J_5'', J_5'J_1''\},$$

which is minimum-weight perfect matching in $G_l - \{J_6', J_2''\}$ with weight $\mathbf{w}(M) = 5$. It corresponds a Hamilton directed path $(J_2, J_3, J_4, J_5, J_1, J_6)$ in D. The sequencing corresponding to this path requires 5 units of adjustment time and, thus, is optimal since it is clear that any sequencing requires at least 5 units of adjustment time.

Note that the equality subgraph also contains a minimum-weight perfect matching

$$M' = \{J_1'J_6'', J_2'J_3'', J_3'J_4'', J_4J_1'', J_6'J_5''\},$$

which corresponds a Hamilton directed path $(J_2, J_3, J_4, J_1, J_6, J_5)$. The sequencing corresponding to this path is also optimal.

Exercises

5.4.1 Prove that (a) $\alpha_l > 0$ and $T \subset N_{G_l}(S)$;

(b) Kuhn-Munkres' algorithm is efficient;

(c) Theorem 5.11.

5.4.2 Find a maximum-weight and a minimum-weight perfect matching in $K_{5,5}$ with weight matrices, respectively,

$$\mathbf{W} = \begin{pmatrix} 9 & 8 & 5 & 3 & 2 \\ 6 & 7 & 8 & 6 & 9 \\ 5 & 8 & 1 & 4 & 7 \\ 7 & 7 & 0 & 3 & 6 \\ 9 & 8 & 6 & 4 & 5 \end{pmatrix} \qquad \mathbf{W}' = \begin{pmatrix} 3 & 2 & 1 & 2 & 3 \\ 1 & 4 & 2 & 1 & 2 \\ 5 & 1 & 2 & 3 & 1 \\ 3 & 2 & 6 & 4 & 1 \\ 1 & 2 & 3 & 1 & 2 \end{pmatrix}$$

5.4.3 Let $\mathbf{A}$ be an n-square matrix. A *diagonal line* of $\mathbf{A}$ is a set of n entries from different rows and columns of $\mathbf{A}$, and its weight is the sum of these n entries.

(a) Find a maximum-weight and a minimum-weight diagonal line of the following matrices and their weights, respectively,

$$\mathbf{A} = \begin{pmatrix} 4 & 5 & 8 & 10 & 11 \\ 7 & 6 & 5 & 7 & 4 \\ 8 & 5 & 12 & 9 & 6 \\ 6 & 6 & 13 & 10 & 7 \\ 4 & 5 & 7 & 9 & 8 \end{pmatrix} \qquad \mathbf{B} = \begin{pmatrix} 9 & 8 & 0 & 0 & 0 \\ 0 & 0 & 8 & 6 & 9 \\ 0 & 8 & 0 & 4 & 7 \\ 0 & 7 & 0 & 3 & 6 \\ 9 & 8 & 6 & 4 & 0 \end{pmatrix}$$

(b) Prove that all diagonal lines of $\mathbf{B}$ admit the same weight.

5.4.4 Six jobs $J_1, J_2, \cdots, J_6$ need to be processed, the time t_{ij} of adaptation from job J_i to job J_j is as follows:

$$\mathbf{T} = (t_{ij}) = \begin{pmatrix} 0 & 3 & 2 & 5 & 1 & 3 \\ 2 & 0 & 4 & 5 & 4 & 1 \\ 1 & 3 & 0 & 1 & 2 & 2 \\ 4 & 2 & 2 & 0 & 1 & 3 \\ 3 & 1 & 4 & 5 & 0 & 2 \\ 2 & 5 & 3 & 1 & 2 & 0 \end{pmatrix}$$

Find a sequencing of jobs as optimal as possible.

5.5 The Travelling Salesman Problem

A travelling salesman, starting in his own town, has to visit each of towns where he should go to precisely or at least once and return his home by the shortest route. This is known as the *travelling salesman problem.*

At first glance, this problem is extremely similar to the Chinese postman problem. However, no efficient algorithm for solving the travelling salesman problem is known as far. We will here describe an algorithm of Christofides [32] for approximately solving the travelling salesman problem, following a comprehensive treatment of Gibbons [74].

We use a connected undirected weighted graph $(G, \mathbf{w})$ to model the traffic system that the travelling salesman has to visit. We call a closed walk that contains each vertex of G at least once to be a *salesman route.* In the graph theoretic language, the travelling salesman problem can be stated as finding a minimum weight Hamilton cycle or a minimum weight salesman route in $(G, \mathbf{w})$. Such a Hamilton cycle, if exists, is called an *optimal cycle*; such a salesman route, which exists certainly, is called an *optimal route.* In general, the two definitions of the travelling salesman problem may have two different solutions even if optimal cycles exist. For example, for the weighted graph $(G, \mathbf{w})$ in Figure 5.15, it clearly contains an optimal cycle (x, y, z, x) of weight 5 and an optimal route (x, z, x, y, x) of weight 4. As his travelling route, any wise salesman would choose the latter rather than the former.

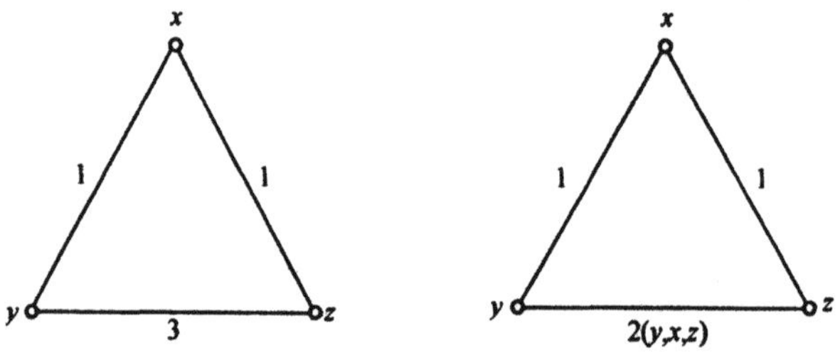

Figure 5.15: (a) An optimal cycle; (b) an optimal route

Generally speaking, a connected weighted graph contains no optimal cycle perhaps, but contains an optimal route certainly. In the following discussion of the travelling salesman problem, we always adopt the second definition, that is, to find an optimal route in a given weighted graph.

Let $(G, \mathbf{w})$ be a connected undirected weighted graph. If for any two distinct vertices x and y of G, the weighted distance $\mathbf{w}$(x,y), the minimum-weight of any xy-path in $(G, \mathbf{w})$, satisfies

$$\mathbf{w}(x,y) \leq \mathbf{w}(x,z) + \mathbf{w}(z,y), \quad \forall\, z \in V(G) \setminus \{x,y\},$$

then we will say that the triangle inequality is satisfied in $(G, \mathbf{w})$.

For an undirected graph G, we can construct a weighted complete graph $(K_v, \mathbf{w}')$ with vertex-set as the same as G, where the weight $\mathbf{w}'(xy)$ of the edge xy of K_v is the weighted distance $\mathbf{w}(x,y)$ between x and y in $(G, \mathbf{w})$. It is clear that the triangle inequality is satisfied in $(K_v, \mathbf{w}')$, and each edge xy of K_v corresponds to an xy-path P in G with $\mathbf{w}(P) = \mathbf{w}'(xy)$ or an edge xy in G with $\mathbf{w}(xy) = \mathbf{w}'(xy)$. For example, see Figure 5.15, $(K_3, \mathbf{w}')$ in (b) is constructed from $(G, \mathbf{w})$ in (a) by the above way.

The following observations are simple, but very useful in the proof of Theorem 5.12.

Lemma 5.12 (i) For any Hamilton cycle C in $(K_v, \mathbf{w}')$, there exists a salesman route R in $(G, \mathbf{w})$ with $\mathbf{w}(R) = \mathbf{w}'(C)$.

(ii) For any optimal route R in $(G, \mathbf{w})$, there exists a Hamilton cycle C in $(K_v, \mathbf{w}')$ with $\mathbf{w}'(C) = \mathbf{w}(R)$.

Proof (i) A required salesman route R in $(G, \mathbf{w})$ can be constructed as follows. For $xy \in E(C)$, if $xy \in E(G)$, then $\mathbf{w}(xy) = \mathbf{w}'(xy)$, and set $xy \in R$; if $xy \notin E(G)$, then G contains an xy-path P with $\mathbf{w}(P) = \mathbf{w}'(xy)$, and set $P \subseteq R$.

(ii) A required Hamilton cycle in K_v can be constructed as follows. Trace R, starting at a vertex x, and delete the vertex that has visited before in turn. Then remaining vertices, in the original order in R, institute a Hamilton cycle C in $(K_v, \mathbf{w}')$, which satisfies $\mathbf{w}'(C) = \mathbf{w}(R)$. ∎

Theorem 5.12 An optimal route R in $(G, \mathbf{w})$ corresponds to an optimal cycle C in $(K_v, \mathbf{w}')$ with the same weight as R, and vice versa.

Proof Suppose that R is an optimal route in $(G, \mathbf{w})$. By Lemma 5.12(ii), there exists a Hamilton cycle C in $(K_v, \mathbf{w}')$ with $\mathbf{w}'(C) = \mathbf{w}(R)$. If C is not optimal, let C^* be an optimal cycle in $(K_v, \mathbf{w}')$, then $\mathbf{w}'(C^*) < \mathbf{w}'(C)$. Thus, by Lemma 5.12(i), there exists a salesman route R' in $(G, \mathbf{w})$ such that$\mathbf{w}(R') = \mathbf{w}'(C^*)$. Thus,

$$\mathbf{w}(R) \leq \mathbf{w}(R') = \mathbf{w}'(C^*) < \mathbf{w}'(C) = \mathbf{w}(R).$$

This contradiction implies that C is an optimal cycle in $(K_v, \mathbf{w}')$.

Conversely, suppose that C is an optimal cycle in $(K_v, \mathbf{w}')$. Then, by Lemma 5.12(i), there exists a salesman route R in $(G, \mathbf{w})$ with $\mathbf{w}(R) = \mathbf{w}'(C)$. If R is not optimal, and let R' be an optimal route in $(G, \mathbf{w})$. Then, by Lemma 5.12(ii), there is a Hamilton cycle C' in $(K_v, \mathbf{w}')$ with $\mathbf{w}'(C') = \mathbf{w}(R')$. Thus,

$$\mathbf{w}'(C') = \mathbf{w}(R') < \mathbf{w}(R) = \mathbf{w}'(C) \leq \mathbf{w}'(C').$$

This contradiction implies that R be an optimal route in $(G, \mathbf{w})$. ∎

By Theorem 5.12, finding an optimal route in $(G, \mathbf{w})$ can be referred to finding an optimal cycle in $(K_v, \mathbf{w}')$ within which the triangle inequality is satisfied. If the triangle inequality is satisfied in $(G, \mathbf{w})$, then G is a spanning subgraph of K_v, and $\mathbf{w} = \mathbf{w}'|E(G)$. Thus, if C is an optimal cycle in $(G, \mathbf{w})$, then C is certainly a Hamilton cycle in $(K_v, \mathbf{w}')$, and $\mathbf{w}(C) = \mathbf{w}'(C)$. Conversely, suppose that C is an optimal cycle in $(K_v, \mathbf{w}')$. If $C \subseteq G$, then C is an optimal cycle in $(G, \mathbf{w})$; if $C \not\subseteq G$, then, by Theorem 5.12, there is an optimal route R in $(G, \mathbf{w})$ such that $\mathbf{w}(R) = \mathbf{w}'(C)$. This shows that if the triangle inequality is satisfied in $(G, \mathbf{w})$, then an optimal cycle in $(K_v, \mathbf{w}')$ corresponds to either an optimal cycle or an optimal route in $(G, \mathbf{w})$. However, if the triangle inequality is not satisfied in $(G, \mathbf{w})$, then an optimal cycle in $(K_v, \mathbf{w}')$ would correspond to an optimal route rather than an optimal cycle in $(G, \mathbf{w})$.

Thus, we need to only find an optimal cycle in $(K_v, \mathbf{w}')$ within which the triangle inequality is satisfied. An immediately obvious method is to enumerate all Hamilton cycles and then by comparison to find the minimum. This approach, although straightforward, presents us with an unacceptably large amount of computation. For a complete undirected graph K_v, there are $\frac{1}{2}(v-1)!$ essentially different Hamilton cycles. Unfortunately, as far no efficient algorithm is known for finding an optimal cycle in a weighted complete graph. In fact, it has been proved to be an NP-hard problem (see, for example, Garey and Johnson [76]), even if weight of every edge is restricted to one and two, see Papadimitriou and Yannakakis [142].

For the travelling salesman problem, as indeed for any other intractable problem, it is useful to have a polynomial-time algorithm which will produce, within known bound, an approximation to the required result. Such algorithms are called *approximation algorithms*. We now describe the best one in these known algorithms for solving the travelling salesman problem, discovered by Christofides [32] in 1976.

Christofides' Approximation Algorithm

1. Find the weighted distance matrix $\mathbf{W}'$ of $(G, \mathbf{w})$ and construct $(K_v, \mathbf{w}')$.

2. Find a minimum tree T in $(K_v, \mathbf{w}')$.

3. Find the set V' of vertices of odd degree in T and a minimum weight perfect matching M in $G' = K_v[V']$.

4. Find an Euler circuit $C_0 = (x, y, z, \cdots, x)$ in $G^* = T \oplus M$.

5. Starting at vertex x, we trace C_0 and delete the vertex that has visited before in turn. Then remaining vertices, in the original order in C_0, determine a Hamilton cycle C in K_v, which is a required approximation optimal cycle.

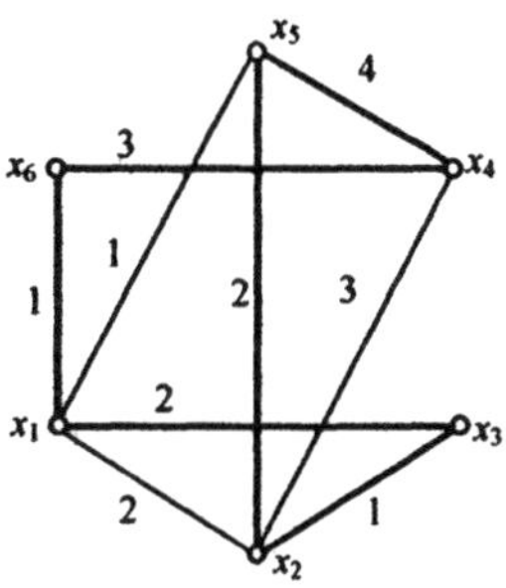

$$\mathbf{w}' = \begin{pmatrix} 0 & 2 & 2 & 4 & 1 & 1 \\ 2 & 0 & 1 & 4 & 2 & 3 \\ 2 & 1 & 0 & 4 & 3 & 3 \\ 4 & 3 & 4 & 0 & 4 & 3 \\ 1 & 2 & 3 & 4 & 0 & 2 \\ 1 & 3 & 3 & 3 & 2 & 0 \end{pmatrix}$$

(a) $(G, \mathbf{w})$ and an optimal cycle C (b) the weighted matrix of $(G, \mathbf{w})$

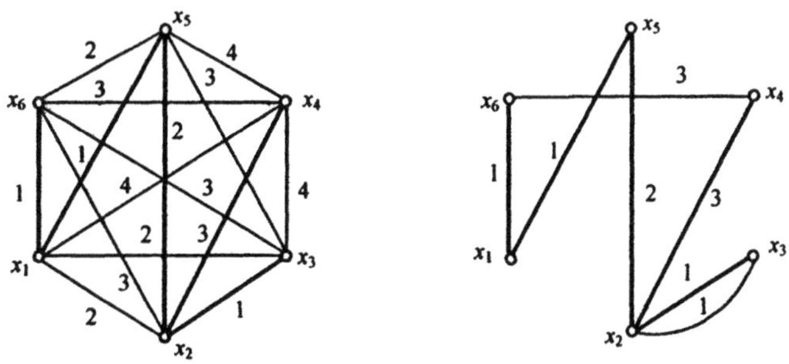

(c) $(K_6, \mathbf{w}')$ and a minimum tree (d) $G^* = T \oplus M$

Figure 5.16: An application of Christofides' algorithm

In the algorithm, the Dijsktra's (see Section 2.5) algorithm can be used in Step 1; the Prim's algorithm (see Section 2.4) can be used in Step 2. In Step 3, V' is nonempty certainly, and $|V'|$ is even by Corollary 1.1. Since $G' = K_v[V']$ is a complete graphs of even order, it must contain a perfect matching M by Corollary 5.2.1. Edmonds and Johnson [48] have presented an efficient algorithm for finding minimum weight perfect in any weighted graph. In Step 4, every vertex of G^* is even degree and, hence, G^* is eulerian by Corollary 1.7.2. Using the Edmonds and Johnson's algorithm can find an Euler circuit C_0 in G^*. All of the above-mentioned algorithms are efficient, thus, Christofides' approximation algorithm is efficient.

Example 5.5.1 Consider the weighted graph $(G, \mathbf{w})$ in Figure 5.16 (a), within which the triangle inequality is satisfied. (b) shows its weighted distance matrix $\mathbf{W}'$; (c) shows $(K_6, \mathbf{w}')$ and a minimum tree T. $V' = \{x_2, x_3, x_4, x_6\}$ is the set of vertices of odd degree in T, $M = \{x_2x_3, x_4x_6\}$ is a minimum weight perfect matching of $K_6[V']$. $G^* = T \oplus M$, shown in (d), has an Euler circuit $C_0 = (x_2, x_5, x_3, x_2, x_4, x_6, x_1)$. Deleting a repeated vertex x_2 from C_0 results in a Hamilton cycle $C = (x_2, x_5, x_3, x_4, x_6, x_1)$ in $(K_6, \mathbf{w}')$ with $\mathbf{w}'(C) = 12$. Because the edge x_3x_4 of C is not in G, C corresponds a salesman route $P = (x_1, x_5, x_2, x_3, x_2, x_4, x_6, x_1)$ with $\mathbf{w}(P) = 12$ which visits each vertex of G at least once.

Notice that G contains only one Hamilton cycle C shown in (a), and so is optimal. However, $\mathbf{w}(C) = 13 > 12 = \mathbf{w}(P)$. ∎

A quality measurement of an approximation algorithm is the *performance ratio*. Let L be a value obtained by an approximation algorithm and L_0 be an exact value. We require a quality guarantee for the approximation algorithm which could, for a minimization (resp. maximization) problem, be stated in the form:

$$1 \leq \frac{L}{L_0} \leq \alpha \quad (\text{resp. } 1 \leq \frac{L_0}{L} \leq \alpha). \tag{5.10}$$

We would, of course, like α to be as close to one as possible. For an approximation algorithm, if there exists a constant α such that (5.10) holds, then the approximation algorithm is called an *α-approximation algorithm.*

Theorem 5.13 For any weighted complete graph within which the triangle inequality is satisfied, Christofides's algorithm for finding a Hamilton cycle is a $\frac{3}{2}$-approximation algorithm.

Proof Suppose that the triangle inequality is satisfied in a weighted complete graph $(K_v, \mathbf{w})$, C_0 and C are an Euler circuit and a Hamilton cycle, respectively, obtained by Christofides' approximation algorithm. Then

$$\mathbf{w}(C) \le \mathbf{w}(C_0) = \mathbf{w}(T) + \mathbf{w}(M), \tag{5.11}$$

where T is a minimum spanning tree of $(K_v, \mathbf{w})$, M is a minimum weight perfect matching in $G' = K_v[V']$, and V' is the set of vertices of odd degree in T.

Let C^* be an optimal cycle in $(K_v, \mathbf{w})$ and T' be a spanning tree of K_v obtained by deleting any one edge from C^*. Then

$$\mathbf{w}(T) \le \mathbf{w}(T') < \mathbf{w}(C^*) \tag{5.12}$$

Let C' be a Hamilton cycle in $G' = K_v[V']$ obtained by following C^*. Because the triangle inequality is satisfied in $(K_v, \mathbf{w})$, $\mathbf{w}(C') \le \mathbf{w}(C^*)$. Since C' is an even cycle, the set of edges of C' can be divided into two edge-disjoint perfect matchings M_1 and M_1. Without loss of generality, suppose $\mathbf{w}(M_1) \le \mathbf{w}(M_2)$. Thus, M_1 is a perfect matching in G' and

$$\mathbf{w}(M) \le \mathbf{w}(M_1) \le \frac{1}{2}\mathbf{w}(C') \le \mathbf{w}(C^*). \tag{5.13}$$

By (5.11), (5.12) and (5.13), we have

$$\mathbf{w}(C) < \mathbf{w}(C^*) + \frac{1}{2}\mathbf{w}(C^*) = \frac{3}{2}\mathbf{w}(C^*),$$

that is,

$$\frac{L_0}{L} = \frac{\mathbf{w}(C)}{\mathbf{w}(C^*)} < \frac{3}{2}$$

as desired. ∎

No such an approximation algorithm for the travelling salesman problem is as far found whose performance ratio is smaller than one of Christofides' algorithm. Although this algorithm can efficiently solve one class of travelling salesman problem, Sahni and Gonzalez [156] have proved that unless the NP-complete problems have solutions in polynomial-time, there is no algorithm with a constant of performance ratio for the optimal cycle problem in $(K_v, \mathbf{w})$ within which the triangle inequality is not satisfied.

There is a great volume of literature associated with the travelling salesman problem. See, for example, Bellmore and Nemhauser [8] for a survey of earlier works and Lawler *et al.* [117] for more.

Exercises

5.5.1 Find an approximately optimal route in the following weighted graph by Christofides' algorithm.

5.5.2 Suppose that the distance matrix of six cities is as follows. Find an approximation solution for the travelling salesman problem by the Christofides' algorithm.

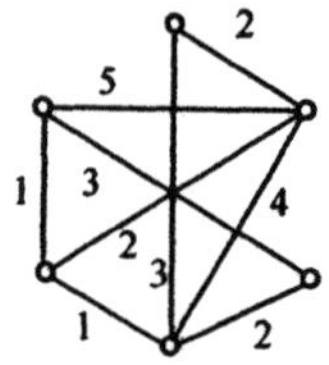

(the exercise 5.5.1)

–	56	32	2	51	60
56	–	21	57	78	70
35	21	–	36	68	68
2	57	36	–	51	61
51	78	68	51	–	13
60	70	68	61	13	–

(the exercise 5.5.2)

5.5.3 Suppose that the triangle inequality is satisfied in a weighted complete graph $(K_v, \mathbf{w})$. Prove that if C is an optimal cycle and T is a minimum tree in $(K_v, \mathbf{w})$ then $\mathbf{w}(C) \leq 2\mathbf{w}(T)$.

5.5.4 Solve the travelling salesman problem in the following traffic system (the minimum weight is 8117).

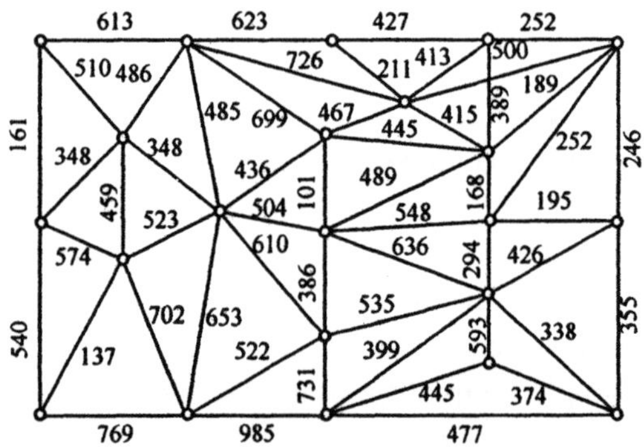

(the exercise 5.5.4)

Chapter 6

Coloring Theory

In the history of graph theory, the problems involving the coloring of graphs have received considerable attention – mainly because of one problem, the four-color problem proposed in 1852: whether four colors will be enough to color the countries of any map so that no two countries which have a common boundary are assigned the same color.

Since more than 150 years, in the process of attempt at the four-color problem, one has greatly developed and enriched coloring theory of graphs. In this chapter, we will introduce basic concepts of vertex-coloring and edge-coloring of a graph and two graphic parameters, chromatic number and edge-chromatic number, closely related to the two types of colorings. We will present two classical results on coloring theory of graphs, Brooks' theorem and Vizing's theorem. We will also present the equivalence of certain problems concerning vertex-coloring and edge-coloring with the four-color problem by Tait's theorem.

We will find the so-called chromatic number (rep. edge-chromatic number) of a graph is the least number of independent subsets (resp. matchings) into which the vertex-set (resp. edge-set) of the graph can be partitioned. In view of this, the coloring theory provided in this chapter is a continuation and extensions of theory concerning independent sets and matchings.

Apart from its own theoretical interest, the study of coloring of graphs is also motivated by its increasing importance in applications of the real-world problems. Unfortunately, as far no efficient algorithm is known for solving these problems.

6.1 Vertex Colorings

Let G be a loopless graph. A *k-vertex coloring* of G is an assignment of k colors, $1, 2, \cdots, k$, to the vertices of G such that adjacent vertices are assigned different colors. In other words, a k-vertex coloring of G is a mapping

$$\pi : V(G) \to \{1, 2, \cdots, k\}$$

such that for each $i = 1, 2, \cdots, k$,

$$V_i = \{x \in V(G) : \pi(x) = i\}$$

is an independent set of G or an empty set. The subset V_i is called a *color class* of π. We often write $\pi = (V_1, V_2, \cdots, V_k)$ for a k-vertex coloring.

The concept of coloring bears no relation to orientations of edges, loops and parallel edges. In discussing vertex colorings, therefore, we will restrict ourselves to simple undirected graphs. Figure 6.1 illustrates a 3-vertex coloring of C_5 and a 3-vertex coloring of Petersen graph, respectively.

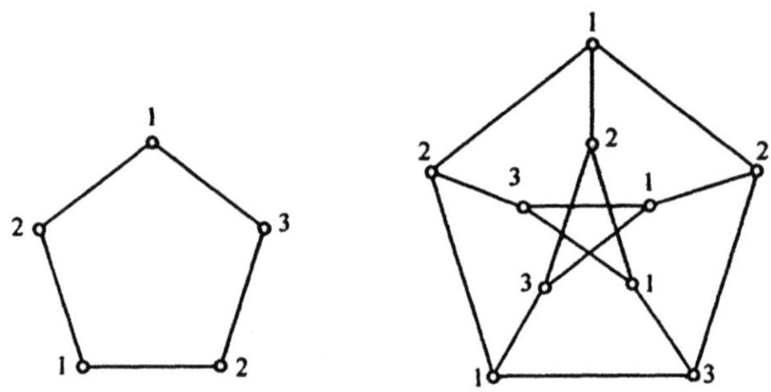

Figure 6.1: Two illustrations of 3-vertex colorings

G is said to be *k-vertex-colorable* if it has a k-vertex coloring. Clearly, every graph is v-vertex-colorable, and if G is k-vertex-colorable then it is necessarily l-vertex-colorable for any $l \geq k$. The *(vertex-)chromatic number*, $\chi(G)$, of G is the minimum number k for which G is k-vertex-colorable.

From definition, for a loopless graph G, if $\chi(G) = k$ then each color class V_i of a k-vertex coloring $\pi = (V_1, V_2, \cdots, V_k)$ is a nonempty independent set of G and $\{V_1, V_2, \cdots, V_k\}$ is a partition of $V(G)$. In other words, the chromatic number of G may be defined alternatively as the minimum number k of independent subsets into which $V(G)$ can be partitioned. Each such independent set is then a color class in the k-vertex coloring of G so defined.

A number of real-world problems show that determining chromatic number is of great importance. For example, suppose that a school assigns end-of-term examinations for several subjects for its students. It is imperative, of course, that two subjects should not be scheduled at the same time if some student is to attend both subjects. Furthermore, it is more efficient to minimize the number of periods used for examinations. This situation can be represented by a simple undirected graph G whose vertices represent the subjects and two vertices are linked by an undirected edge if and only if there is at least one student who is to attend both of the corresponding subjects. The least number of the periods required is then $\chi(G)$.

For several special classes of graphs, the chromatic number is quite easy to determine. For example,

$\chi(G) = 1 \Longleftrightarrow G \cong K_v^c$;
$\chi(G) = 2 \Longleftrightarrow G$ is a nonempty bipartite graph;
$\chi(G) = v \Longleftrightarrow G \cong K_v$, and
$\chi(C_{2n+1}) = 3$ for $n \geq 1$.

A graph G is called to be *k-chromatic* if $\chi(G) = k$. For example, K_v^c is 1-chromatic; a nonempty bipartite graph is 2-chromatic and an odd cycle is 3-chromatic and K_v is v-chromatic.

A graph G is said to be *critical k-chromatic* if $\chi(G) = k$ and $\chi(H) < k$ for every proper subgraph H of G. Clearly, from definition, any critical k-chromatic graph is simple certainly, and a graph

G is critical 1-chromatic $\Longleftrightarrow G \cong K_1$;
G is critical 2-chromatic $\Longleftrightarrow G \cong K_2$;
G is critical 3-chromatic $\Longleftrightarrow G \cong K_3$.

Generally, no necessary and sufficient condition for a graph to be k-chromatic or critical k-chromatic has been known so far. However, it is clear that any k-chromatic graph contains a critical k-chromatic subgraph.

The concept of critical k-chromatic graphs, proposed first by Dirac [39], plays an important role to study vertex colorings. The following theorem gives a necessary condition for a graph to be critical k-chromatic, by means of edge-connectivity, due to Dirac [39], which is here deduced from König's theorem (5.3).

Theorem 6.1 $\lambda(G) \geq k-1$ for any critical k (≥ 2)-chromatic graph G.

Proof Suppose that G is a critical k-chromatic graph with $k \geq 2$. If $k = 2$ then theorem holds clearly since, in this case, $G \cong K_2$, and so $\lambda(G) = 1$. Suppose below $k \geq 3$ and suppose to the contrary that $\lambda(G) < k - 1$. There exists a subset $S \subset V(G)$ such that $|[S, \overline{S}]| = \lambda(G) < k-1$. Since G is critical

k-chromatic, both $G_1 = G[S]$ and $G_2 = G[\overline{S}]$ are $(k-1)$-vertex-colorable. Suppose that

$$\pi_1 = (U_1, U_2, \cdots, U_{k-1}) \quad \text{and} \quad \pi_2 = (W_1, W_2, \cdots, W_{k-1})$$

are $(k-1)$-vertex colorings of G_1 and G_2, respectively. Construct a bipartite simple graph H with bipartition $\{X, Y\}$ as follows. $X = \{x_1, x_2, \cdots, x_{k-1}\}$ and $Y = \{y_1, y_2, \cdots, y_{k-1}\}$, $x_i y_j \in E(H) \iff E_G(U_i, W_j) = \emptyset$. It follows from $|E_G(S, \overline{S})| = \lambda(G) < k-1$ that

$$\varepsilon(H) > (k-1)^2 - (k-1) = (k-1)(k-2).$$

By Corollary 5.3 of König's theorem, H contains a perfect matching, say $M = \{x_i y_{j_i} : i = 1, 2, \cdots, k-1\}$. Thus, $V_i = U_i \cup W_{j_i}$ is an independent set of G for each $i = 1, 2, \cdots, k-1$. Therefore, $\pi = (V_1, V_2, \cdots, V_{k-1})$ is a $(k-1)$-vertex coloring of G, which contradicts the hypothesis of $\chi(G) = k$. Thus, $\lambda(G) \geq k-1$, and theorem follows. ∎

Corollary 6.1.1 $\delta(G) \geq k-1$ for any critical k-chromatic graph G.

Proof It is immediate from Theorem 4.4 and Theorem 6.1. ∎

Corollary 6.1.2 $\chi(G) \leq \Delta(G) + 1$ for any simple graph G.

Proof Suppose that $\chi(G) = k$ and H is a critical k-chromatic subgraph of G. By Corollary 6.1.1, $\delta(H) \geq k-1$. Thus

$$\Delta(G) \geq \Delta(H) \geq \delta(H) \geq k-1 = \chi(G) - 1,$$

and the corollary follows. ∎

We have seen $\chi = \Delta + 1$ for any odd cycle and any complete graph. In fact, it has been prove that odd cycles and complete graphs are only two types of graphs for which $\chi = \Delta + 1$. This is the following classical theorem, known as Brooks' theorem, the proof given here is due to Lovász [121].

Theorem 6.2 (Brooks [21]) If G is a connected simple graph and is neither an odd cycle nor a complete graph, then $\chi(G) \leq \Delta(G)$.

Proof Suppose that $\chi(G) = k$ and H is a critical k-chromatic subgraph of G. If $H \cong K_k$, then $\Delta(H) = \Delta(K_k) = k-1$. Thus $\chi(G) = k = \Delta(H)+1 \leq \Delta(G)$. If H is an odd cycle, then $\chi(G) = \chi(H) = 3 \leq \Delta(G)$ since G is not an odd cycle.

Suppose now that H is neither an odd cycle nor a complete graph. Then $k \geq 4$ and $v(H) = p \geq 5$. Moreover, $\delta(H) \geq 3$ by Corollary 6.1.1. Subject to these hypotheses, we will prove that H is $\Delta(H)$-vertex-colorable.

Since H is not a complete graph, there are $x, y, z \in V(H)$ such that $xy \notin E(H)$, but $xz, yz \in E(H)$. Let $x = x_1, y = x_2$ and let $x_3, x_4, \cdots, x_p$ be an ordering of the vertices in $H - \{x, y\}$ such that $d_H(x_i, z) \geq d_H(x_{i+1}, z)$ for each $i = 3, 4, \cdots, p$. Then $z = x_p$. Let $h = \Delta(H)$. Then $h \leq \Delta(G)$.

We can now describe an h-vertex coloring of G: assign color 1 to both x_1 and x_2; then successively color $x_3, x_4, \cdots, x_p$, each with the first available color in the list $1, 2, \cdots, h$. By the construction of the sequence $x_1, x_2, \cdots, x_p$, each vertex x_i $(i = 1, 2, \cdots, p-1)$ is adjacent to some vertex x_j with $j > i$, and therefore to at most $h-1$ vertices x_l with $l < i$. It follows, when its turn comes to be colored, x_i is adjacent to at most $h-1$ colors, and thus that one of the colors $1, 2, 3, \cdots, h$ will be available. Finally, since x_p is adjacent to both x_1 and x_2 that have been assigned the color 1, it is adjacent to at most $h-2$ other vertices and can be assigned one of the colors $2, 3, \cdots, h$, and so theorem follows. ∎

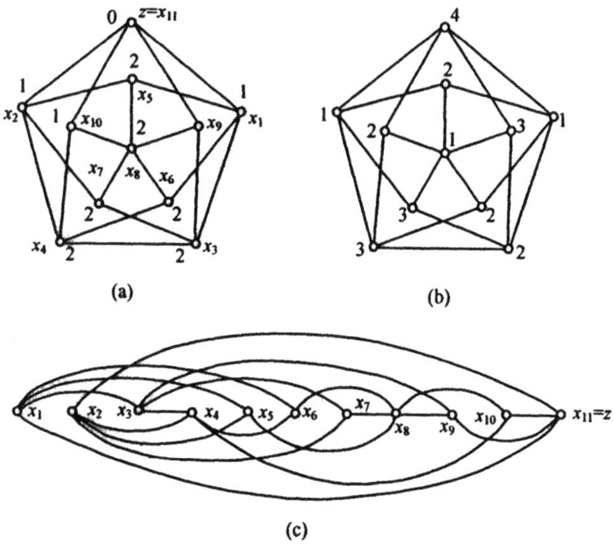

Figure 6.2: The sequential coloring of Grótzsch graph

Example 6.1.1 Consider Petersen graph G. Since it contains odd cycles, G is not bipartite, and so $\chi(G) \geq 3$. On the other hand, G is neither an odd cycle nor a complete graph, then, by Theorem 6.2, $\chi(G) \leq \Delta(G) = 3$. Thus, Petersen graph is a 3-chromatic graph. ∎

The proof of Theorem 6.2 obviously provides a heuristic algorithm for Δ-vertex coloring of a graph. This algorithm is called *sequential coloring.*

Example 6.1.2 Consider the Grótzsch graph G, shown in Figure 6.2, neither an odd cycle nor a complete graph, by Theorem 6.2, then $\chi(G) \leq \Delta(G) = 5$. In fact, we can give a 4-vertex coloring by the sequential coloring.

Arbitrarily choose a vertex z of the Grótzsch graph G and then count the distance $d_G(z, x)$ for every $x \in V(G)$, indicated by the digits nearby vertices in Figure 6.2 (a). Label each vertex of G by $x_1, x_2, \cdots, x_{11}$ such that $x_1, x_2 \in N_G(z)$ and $x_1x_2 \notin E(G)$ (such two vertices must exist since $d_G(z) \geq 3$ and G contains no triangle) and $d_G(z, x_i) \geq d_G(z, x_{i+1})$ for each $i = 3, 4, \cdots, 11$, and so $z = x_{11}$, see Figure 6.2 (a) or (c).

First assign the color 1 to both x_1 and x_2; then successively color $x_3, x_4, \cdots, x_{11}$, each with the first available color in the list $1, 2, 3, 4, 5$. The resulting coloring, a 4-vertex coloring, is shown in (b). ∎

Example 6.1.3 (Roy [155]; Gallai [71]) Let G be a digraph with $\chi(G) = \chi$. Then G contains a directed path of length at least $\chi - 1$.

Proof Let E' be a minimal subset of $E(G)$ such that $G' = G - E'$ contains no directed cycle, which implies that $G' + e$ contains a directed cycle for any $e \in E'$. Suppose that the length of a longest directed path in G' is k. It is sufficient to prove $\chi \leq k+1$. We can do this by constructing a $(k+1)$-vertex coloring of G.

For each $i = 1, 2, \cdots, k+1$, let V_i be a subset of $V(G)$: $x \in V_i$ if and only if the length of a longest directed path in G' with origin x is $i - 1$. Then $\{V_1, V_2, \cdots, V_{k+1}\}$ is a partition of $V(G)$. We first prove that V_i satisfies the following properties.

(i) G' contains no directed path whose origin and terminus both are in V_i for any $i\,(1 \leq i \leq k+1)$. For otherwise, consider a directed (x, y)-path P in G' with $x, y \in V_i$. Then G' contains a directed path Q of length $i - 1$ with origin y. Since G' contains no directed cycle, $P \cup Q$ is a directed path in G' whose length is at least i. This contradicts the choice of $x \in V_i$.

(ii) V_i is an empty set or an independent set of G for each $i = 1, 2, \cdots, k+1$. By contradiction. Suppose that x and y are two distinct vertices in some V_i and are adjacent in G. Thus there exists $e \in E(G)$ such that $\psi(e) = (x, y)$. Then $e \in E'$ since G' contains no (x, y)-path by (i). Thus $G' + e$ contains a directed cycle, say C. However, $C - e$ is a (y, x)-path, which contradicts (i).

By (ii), $\pi = (V_1, V_2, \cdots, V_{k+1})$ is a $(k+1)$-vertex coloring of G, that is, $\chi(G) \leq k+1$. ∎

Example 6.1.4 (V. Chvátal and J. Komlós, 1971) Let G be a simple digraph with $\chi(G) > mn$ and $\mathbf{f} \in \mathcal{V}(G)$. Then G contains either a directed path $(x_0, x_1, \cdots, x_m)$ such that $\mathbf{f}(x_0) \leq \mathbf{f}(x_1) \leq \cdots \leq \mathbf{f}(x_m)$ or a directed path $(y_0, y_1, \cdots, y_n)$ such that $\mathbf{f}(y_0) > \mathbf{f}(y_1) > \cdots > \mathbf{f}(y_n)$.

Proof Construct two spanning subgraphs G_1 and G_2 of G as follows. For $(x, y) \in E(G)$,

$$(x, y) \in E(G_1) \Longleftrightarrow \mathbf{f}(x) \leq \mathbf{f}(y) \text{ or}$$

$$(x, y) \in E(G_2) \Longleftrightarrow \mathbf{f}(x) > \mathbf{f}(y).$$

Clearly, $G = G_1 \oplus G_2$. Suppose that $\chi(G_1) \leq m$ and $\chi(G_2) \leq n$, and let $\pi_1 = (V_1, V_2, \cdots, V_m)$ and $\pi_2 = (V_1', V_2', \cdots, V_m')$ be an m-vertex coloring of G_1 and an n-vertex coloring of G_2, respectively. Set

$$V_{ij} = \{x \in V(G) : \ x \in V_i \cap V_j'\}, \quad i \leq i \leq m, 1 \leq j \leq n.$$

V_i and V_j' are either an independent set or an empty set of G, so is V_{ij}. Thus,

$$\pi = \{V_{ij} : \ i \leq i \leq m, 1 \leq j \leq n\}$$

is an mn-vertex coloring of G. This implies $\chi(G) \leq mn$, which contradicts the hypothesis. Therefore, we have either $\chi(G_1) > m$ or $\chi(G_2) > n$.

If $\chi(G_1) > m$, then, by Example 6.1.3, G_1 contains a directed path P of length at least $\chi(G) - 1 \geq m$. Let $(x_0, x_1, \cdots, x_m)$ be a section of P of length m. By construction of G_1, we have $\mathbf{f}(x_0) \leq \mathbf{f}(x_1) \leq \cdots \leq \mathbf{f}(x_m)$.

Similarly, if $\chi(G_2) > n$, then G_2 contains a directed path $(y_0, y_1, \cdots, y_n)$ such that $\mathbf{f}(y_0) > \mathbf{f}(y_1) > \cdots > \mathbf{f}(y_n)$. ∎

Example 6.1.5 (P. Erdös and G. Szekeres, 1935) Any sequence of $mn+1$ distinct integers contains either an increasing subsequence of m terms or a decreasing subsequence of n terms.

Proof Let $(a_1, a_2, \cdots, a_{mn+1})$ be any sequence of $mn+1$ distinct integers. Construct a simple digraph $G = (V, E)$ as follows. $V(G) = \{a_1, a_2, \cdots, a_{mn+1}\}$, and $(a_i, a_j) \in E(G) \Longleftrightarrow a_i < a_j$. It is easy to see that G is a tournament and $\chi(G) = mn+1$. Let $\mathbf{f} \in \mathcal{V}(G)$ such that $\mathbf{f}(a_i) = a_i$ for each $i = 1, 2, \cdots, mn+1$. The conclusion follows immediately from Example 6.1.4. ∎

Exercises

6.1.1 Let G be a critical k-chromatic graph with $k \geq 3$. Prove that

(a) for any separating set S of G, the induced subgraph $G[S]$ is not a complete graph;

(b) G is 2-connected.

6.1.2 Prove that if G is a critical k-chromatic graph then $v(G) \neq k+1$.

6.1.3 Prove that

(a) any k-chromatic graph contains a critical k-chromatic subgraph;

(b) any k-chromatic simple graph with minimum order must be a critical k-chromatic graph;

(c) $\chi(G) \leq 1 + \max\{\delta(H) : H \subseteq G\}$ for any graph G;

(d) $\chi(G) \leq 1 + l(G)$ for any graph G, where $l(G)$ is the length of longest path in G.

6.1.4 Prove that

(a) $\left\lceil \dfrac{v}{\alpha} \right\rceil \leq \chi(G) \leq v + 1 - \alpha$ for any graph G;

(b) $\dfrac{v^2}{v^2 - 2\varepsilon} \leq \chi(G) \leq \dfrac{1}{2} + \sqrt{2\,\varepsilon + \dfrac{1}{4}}$ for any simple graph G.

6.1.5 Prove that if any two odd cycles in a graph G have a vertex in common then $\chi(G) \leq 5$.

6.1.6 Prove that Brooks' theorem is equivalent to the following statement: if G is k-critical ($k \geq 4$) and not complete, then $2\,\varepsilon \geq v(k-1) + 1$.

6.1.7 (a) Prove that a generalized Brooks' theorem: if a graph G has either $\Delta(G) = 2$ and no odd component or $\Delta(G) \geq 3$ and no component that contains $K_{\Delta+1}$, then $\chi(G) \leq \Delta(G)$.

(b) Prove that for any integers k and m with $2 \leq k \leq m$, there exists a graph G such that $\Delta(G) = m$ and $\chi(G) = k$.

6.1.8 Prove that an undirected graph G is k-vertex colorable if and only if G has an orientation D in which each directed path is of length at most $k-1$.

6.2 Edge Colorings

Let G be a loopless graph. A *k-edge coloring* of G is an assignment of k colors, $1, 2, \cdots, k$, to the edges of G such that adjacent edges are assigned different colors. In other words, a k-edge coloring of G is a mapping

$$\pi' : E(G) \to \{1, 2, \cdots, k\}$$

such that for each $i = 1, 2, \cdots, k$,

$$E_i = \{e \in E(G) : \pi'(e) = i\}$$

is a matching of G or empty set. We often write $\pi' = (E_1, E_2, \cdots, E_k)$ for a k-edge coloring, where E_i is called an *edge-color class* of π'.

The concept of edge-coloring bears no relation to orientations of edges. Therefore, in discussing edge colorings, we will restrict ourselves to undirected graphs. Figure 6.3 illustrate a 3-edge coloring of odd cycle C_5 and a 4-edge coloring of Petersen graph, respectively.

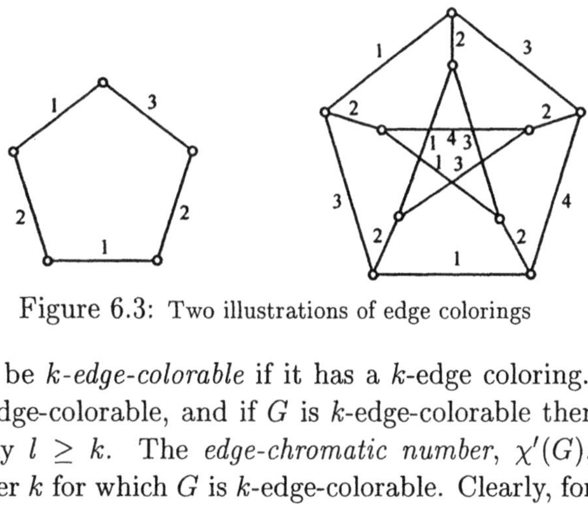

Figure 6.3: Two illustrations of edge colorings

G is said to be *k-edge-colorable* if it has a k-edge coloring. Clearly, every graph is ε-edge-colorable, and if G is k-edge-colorable then it is l-edge-colorable for any $l \geq k$. The *edge-chromatic number*, $\chi'(G)$, of G is the minimum number k for which G is k-edge-colorable. Clearly, for any loopless graph G

$$\chi'(G) \geq \Delta(G). \tag{6.1}$$

From definition, if $\chi'(G) = k$ then each edge-color class E_i of any k-edge coloring $\pi = (E_1, E_2, \cdots, E_k)$ is a nonempty matching of G. In other words, the edge-chromatic number k of G may be defined alternatively as the minimum number of matchings into which $E(G)$ may be partitioned. Each such matching is then an edge-color class in the k-edge coloring of G so defined.

A number of real-world problems can be described by edge coloring of a graph. For example, in a school, there are m teachers $x_1, x_2, \cdots, x_m$ and n classes $y_1, y_2, \cdots, y_n$. Given that the teacher x_i is required to teach the class y_j for p_{ij} periods, schedule a complete timetable in the minimum possible number of periods. We construct a bipartite graph G with bipartition $\{X, Y\}$, where $X = \{x_1, x_2, \cdots, x_m\}$ and $Y = \{y_1, y_2, \cdots, y_n\}$, and vertices x_i and y_j are jointed by p_{ij} undirected edges. Now, in any one period, each teacher can teach at most one class, and each class can be taught by at most one teacher. Thus a teaching schedule for one period corresponds to a matching of G and, conversely, each matching of G corresponds to a possible assignment of teachers to classes for one period. Our problem, therefore, is to partition the edges of G into as few matchings as possible or, equivalently, to color the edges of G with as few colors as possible. This the minimum number is $\chi'(G)$.

For several special classes of graphs, the edge-chromatic number is quite easy to determine. For example, for a cycle C_n, $\chi'(C_n)$ is equal to 2 if n is even, and 3 if n is odd; for a complete graph K_n, $\chi'(K_n)$ is equal to $n-1$ if n is even, and n if n is odd; and for a bipartite graph H, $\pi'(H) = \Delta(H)$ by Corollary 5.9.1.

From definition, the problem of determining the edge-chromatic number of a graph G can be immediately transformed into that of dealing with chromatic number by considering its line graph $L(G)$, namely, if G is nonempty,

$$\pi'(G) = \pi(L(G)).$$

This observation appears to be of little value in computing edge-chromatic number, however, since chromatic numbers are extremely difficult to evaluate in general.

In what must be introduced the fundamental result on edge colorings is the following theorem, known as Vizing's theorem in the literature, first found by Vizing [166]. The proof given here is due to Xu [180]

Theorem 6.3 (Vizing's theorem) For any loopless nonempty graph G,

$$\Delta(G) \leq \chi'(G) \leq \Delta(G) + \mu(G). \tag{6.2}$$

In particular, if G is a simple nonempty graph, then

$$\Delta(G) \leq \chi'(G) \leq \Delta(G) + 1. \tag{6.3}$$

Proof By the inequality (6.1), we need to only prove the right-hand inequality in (6.2). By contradiction. Suppose that there exist a graph G with edge-chromatic number $\chi'(G) = k > \Delta(G) + \mu(G)$ and $e_0 \in E(G)$ such that $G - e_0$ has a $(k-1)$-edge coloring $\pi' = (E_1, E_2, \cdots, E_{k-1})$.

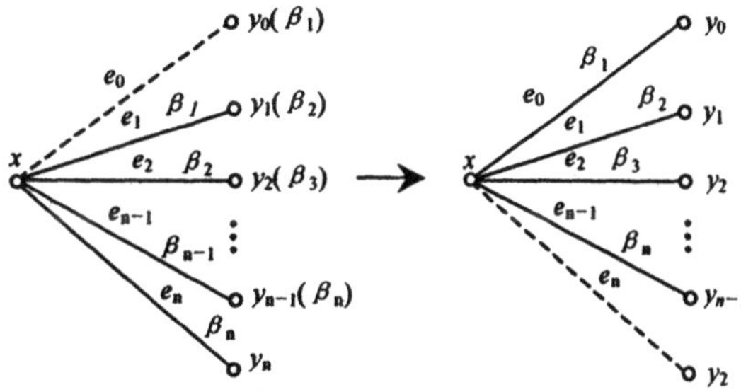

Figure 6.4: A recoloring procedure

For $u \in V(G)$, denote by $C_{\pi'}(u)$ (resp. $C'_{\pi'}(u)$) the set of colors appearing (resp. not appearing) at u under π'. Then $C'_{\pi'}(u) \neq \emptyset$ because $d_{G-e_0}(u) \leq \Delta(G) < k - \mu(G)$.

Let $\psi_G(e_0) = xy_0$. Choose $\beta_1 \in C'_{\pi'}(y_0)$. If $\beta_1 \in C'_{\pi'}(x)$, then we can obtain a $(k-1)$-edge coloring of G by assigning the color β_1 to e_0, which contradicts $\pi'(G) = k$. Thus, $\beta_1 \in C_{\pi'}(x)$ and so there is some $e_1 \in E_G(x)$ such that $\pi'(e_1) = \beta_1$. Let $\psi_G(e_1) = xy_1$ and $F_x(1, \pi') = \{e_0, e_1\}$. Choose $\beta_2 \in C'_{\pi'}(y_1)$. If $\beta_2 \in C'_{\pi'}(x)$, then we can obtain a $(k-1)$-edge coloring of G by assigning the color β_1 to e_0 and the color β_2 to e_1, which contradicts $\pi'(G) = k$. Thus, $\beta_2 \in C_{\pi'}(x)$ and so there is some $e_2 \in E_G(x)$ such that $\pi'(e_2) = \beta_2$. Let $\psi_G(e_2) = xy_2$ (maybe $y_2 = y_0$) and $F_x(2, \pi') = \{e_0, e_1, e_2\}$. Generally, we can obtain a subset of $E(G)$

$$F_x(n, \pi') = \{e_0, e_1, \cdots, e_n\},$$

where $n \geq 1$, and

$$\begin{aligned} &\psi_G(e_i) = xy_i, \quad i = 0, 1, 2, \cdots, n, \\ &\pi'(e_i) \in C'_{\pi'}(y_{i-1}), \quad i = 1, 2, \cdots, n, \end{aligned}$$

where $y_0, y_1, \cdots, y_n$ do not need different. Recoloring $F_x(n, \pi')$ implies such a coloring procedure that assigns the color $\pi'(e_i)$ to the edge e_{i-1} for $i =$

$1, 2, \cdots, n$ and makes e_n uncolored. Notice that recoloring $F_x(n, \pi')$ gives a $(k-1)$-edge coloring $G - e_n$. Figure 6.4 illustrates such a recoloring procedure.

We now show that $F_x(n, \pi')$ satisfies the following two properties.

(i) $C'_{\pi'}(x) \cap C'_{\pi'}(y_i) = \emptyset$ for each $i = 0, 1, 2, \cdots, n$.

Suppose that there are a color α and some vertex y_i such that $\alpha \in C'_{\pi'}(x) \cap C'_{\pi'}(y_i)$. Then a $(k-1)$-edge coloring of G can be obtained by recoloring $F_x(i, \pi')$ and assigning e_i the color α, which contradicts the hypothesis that $\chi'(G) = k$.

(ii) $C'_{\pi'}(y_i) \cap C'_{\pi'}(y_j) = \emptyset$ for any pair of different vertices y_i and y_j.

Suppose that there are a color β and two distinct vertices y_i and y_j such that $\beta \in C'_{\pi'}(y_i) \cap C'_{\pi'}(y_j)$. Choose y_i and y_j such that both i and j are as small as possible. Without loss of generality, suppose that $i < j$.

Let $\alpha \in C'_{\pi'}(x)$. Then $\alpha \neq \beta, \beta \in C_{\pi'}(x), \alpha \in C_{\pi'}(y_i) \cap C_{\pi'}(y_j)$ by (i). Let $H = G[E_\alpha \cup E_\beta]$. Then $d_H(x) = d_H(y_i) = d_H(y_j) = 1$. Let H' be the connected component containing x in H. Then H' is a path and at least one of y_i and y_j is not in H'. A $(k-1)$-edge coloring of G can be obtained either by interchanging the colors α and β in the component containing y_i in H if y_i is not in H' and recoloring $F_x(i, \pi')$ and coloring e_i the color α, or by interchanging the colors α and β in the component containing y_2 in H if y_i is in H' and recoloring $F_x(j, \pi')$ and giving e_j the color α. This contradicts the hypothesis that $\chi'(G) = k$.

Choose $F_x(n, \pi')$ such that n is as large as possible. Let $A(x) = \{y_{n_0}, y_{n_1}, y_{n_2}, \cdots, y_{n_m}\}$ be the end-vertices of edges in $F_x(n, \pi')$ other than x. By (ii) and the choice of n each color in $C'(y_{n_i})$, $i = 0, 1, \cdots, m$, must be used in some edge in $F_x(n, \pi')$. Thus there are at least

$$\begin{aligned} n &\geq |C'(y_{n_0})| + |C'(y_{n_1})| + \cdots + |C'(y_{n_m})| + |\{e_0\}| \\ &\geq (m+1)(k-1-\Delta) + 1, \end{aligned}$$

edges from x to $A(x)$ in G, at least $k - \Delta$ of them must share the same end-vertex in A, a contradiction, since $k - \Delta > \mu$. The proof is completed. ∎

Vizing's theorem, or the inequality (6.3), gives us a simple way of classifying simple graphs into two classes. A simple graph G is said to belong to *class one* if $\pi'(G) = \Delta(G)$, and to *class two* if $\pi'(G) = \Delta(G) + 1$. The problem of deciding which graphs belong to which class is the so-called *classification problem*. We have seen that a complete graph K_{2n} and a bipartite graph belong to class one, and an odd cycle C_{2n+1} and a complete graph K_{2n+1}

belong to class two, but the general classification problem is unsolved. The importance and difficulty of this problem have become apparent since we will realized that its solution would imply that the four-color problem solves, in view of Tait's theorem (see the next section).

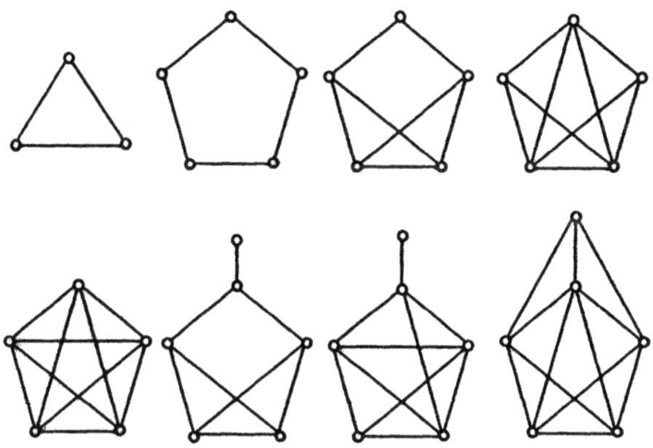

Figure 6.5: All connected simple graphs of order at most six of class two

Although the classification problem has been proved to be NP-hard by Holyer [98], it seems that graphs of class two are relatively scarce. For example, of the 143 connected simple graphs of order at most six, only eight belong to class two (see Figure 6.5). A more general result of this kind is due to Erdös and Wilson [56], who proved that almost all nonempty simple graphs belong to class one, that is,

$$\lim_{v\to\infty} \frac{|C^1(v)|}{|C^1(v) \cup C^2(v)|} = 1,$$

where $C^1(v)$ and $C^2(v)$ are the set of all nonempty simple graphs of order v belonging to class one and two, respectively.

No progress has been made on the more difficult problem of deciding which class contains almost all graphs with a given maximum degree Δ, even for $\Delta = 3$ this is unknown. However, there exist planar graphs of class one with the maximum degree Δ for any $\Delta \geq 2$. For example, the star graph $K_{1,\Delta}$ is such a graph. There exist also planar graphs of class two with the maximum degree Δ for $\Delta = 2, 3, 4, 5$. For example, $\chi'(K_3) = 3 = \Delta(K_3) + 1$; other three planar graphs of class two are shown in Figure 6.6. Vizing [167] has proved that there is no planar graph of class two with the maximum degree $\Delta \geq 8$. In his another paper, Vizing [168] conjectured that there is no planar graph of class two with the maximum degree $\Delta = 6$ or 7. Zhang [184] has proved that Vizing's conjecture is true for $\Delta = 7$. We have not, however, known whether or not Vizing's conjecture is true for $\Delta = 6$.

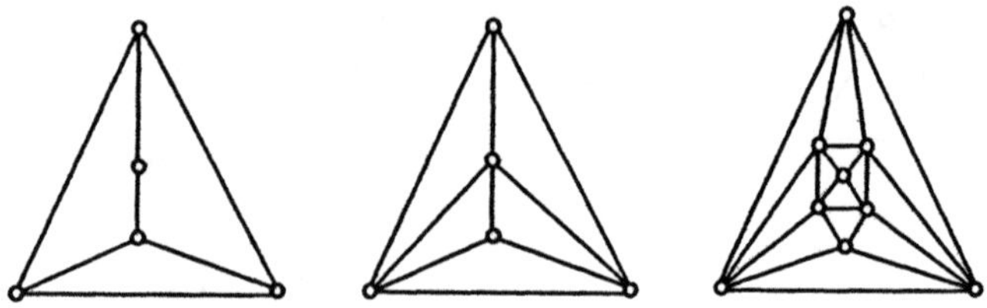

Figure 6.6: Three planar graphs of class two

Exercises

6.2.1 Prove that

(a) each of bipartite graphs, 3-regular hamiltonian graphs and even complete graphs belongs to class one;

(b) each of odd cycles and odd complete graphs belongs to class two.

6.2.2 Prove that a simple graph G belongs to class two if it satisfies one of the following conditions:

(a) $\varepsilon > \Delta\alpha'$;

(b) $\varepsilon > \Delta \lfloor \frac{v}{2} \rfloor$; (L.W.Beineke and R.J.Wilson, 1973)

(c) G has odd order and is nonempty, regular;

(d) G is regular and contains a cut-vertex. (V.G.Vizing, 1965)

6.2.3 Prove that $\chi'(G) = \chi(L(G))$ if G is nonempty.

6.2.4 Prove that a complete k-partite graph $K_n(k)$ belongs to class one if nk is even; to class two otherwise. (R. Laskar and W.Hare, 1971)

6.2.5 Prove that $\chi'(G) \leq \left\lfloor \frac{3}{2}\Delta \right\rfloor$ if G is loopless. (C.E.Shannon, 1949)

6.2.6 Prove that $\chi'(G) \leq 3\Delta - 2$ if G is simple and $\Delta \geq 3$ by using Brooks' theorem (6.2) and the exercise 6.2.3.

Applications

6.3 The Four-Color Problem

This problem can be traced back to 1852. While coloring a map of the counties of England, Francis Guthrie, a London student, noticed that four colors are sometimes needed, and then proposed a conjecture to his brother Frederick that four colors are always sufficient to color the countries of any map so that no two countries which have a common boundary are assigned the same color. But Frederick was unable to prove this and brought the problem to the attention of Augustus De Morgan, Professor of Mathematics at University College, London, who mentioned it on a number of occasion, giving credit to Francis Guthrie for proposing it. However, it was not until 1878 (after De Morgan's death) that the problem became widely known. At a meeting of the London Mathematical Society in that year, Arthur Cayley asked whether the problem had been solved, and shortly afterwards wrote a note [25] in which he attempted to explain where the difficulty lies.

Note that a map can be regarded as a plane graph with its countries as the faces of the graph. This leads to a concept of face-colorings of a plane graph.

A *k-face coloring* π^* of a plane graph G is an assignment of k colors, $1, 2, \cdots, k$, to the faces of G such that no two faces which have a common boundary are assigned the same color. We write $\pi^* = (F_1, F_2, \cdots, F_k)$ for a k-face coloring, where $F_i = \{f \in F(G) : \pi^*(f) = i\}$. A plane graph G is said to be *k-face-colorable* if it has a k-face coloring. The parameter $\chi^*(G) = \min\{k : G \text{ is } k\text{-face-colorable}\}$ is called the *face-chromatic number* of G. For example $\chi^*(K_4) = 4$.

From definition, by making use of the geometric dual G^* of a plane graph G, we immediately obverse the following relationship between the face-chromatic number of G and the vertex-chromatic number of G^*:

$$\chi^*(G) = \chi(G^*). \tag{6.4}$$

By Corollary 3.3.2, all planar embeddings of a given connected planar graph have the same number of faces. By (6.4), therefore, the four-color conjecture can be equivalently stated as the following formula.

Four-Color Conjecture Every plane graph is 4-face-colorable or every planar graph is 4-vertex-colorable.

The four-color conjecture is one of the best-known conjecture in the whole of mathematics. The problem of deciding whether the four-color conjecture is true or not is called the *four-color problem.*

The first serious attempt at a proof of the four-color conjecture seems to have been made by Kempe [104], a barrister and keen amateur mathematician who was Treasurer, and later President, of the London Mathematical Society. In 1879, he published a "proof" of the four-color conjecture. In order to describe Kempe's ideas in his proof in modern terminology, we need a few definitions.

Clearly, it is sufficient to consider plane triangulations for the four-color conjecture. A plane graph is called a *configuration* if each of its bounded faces is a triangle. The four graphs shown in Figure 6.7, for example, are configurations, denoted by O, P, Q, R, respectively. A set $\mathcal{F}$ consisting of finite configurations is called *unavoidable complete* if every plane triangulation must contain at least one element of $\mathcal{F}$. By Corollary 3.4.3, it is clear that the set $\mathcal{F} = \{O, P, Q, R\}$ is an unavoidable complete set.

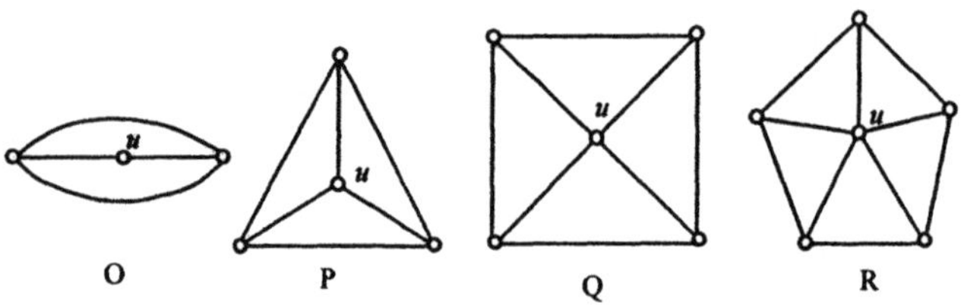

Figure 6.7: An unavoidable complete set $\mathcal{F}$

Suppose that there exists a counterexample to the four-color conjecture. We may choose a counterexample with order as small as possible, such a plane triangulation G is called a *minimal graph.* So $\chi(G) = 5$ and $\chi(H) \leq 4$ for any plane graph H with fewer vertices than G. Kempe attempted to prove there exists no minimal graph.

Suppose that G is a minimal graph. So G must contain at least one of configurations in $\mathcal{F} = \{O, P, Q, R\}$. If G contains either O or P, then $\chi(G - u) \leq 4$. Since $d_G(u) \leq 3$, there is always a spare color that can be used to color u for any 4-vertex coloring of $G - u$. This implies $\chi(G) \leq 4$, a contradiction. Therefore, G contains neither O nor P.

Suppose that G contains the configuration Q. Let $N_G(u) = \{u_1, u_2, u_3, u_4\}$, and let $\pi = \{V_1, V_2, V_3, V_4\}$ be a 4-vertex coloring of $G-u$. Without loss of generality, we may suppose $u_i \in V_i$ for each $i = 1, 2, 3, 4$. Then either u_1 and u_4 are not connected in $G_{14} = G[V_1 \cup V_4]$ or u_2 and u_3 are not connected in $G_{23} = G[V_2 \cup V_3]$; for otherwise, a u_1u_4-path in G_{14} and a u_2u_3-path in G_{23} have a common vertex with two different colors, see Figure 6.8, where the digit i nearby a vertex indicates the color used in the vertex. Without loss of generality, suppose that u_1 and u_4 are not connected in G_{14}. By interchanging the colors 1 and 4 in the component containing u_1 of G_{14}, we can obtain a spare color 1 that can be used to color u, which results is a 4-vertex coloring of G, a contradiction. Therefore, G can not contain the configuration Q.

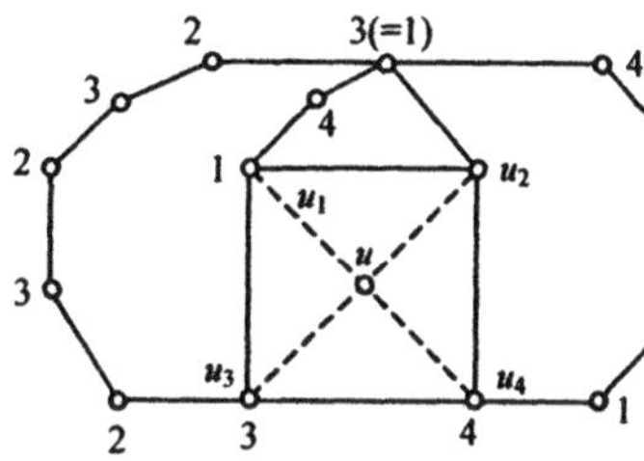

Figure 6.8: An illustration of Kempe's argument

Using the same way, Kempe "proved" that G can not contain the configuration R. Thus, G contains none of elements of $\mathcal{F}$, which contradicts to the fact that $\mathcal{F} = \{O, P, Q, R\}$ is an unavoidable complete set. So Kempe regarded he had proved the four conjecture.

Eleven years later, in 1890, Heawood [91] gave a counterexample, shown in Figure 6.9, showing that Kempe's discussion of the configuration R is incorrect.

Let $N_G(u) = \{u_1, u_2, u_3, u_4, u_5\}$, and let π be a 4-vertex coloring of $G-u$, see Figure 6.9, where the digit i nearby a vertex indicates the color used in the vertex. Observing that u_2 and u_4 are connected in G_{24}, and u_2 and u_5 are connected in G_{23}, and interchanging the two colors in G_{24} or G_{23}, we can not obtain any spare color to color u. Noting that u_1 and u_4 are not connected in G_{24}, we may consider to interchange the two colors in the component containing u_1 of G_{24} (indicated by the digits in the brackets) to spare the color 2. However, the color 2 can not be used to color the vertex

u since it has been used to color the vertex u_1. Noting that u_3 and u_5 are not connected in G_{23}, we may consider to interchange the two colors in the component containing u_3 of G_{23} to spare the color 2, in this case, however, the vertices x and y will obtain the same color 2. Therefore, Kempe's proof is invalid. Making use of Kempe's argument, Heawood proved the following theorem, known as the *five-color theorem* on planar graphs.

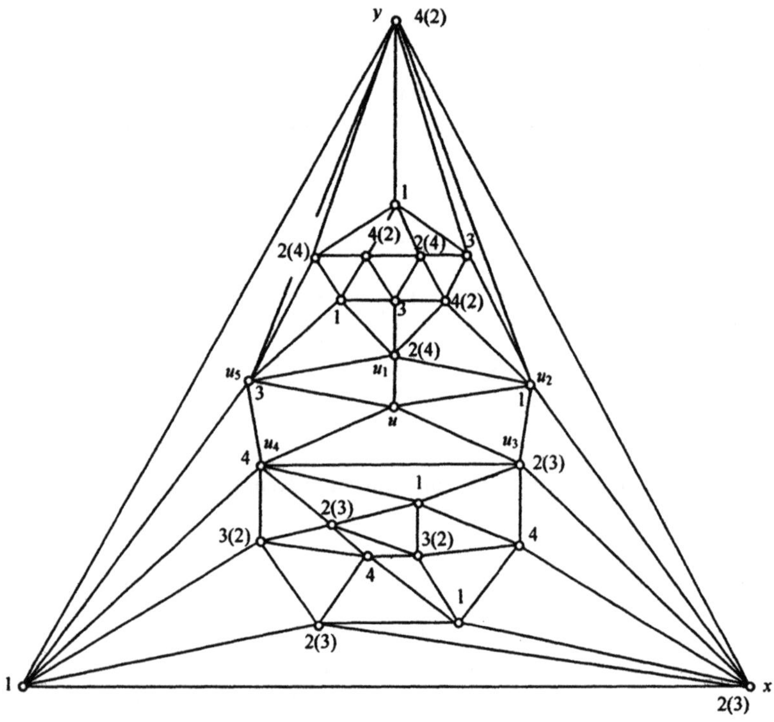

Figure 6.9: A counterexample to Kempe's proof

Theorem 6.4 $\chi(G) \leq 5$ for any planar graph G.

Proof Suppose to the contrary that the theorem is false. Then there exists a critical 6-chromatic plane graph G. Since a critical graph is simple, we have $\delta(G) = 5$ by Corollary 3.4.2 and Theorem 6.2. Let $u \in V(G)$ with $d_G(u) = 5$ and $N_G(u) = \{u_1, u_2, u_3, u_4, u_5\}$. Suppose that $\pi = (V_1, V_2, V_3, V_4, V_5)$ is a 5-vertex coloring of $G - u$. Noting $\chi(G) = 6$, we can, without loss of generality, that $u_i \in V_i$ for $i = 1, 2, 3, 4, 5$.

Let $G_{ij} = G[V_i \cup V_j]$. If there exist i and j such that u_i and u_j are not connected in G_{ij}, then by interchanging the colors i and j in the component containing u_i, we obtain a spare color i, which can be used to color u, and so obtain a 5-vertex coloring of G. This contradicts the hypothesis of $\chi(G) = 6$.

Therefore, there exists a u_iu_j-path P_{ij} in G_{ij} for any i, j $(1 \leq i \neq j \leq 5)$. Consider the cycle $C = uu_1 + P_{13} + u_3u$. Then the path P_{24} must meet a vertex of C. But this is impossible, since the vertices of P_{24} have colors 2 and 4, whereas no vertex of C has either of these colors. ∎

For some years afterwards, the flaw in Kempe's proof seems not to have been recognized as serious, but as the years went by and nobody found a satisfactory way around the difficulty, it gradually became realized that the problem was much harder than originally supposed. Since then, many mathematicians have intended to prove the conjecture. Although Kempe's proof was fallacious, his several important ideas contributed in the proof has became the foundation for almost all subsequent attempts on the problem.

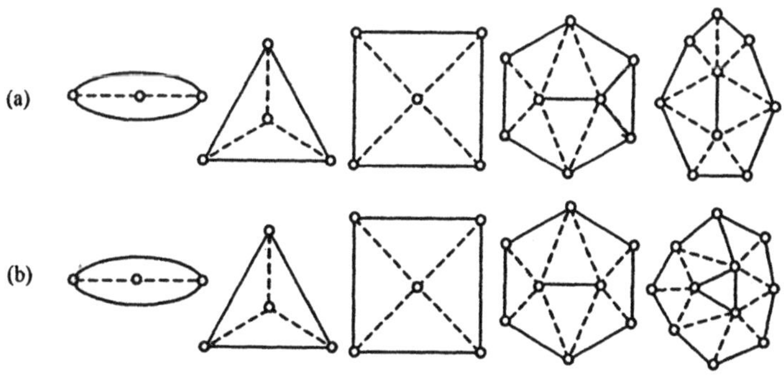

Figure 6.10: Two unavoidable complete sets

A configuration is called to be *reducible* if it can not be contained in any minimal graph. Kempe only proved that the configurations O, P, Q all are reducible, but he was unable to prove that the configuration R is reducible. Kempe's ideas showed that in order to prove the four-color conjecture, it suffices to find an unavoidable complete set of reducible configurations. Since the last of the configurations in $\mathcal{F} = \{O, P, Q, R\}$ has not been shown to be reducible, it is natural to ask whether it can be replaced by any other configurations to form another unavoidable complete set. In 1904, Wernicke [172] found an unavoidable complete set, and, in 1913, Birkhoff [12], shown in Figure 4.9 (a) and (b), respectively. Unfortunately, they were unable to show that the last two configurations are reducible. Since then, many mathematicians have joined the search for reducible configuration, and thousand of such configurations are found.

We should specifically mention Heesch [92] who developed the two main ingredients needed for the ultimate proof - reducibility and discharging. It was he who conjectured that a suitable development of this method would solve the four-color problem.

This was confirmed by Appel and Haken [3, 4, 5, 6] in 1976, when they announced a proof of the four-color conjecture by using a computer. Another similar proof of the conjecture, but simpler than Appel and Haken's in several respects, in 1997, was given by Robertson *et al.* [153] who exhibited an unavoidable set of 633 reducible configurations. However, their proof has not been fully accepted. There has remained a certain amount of doubt about it validity, basically for a main reason: part of the proof uses a computer and can not be verified by hand. Thus, the proof of the four-color conjecture by hand is still necessary.

We should also mention another failed proof, due to Tait [158] in 1880. To prove the four-color conjecture, he observed that it would be sufficient to show that every 3-regular 3-connected planar graph is 3-edge-colorable (the exercise 6.3.8). By mistakenly assuming that every such graph is hamiltonian, he gave a "proof" of the four-color conjecture (the exercise 6.3.3 (b)). In 1946, Tutte [162] pointed out Tait's proof to be invalid by constructing a nonhamiltonian 3-regular 3-connected planar graph, depicted in the exercise 3.2.7. In fact, in 1968, Grinberg [78] discovered a necessary condition for a plane graph to be hamiltonian (see the exercise 3.2.7). His discovery has led to the construction of many nonhamiltonian 3-regular 3-connected planar graphs.

Although Tait's proof was incorrect, he found an equivalent formulation of the four-color conjecture in terms of 3-edge-coloring.

Theorem 6.5 (Tait's theorem) The four-color conjecture is equivalent to as $\chi'(G) = 3$ for every 2-edge-connected 3-regular simple planar graph G.

Proof Suppose that the four-color conjecture holds. Let G be a simple 2-edge-connected 3-regular planar graph. By the inequality (6.1), we need to only show that G is 3-edge-colorable. To the end, let $\widetilde{G}$ be a planar embedding of G. By the hypothesis, $\widetilde{G}$ is 4-face-colorable. It is, of course, immaterial which symbols are used as the "colors", and in this case we denote the four colors by the vectors

$$\mathbf{C}_0 = (0,0),\ \mathbf{C}_1 = (1,0),\ \mathbf{C}_3 = (0,1),\ \mathbf{C}_4 = (1,1)$$

over the field of integers modulo 2. Note that at most one edge is in a common boundary of two different faces of $\widetilde{G}$ since $\widetilde{G}$ is 3-regular. We define an edge-coloring π' of $\widetilde{G}$ according to the following rules:

$$\pi'(e) = \mathbf{C}_i + \mathbf{C}_j \Longleftrightarrow B_G(f) \cap B_G(g) = \{e\},\ \forall\ e \in E(\widetilde{G}),$$

where $f, g \in F(\widetilde{G})$ with colors $\mathbf{C}_i$ and $\mathbf{C}_j$, respectively (see Figure 6.11 (a)). If $\mathbf{C}_i$, $\mathbf{C}_j$, $\mathbf{C}_k$ are the three colors assigned to the three face incident with a vertex x, then $\mathbf{C}_i + \mathbf{C}_j$, $\mathbf{C}_j + \mathbf{C}_k$ and $\mathbf{C}_k + \mathbf{C}_i$ are the colors assigned to the three edges incident with x under π'. Since $\widetilde{G}$ is 2-connected, each edges is in a common boundary of two different faces and it follows that no edge is assigned the color $\mathbf{C}_0$ under π'. It is also clear that the three edges incident with a given vertex are assigned different colors under π'. Thus π' is a 3-edge coloring of $\widetilde{G}$.

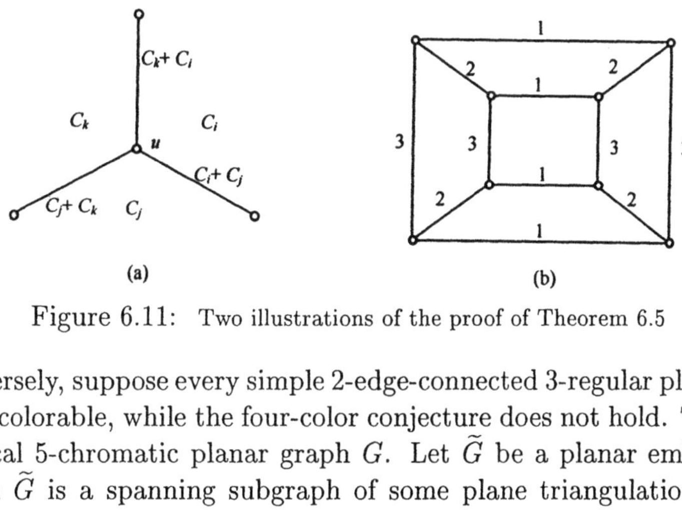

Figure 6.11: Two illustrations of the proof of Theorem 6.5

Conversely, suppose every simple 2-edge-connected 3-regular planar graph is 3-edge-colorable, while the four-color conjecture does not hold. Then there is a critical 5-chromatic planar graph G. Let $\widetilde{G}$ be a planar embedding of G. Then $\widetilde{G}$ is a spanning subgraph of some plane triangulation H. The geometric dual H^* of H is a 2-edge-connected 3-regular simple planar graph (see the exercise 4.3.8). Thus, by the hypothesis, H^* has a 3-edge coloring $\pi' = (E_1, E_2, E_3)$. For $i \neq j$, let $H^*_{ij} = H^*[E_i \cup E_j]$. Since each vertex of H^* is incident with one edge of E_i and one edge of E_j, by Lemma 5.8, H^*_{ij} is a union of disjoint even-cycles (see Figure 6.11 (b)) and is therefore 2-face-colorable (see the exercise 6.3.1). Since H^* is connected, therefore without loss of generality, suppose that $H^* = H^*_{12} \cup H^*_{13}$. Let

$$\pi^*_{12} : F(H^*_{12}) \to \{\alpha, \beta\}, \quad \text{and} \quad \pi^*_{13} : F(H^*_{13}) \to \{\gamma, \delta\}$$

be 2-face colorings of H^*_{12} and H^*_{13}, respectively, and let f be any a face of H^*. Then $f \in F(H^*) = F(H^*_{12}) \cap F(H^*_{13})$. Let $\pi^*_{12}(f) = x \in \{\alpha, \beta\}$ and $\pi^*_{13}(f) = y \in \{\gamma, \delta\}$. It is easily verified that

$$\begin{aligned} \pi^* : F(H^*) &\to \{(\alpha,\gamma), (\alpha,\delta), (\beta,\gamma), (\beta,\delta)\} \\ f &\mapsto (x, y) \end{aligned}$$

is a 4-face coloring of H^*. Since $\widetilde{G}$ is a spanning subgraph of H, we have $5 = \chi(G) \le \chi(\widetilde{G}) \le \chi(H) = \chi^*(H^*) \le 4$. This contradiction shows that the four-color conjecture holds. ∎

Exercises

6.3.1 Prove that a 2-edge-connected plane graph G is 2-face-colorable if and only if and G contains no vertex of odd degree.

6.3.2 Prove that a plane triangulation G is 3-face-colorable if and only if and G contains no vertex of odd degree.

6.3.3 Prove that

(a) every Hamiltonian plane graph is 4-face-colorable;

(b) every Hamiltonian 3-regular graph is 3-edge-colorable.

6.3.4 Prove that every plane graph is 4-face-colorable if and only if every simple 2-edge-connected 3-regular plane graph is 4-face-colorable.

6.3.5 Suppose that the plane is divided into several regions by $n\ (\ge 1)$ lines. Prove that these regions can be colored with two colors so that no two regions that share a length of common border are assigned the same color.

6.3.6 Prove that every 3-regular plane graph G is 3-face-colorable if and only if G contains no face of odd degree.

6.3.7 Prove that if every 3-regular plane graph is 4-face-colorable, then the four-color conjecture holds.

6.3.8 Prove that the four-color conjecture is equivalent to *Tait's conjecture*: every 3-connected 3-regular simple planar graph is 3-edge-colorable.

Chapter 7

Graphs and Groups

To aim to explore mathematical essence of graph theory further, the closed connection between graphs and groups will be discussed in this chapter. The reader who will read this chapter is supposed to familiarize himself with some basic concepts and methods of group theory.

As we have known from group theory that every finite set with an appropriate relation or operation there exists a group of permutations preserving the relation or operation. As a set with a given binary relation, of course, graph is no exception. We will see that every graph is associated with a group of permutations preserving adjacency on its vertex-set, called an automorphism group of the graph. On the other hand, we will also see, for every finite abstract group Γ there exists a digraph, called Cayley graph, its color-preserving automorphism group, a subgroup of the automorphism group of the Cayley graph, is isomorphic to Γ; there also exists an undirected graph, called Frucht graph, its automorphism group is isomorphic to Γ, the latter answers a question proposed by Köning in 1936.

With the aid of the automorphism group, we will discuss an important class of graphs, vertex-transitive graphs, and investigate their structural properties. The Cayley graph, a vertex-transitive graph, can be constructed for a given finite group and a set of the group without the identity. The Frucht graph can be also constructed by making use of a Cayley digraph.

Vertex-transitive graphs are high symmetric. We will introduce their applications in designs of interconnection networks.

7.1 Group Presentation of Graphs

Recall concept of isomorphism of graphs defined in Section 1.2. Two simple graphs G and H are *vertex-isomorphic* if there exists a bijective mapping $\theta : V(G) \to V(H)$ which satisfies the *adjacency-preserving condition*, that is,

$$(x, y) \in E(G) \Longleftrightarrow (\theta(x), \theta(y)) \in E(H).$$

The mapping θ is called a *vertex-isomorphic mapping* from G to H.

Similarly, we can define concept of edge-isomorphism of simple graphs. Two simple graphs G and H are *edge-isomorphic* if there exists a bijective mapping $\phi : E(G) \to E(H)$ such that for any two edges a and b of G, the head of a is the tail of b if and only if the head of $\phi(a)$ is the tail of $\phi(b)$. The mapping ϕ is called an *edge-isomorphic mapping* from G to H. Two digraphs shown in Figure 7.1 are edge-isomorphic, but not vertex-isomorphic.

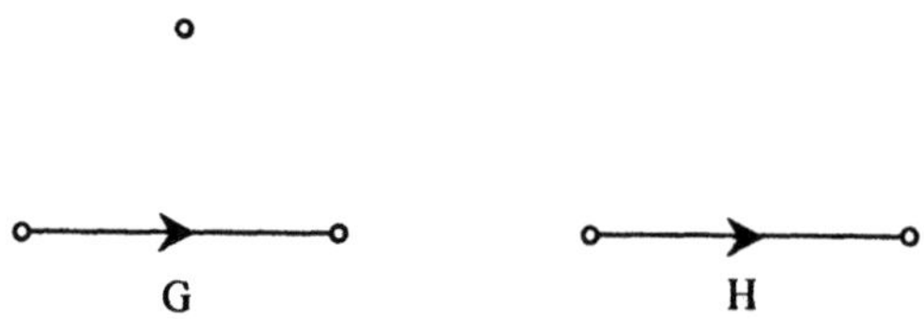

Figure 7.1: Two edge-isomorphic, but not vertex-isomorphic digraphs

Lemma 7.1 If θ is a vertex-isomorphism of two nonempty simple graphs G and H, then the following mapping is an edge-isomorphism from G to H

$$\begin{aligned} \theta^* : \quad & E(G) \to E(H) \\ a = (x, y) & \mapsto \theta^*(a) = (\theta(x), \theta(y)). \end{aligned} \tag{7.1}$$

Proof It is clear that $\theta^* : E(G) \to E(H)$ is a bijective mapping since $\theta : V(G) \to V(H)$ is bijective. Let $a, b \in E(G)$, and let $a = (x, y)$ and $b = (y, z)$. Then $\theta^*(a) = (\theta(x), \theta(y)) \in E(H)$ and $\theta^*(b) = (\theta(y), \theta(z)) \in E(H)$. Thus, θ^* is an edge-isomorphism from G to H. ∎

The mapping θ^* defined by (7.1) is called an *induced edge-isomorphism* by θ. By Lemma 7.1, we immediately obtain the following result.

Theorem 7.1 Every pair of nonempty vertex-isomorphic simple graphs are edge-isomorphic. ∎

Two digraphs in Figure 7.1 show that the converse of Theorem 7.1 is not always true. However, we have the following result.

Theorem 7.2 Every pair of nonempty edge-isomorphic and connected simple digraphs with $\delta > 0$ is necessarily vertex-isomorphic.

A sketch of the proof Let G and H be two nonempty edge-isomorphic simple connected digraphs, ϕ be an edge-isomorphism from G to H. If there exist two distinct edges a and b of G such that $a = (x, y)$ and $b = (y, x)$, then the tail and the head of $\phi(a)$ are the head and the tail of $\phi(b)$, respectively. This implies that if $\phi(a) = (u, v)$, then $\phi(b) = (v, u)$. Thus, we can without loss of generality that G contains no symmetric edge.

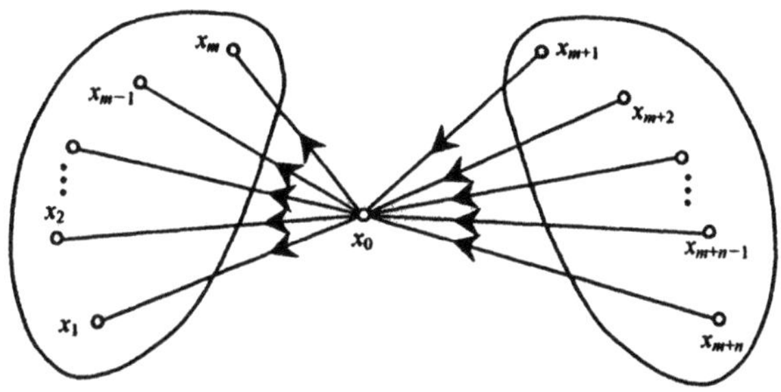

Figure 7.2: An illustration of the proof of Theorem 7.2

Arbitrarily choose $x_0 \in V(G)$, which is not an isolated vertex since G is a nonempty and connected simple digraph. Let $m = d^+_G(x_0)$, $n = d^-_G(x_0)$, and

$$E^+_G(x_0) = \{a_1, a_2, \cdots, a_m\}, \quad E^-_G(x_0) = \{a_{m+1}, a_{m+2}, \cdots, a_{m+n}\},$$

where the head of a_i is x_i for $i = 1, 2, \cdots, m$ and the tail of a_{m+j} is x_{m+j} for $j = 1, 2, \cdots, n$ (see Figure 7.2). Let

$$\phi(a_i) = b_i \in E(H), \quad i = 1, 2, \cdots, m + n.$$

Then there exist $y_0 \in V(H)$ such that it is the tail of b_i and the head of b_j for each $i = 1, 2, \cdots, m$ and each $j = m+1, m+2, \cdots, m+n$. Let y_i be other end-vertex of b_i for each $i = 1, 2, \cdots, m + n$.

If $a_{ij} = (x_i, x_j) \in E(G)$, $i, j \neq 0$ and $i \neq j$, then $\phi(a_{ij}) \in E(H)$ and $\phi(a_i) = b_i$, $\phi(a_j) = b_j$ and $\phi(a_{ij}) = (y_i, y_j)$. Let

$$G_1 = G[\{x_0, x_1, \cdots, x_{m+n}], \quad H_1 = G[\{y_0, y_1, \cdots, y_{m+n}].$$

Define a mapping $\theta: \ V(G_1) \to V(H_1)$

$$x_i \mapsto \theta(x_i) = y_i, \quad i = 1, 2, \cdots, m+n.$$

Then θ is a vertex-isomorphism from G_1 to H_1 clearly and the theorem follows if $G_1 = G$. Suppose $G_1 \subset G$ below.

Since G is connected, there exists a vertex x in G not in G_1 with its neighbors in G_1. Let x_i be a neighbor of x in G_1. Clearly $x_i \neq x_0$. Without loss of generality, suppose $1 \leq i \leq m$ and $(x, x_i) \in E(G)$. Then $\phi((x, x_i)) \notin E(G_1)$. Since x_i is a common head of the edge (x_0, x_i) and the edge (x, x_i), it follows that y_i is a common head of the edge $\phi((x_0, x_i))$ and the edge $\phi((x, x_i))$. Thus there exists a vertex y in H not in H_1 such that $\phi((x, x_i)) = (y, y_i) \in E(H)$. Let

$$G_2 = G[\{x_0, x_1, \cdots, x_{m+n}, x], \quad H_2 = G[\{y_0, y_1, \cdots, y_{m+n}, y].$$

By adding $\theta(x) = y$ to the mapping θ above, we obtain a vertex-isomorphism from G_2 to H_2. If $G_2 = G$, then the theorem holds. Otherwise, continue the above process until the resulting graph is G. ∎

An *automorphism* of a graph G is an isomorphism of G with itself, that is, a permutation on $V(G)$ that preserves adjacency. It is straightforward to verify that under the operation of composition the set of all automorphisms of G forms a group, denoted by Aut (G), and referred to as the *automorphism group*, the *vertex-group*, or simply the *group* of G.

For example, the automorphism group Aut (K_n) of a complete graph K_n is a symmetric group S_n of order $n\,!$, and the automorphism group Aut $(\overrightarrow{C}_n)$ of a directed cycle $\overrightarrow{C}_n$ is a ring group Z_n of order n, while the automorphism group Aut (C_n) of an undirected cycle C_n is a dihedral group D_n of order $2n$. Generally, for a given graph G, deciding the automorphism group Aut G is fairly difficult.

The automorphism group of a graph G is referred to as the *group representation* of G. This kind of representation of a graph is conducive to study properties of the graph by making use of methods and results in group theory.

We can similarly define an edge-automorphism and the edge-automorphism group of a graph.

From Lemma 7.1, every automorphism θ of G can induce an edge-automorphism θ^* of G, called an *induced edge-automorphism.* The set of all induced edge-automorphisms forms a group, called the *induced edge-automorphism group* of G, denoted by Aut$^*(G)$.

Theorem 7.3 Let G be a nonempty simple digraph, then $\text{Aut}(G) \cong \text{Aut}^*(G)$ if and only if G contains at most one isolated vertex.

Proof Suppose $\text{Aut}(G) \cong \text{Aut}^*(G)$ and suppose to the contrary that x and y are two distinct isolated vertices of G. Let $\alpha \in \text{Aut}(G)$ so that $\alpha(x) = y$ and others are fixed. Let e is the identity element of $\text{Aut}(G)$. Then the two elements induced by α and e both are the identity element of $\text{Aut}^*(G)$, a contradiction.

Conversely, Let

$$\begin{aligned} \varphi: \text{Aut}(G) &\to \text{Aut}^*(G) \\ \alpha &\mapsto \varphi(\alpha) = \alpha^*. \end{aligned} \tag{7.2}$$

We now show that φ is a group isomorphism. It is clear from (7.2) that φ is surjective. Thus, we need to prove that φ is injective and operation-preserving.

Let α and β be two distinct elements in $\text{Aut}(G)$. Then there exists $x \in V(G)$ such that $\alpha(x) \neq \beta(x)$. Since G contains at most one isolated vertex, at least one of $\alpha(x)$ and $\beta(x)$ is not an isolated vertex. Without loss of generality, suppose that $\alpha(x) = u$ is not an isolated vertex and $(u, z) \in E(G)$. Then there exists a vertex y of G such that $\alpha(y) = z$ and

$$(u, z) = (\alpha(x), \alpha(y)) \in E(G) \Longleftrightarrow (x, y) \in E(G).$$

Let $\beta(x) = s$ and $\alpha(y) = t$. Then $(s, t) \in E(G)$, and $(u, z) \neq (s, t)$ for $u = \alpha(x) \neq \beta(x) = s$. It follows that if let $a = (x, y)$ then

$$\alpha^*(a) = (\alpha(x), \alpha(y)) = (u, z) \neq (s, t) = (\beta(x), \beta(y)) = \beta^*(a),$$

that is, $\varphi(\alpha) \neq \varphi(\beta)$. This proves that φ is an injection.

We now prove that φ preserves operation of composition, that is, for any $a \in E(G)$, α, $\beta \in \text{Aut}(G)$, we have $\varphi(\alpha\beta)(a) = (\varphi(\alpha))(\varphi(\beta))(a)$. To the end, let $a = (x, y)$ and let

$$\beta(x) = x', \ \beta(y) = y'; \ \ \alpha(x') = x'', \ \alpha(y') = y''.$$

Then we have

$$\begin{aligned} \varphi(\alpha\beta)(a) &= \varphi(\alpha\beta)(x, y) = (\alpha\beta(x), \ (\alpha\beta(y)) \\ &= (\alpha(x'), \ \alpha(y')) = (x'', \ y''), \end{aligned} \tag{7.3}$$

$$\begin{aligned} (\varphi(\alpha))(\varphi(\beta))(a) &= (\varphi(\alpha))(\varphi(\beta))(x, y) = (\varphi(\alpha))(\beta(x), \ \beta(y)) \\ &= (\varphi(\alpha))(x', \ y') = (\alpha(x'), \ \beta(y')) = (x'', \ y''). \end{aligned} \tag{7.4}$$

The equalities (7.3) and (7.4) implies that $\varphi(\alpha\beta)(a) = (\varphi(\alpha))(\varphi(\beta))(a)$ immediately. ∎

It should be noted that the concept of edge-isomorphism of digraphs is called an *arc-isomorphism* in some of the literature and textbook since the same concept can be often defined in a little different form for undirected graphs. Two undirected graphs G and H are *edge-isomorphic* if there exists a bijective mapping $\phi : E(G) \to E(H)$ such that for any two edges a and b of G, a and b are adjacent in G if and only if $\phi(a)$ and $\phi(b)$ are adjacent in H.

In view of this definition, for example, a complete undirected graph K_3 and a star $K_{1,3}$ are edge-isomorphic, but not vertex-isomorphic. $\mathrm{Aut}\, K_2 = S_2$ while $\mathrm{Aut}^*(K_2) = S_1$ for a complete undirected graph K_2. Thus, for the undirected case, Theorem 7.3 can be stated as follows.

Corollary 7.3 Let G be a nonempty simple undirected graph, then $\mathrm{Aut}\,(G) \cong \mathrm{Aut}^*(G)$ if and only if G has neither K_2 as a component nor two or more isolated vertices. ∎

Exercises

7.1.1 Suppose G and H are simple digraphs and θ is an isomorphism of G and H. Prove that for any $S \subset V(G)$

$$N_H^+(\theta(S)) = \theta(N_H^+(S)),\ N_H^-(\theta(S)) = \theta(N_H^-(S)),$$

where $\theta(S) = \{u \in V(H) :\ \theta(x) = u,\, x \in S\}$.

7.1.2 Prove that Corollary 7.3.

7.1.3 Prove that if G is a simple graph then $|\mathrm{Aut}\,(G)| = n!$ if and only if $G \cong K_n$.

7.1.4 Prove that

(a) $\mathrm{Aut}\,(G) \cong \mathrm{Aut}\,(G^c)$;

(b) $\mathrm{Aut}\,(G) \cong \mathrm{Aut}\,(\overleftarrow{G})$ for any simple digraph G.

7.1.5 Prove that if G is a simple graph and the eigenvalues of the adjacency matrix $\mathbf{A}(G)$ of G are distinct, then $\mathrm{Aut}\,(G)$ is an abelian group.

(C.Y.Chao, 1971)

7.2 Transitive Graphs

In this section, we will investigate an important class of graphs, transitive graphs, by making use of the group presentation of a graph.

Let x and y be two vertices of G. We say that x is *similar* to y if there is an element $\sigma \in \mathrm{Aut}(G)$ such that $y = \sigma(x)$. It is easy to verify that the relation "is similar to" is an equivalence relation on $V(G)$.

A graph G is *vertex-transitive* if every pair of vertices of G are similar. In other words, G is vertex-transitive if there is $\sigma \in \mathrm{Aut}(G)$ such that $y = \sigma(x)$ for any pair (x, y) of vertices in G. It can be easily verified that a vertex-transitive graph is necessarily regular (the exercise 7.2.2).

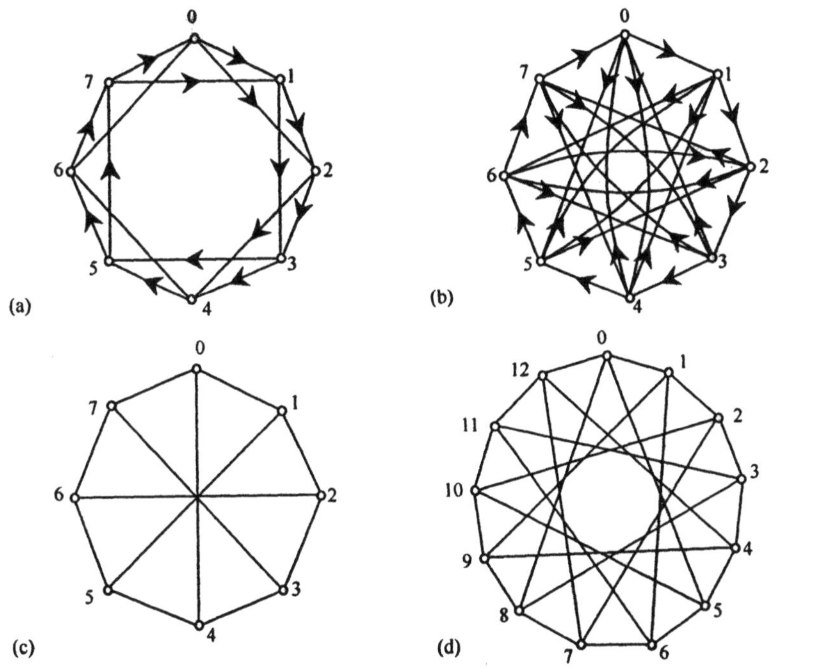

Figure 7.3: (a) $G(8; \{1,2\})$; (b) $G(8; \{1,4,5\})$; (c) $G(8; \pm\{1,4\})$; (d) $G(8; \pm\{1,5\})$

Example 7.2.1 An important class of vertex-transitive graphs is the circulant graphs.

A *circulant digraph*, denoted by $G(n; S)$, where $S \subseteq \{1, 2, \cdots, n-1\}, n \geq 2$, is defined as a digraph consisting of the vertex set $V = \{0, 1, \cdots, n-1\}$

and the edge set $E = \{(i,j)$: there is $s \in S$ such that $j - i \equiv s(\text{mod } n)\}$.

It is clear that $G(n;1)$ is a directed cycle C_n, $G(n;\{1,2,\cdots,n-1\})$ is a complete digraph K_n. The digraph shown in Figure 7.3(a) and (b) are $G(8;\{1,2\})$ and $G(8;\{1,4,5\})$, respectively.

A *circulant undirected graph*, denoted by $G(n;\pm S)$, where $S \subseteq \{1,2,\cdots,\lfloor\frac{1}{2}n\rfloor\}$, $n \geq 3$, is defined as an undirected graph consisting of the vertex set $V = \{0,1,\cdots,n-1\}$ and the edge set $E = \{ij$: there is $s \in S$ such that $|j-i| \equiv s(\text{mod } n)\}$.

It is also clear that $G(n;\pm 1)$ is an undirected cycle C_n and $G(n;\pm\{1,2,\cdots,\lfloor\frac{1}{2}n\rfloor\})$ is a complete graph K_n. The undirected graph shown in Figure 7.3(c) and (d) are $G(8;\pm\{1,4\})$ and $G(8;\pm\{1,5\})$, respectively.

We now show that the circulant graphs are vertex-transitive. In fact, the rotation $\pi = (012\cdots n-1) \in \text{Aut}\,(G(n;S))$ and $\pi^{j-i}(i) = j$ for any $i,j \in V$ $(i<j)$. ∎

Similarly, we have the concept of similar edges. Let G be a simple graph, $a = (x,y)$ and $b = (z,u)$ be two edges of G. We say that a is *similar* to b if there is an element $\sigma \in \text{Aut}\,(G)$ such that $z = \sigma(x)$ and $u = \sigma(y)$. It is easy to verify that this relation is an equivalence relation on $E(G)$.

A graph G is *edge-transitive* if every pair of edges of G are similar.

We now state a theorem, due to Elayne Dauber (see Harary [88]) for the undirected case, whose corollaries describe properties of edge-transitive undirected graphs.

Theorem 7.4 Every edge-transitive graph G with $\delta(G) > 0$ is vertex-transitive or bipartite.

Proof We prove this theorem for directed case.

Let $E(G) = \{e_1, e_2, \cdots, e_m\}$, and let $e_1 = (x,y)$. Then for each $i = 1,2,\cdots,m$, there is an element $\sigma_i \in \text{Aut}\,(G)$ such that

$$(\sigma_i(x), \sigma_i(y)) = e_i \in E(G), \quad i = 1,2,\cdots,m.$$

Let

$$\begin{aligned} V_1 &= \{\sigma_i(x) \in V(G): \ i = 1,2,\cdots,m\}, \quad \text{and} \\ V_2 &= \{\sigma_i(y) \in V(G): \ i = 1,2,\cdots,m\}. \end{aligned}$$

It is clear that $V_1 \neq V_2$ and, moreover, $V(G) = V_1 \cup V_2$ because of $\delta(G) > 0$. There are two possibilities: V_1 and V_2 are disjoint or they are not.

For $V_1 \cap V_2 = \emptyset$, we can show that G is bipartite. Arbitrarily choose

$a_k = (u, w) \in E(G)$. Then there is σ_k such that $\sigma_k(x) = u$ and $\sigma_k(y) = w$. This implies $u \in V_1$ and $w \in V_2$, and, hence, G is bipartite because $V_1 \cap V_2 = \emptyset$.

For $V_1 \cap V_2 \neq \emptyset$, we can show that G is vertex-transitive. Consider any two vertices u and w of G. We wish to show that u is similar to w. Note that $E_G^+(u) \neq \emptyset$ and $E_G^+(w) \neq \emptyset$ since $\delta(G) > 0$. Without loss of generality, we assume that u and w are both in V_1. Since G is edge-transitive, for any $e_i \in E_G^+(u)$ there exists σ_i such that $\sigma_i(x) = u$, and for any $e_j \in E_G^+(w)$ there exists σ_j such that $\sigma_j(x) = w$. Let $\sigma = \sigma_j \sigma_i^{-1}$. Then $\sigma \in \text{Aut}(G)$, and

$$\sigma(u) = \sigma_j \sigma_i^{-1}(u) = \sigma_j(x) = w.$$

This shows that u is similar to w. If u is in V_1 and w is in V_2, then let v be a vertex in $V_1 \cap V_2$. Since u is similar to v and v is similar to w, it follows that u is similar to w. ∎

Corollary 7.4.1 Let G be an edge-transitive graph. If G is not regular, then G is bipartite. ∎

Corollary 7.4.2 Let G be an edge-transitive undirected graph with odd order and $\delta(G) > 0$. If G is regular, then G is vertex-transitive. ∎

Corollary 7.4.3 Let G be an edge-transitive undirected graph with even order v. If G is $d\,(\geq \frac{1}{2}v)$-regular, then G is vertex-transitive. ∎

With these three corollaries, the only edge-transitive undirected graphs not yet characterized are those with even order and regularity $d < \frac{1}{2}v$. The undirected cycle C_6 is an example of such an edge-transitive graph which is both vertex-transitive and bipartite. The icosahedron, the dodecahedron, and the Petersen graph are examples of such edge-transitive graphs which are vertex-transitive but not bipartite. But not all regular edge-transitive graphs are vertex-transitive. In fact, Folkman [63] showed more general result: Whenever $v \geq 20$ is divisible by 4, there exists a regular undirected graph G of order v which is edge-transitive but not vertex-transitive.

We now introduce an important concept of graphs, *atoms*. Let G be a strongly connected digraph, F be a nonempty and proper subset of $V(G)$. If $N_G^+(F)$ is a κ-separating set of G, then F is called a *positive fragment* of G. Similarly, if $N_G^-(F)$ is a κ-separating set of G, then F is called a *negative fragment* of G. $\emptyset \neq F \subset V(G)$ is called a *fragment* of G if F is a positive or negative fragment. A fragment F is called an *atom* of G if F has the minimum cardinality over all fragments of G. The cardinality of an atom of

G is called the *atomic number* of G, denoted by $a(G)$. An atom A of G is positive (resp. negative) if A is a positive (resp. negative) fragment of G.

For undirected graphs, these concepts are simple. For instance, let G be a connected undirected graph, F be a nonempty and proper subset of $V(G)$. If $N_G(F)$ is a κ-separating set of G, then F is called a *fragment* of G.

If F is a positive fragment of a digraph G, then F is a negative fragment of the converse $\overleftarrow{G}$ of G, which is obtained by reversing the orientation of each edge of G. This simple fact is very useful for proving some results on fragments or atoms. For example, assume that some result holds for positive fragments or atoms of any digraph, then this result is valid for negative fragments or atoms by considering the converse of the digraph.

The concept of atoms, proposed first by Watkins [171] for undirected graphs and generalized to digraphs by Chaty [31], plays an important role in investigating properties of a vertex-transitive graph further, particularly connectivity of transitive graphs. These studies are based on the following result, due to Hamidoune [84].

Theorem 7.5 Let G be a strongly connected digraph, A and F be a positive (resp. negative) atom and a positive (resp. negative) fragment of G, respectively. Then $A \subseteq F$, or $A \cap F = \emptyset$.

Proof It is sufficient to prove that $A \subseteq F$ if $A \cap F \neq \emptyset$. Suppose to the contrary that $A \not\subseteq F$, we will derive a contradiction. To this aim, let $S = N_G^+(A)$, $T = N_G^+(F)$, $H = V(G) \setminus (A \cup S)$, $R = V(G) \setminus (F \cup T)$ (see Figure 7.4). Then

$$|A| \leq |F|, \quad |S| = |T| = \kappa(G), \quad |H| \geq |R| \geq |A|.$$

Since $A \cap F \neq \emptyset$, $N_G^+(A \cap F)$ is a separating set of G. Since $|A \cap F| < |A|$, we have $|N_G^+(A \cap F)| > \kappa(G)$, and, thus,

$$|A \cap T| > \kappa - |(S \cap F) \cup (S \cap T)| = |S \cap R|.$$

If $H \cap R \neq \emptyset$, then a contradiction can be derived as follows.

$$\kappa(G) \leq |N_G^-(H \cap R)| \leq |S| + |T| - |N_G^+(A \cap F)| < \kappa(G).$$

Thus $H \cap R = \emptyset$. Note $|A \cap T| > |S \cap R|$, then

$$|A| \leq |R| = |(A \cap R) \cup (S \cap R)| < |A \cap R| + |A \cap T| < |A|,$$

a contradiction. The theorem follows. ∎

Corollary 7.5 (Watkins [171]) Let A and S are an atom and a κ-separating set of a connected undirected graph G, respectively. Then $A \subseteq S$ or $A \cap S = \emptyset$.

Proof Suppose $A \not\subseteq S$. Since S is a κ-separating set of G, there is a fragment F such that $N_G(F) = S$ and $A \cap F \neq \emptyset$. By Theorem 7.5, we have $A \subseteq F$. It follows that $A \cap S = \emptyset$. ∎

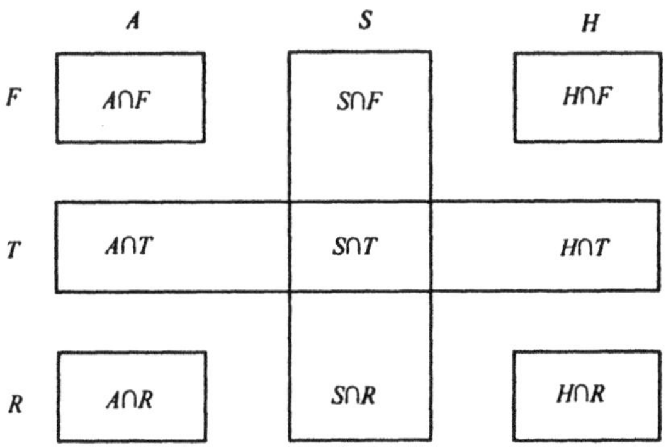

Figure 7.4: An illustration in the proof of Theorem 7.5

By Theorem 7.5 and Corollary 7.5, we have the following important theorem on atoms immediately, due to Mader [123] for the undirected case and to Hamidoune [84] for the directed case, respectively.

Theorem 7.6 Any two distinct positive (or negative) atoms of a strongly connected digraph are disjoint. In particular, any two distinct atoms of a connected undirected graph are disjoint. ∎

We now present the well-known *atomic decomposition theorem* of transitive graphs, due to Watkins [171] for the undirected case and to Hamidoune [84] for the directed case, respectively.

Theorem 7.7 Let G be a connected vertex-transitive graph, a set A be a positive (resp. negative) atom of G. Then

(a) $G[A]$ is vertex-transitive.

(b) There is a partition $\{A_1, A_2, \cdots, A_m\}$ of $V(G)$ such that $G[A_i] \cong G[A]$ for each $i = 1, 2, \cdots, m$, $m \geq 2$.

Proof There is nothing to do for $|A| = 1$. Assume $|A| > 1$ below and, without loss of generality, assume that A is a positive atom of G.

(a) For any $x, y \in V(G)$, there is $\sigma \in \mathrm{Aut}\,(G)$ such that $y = \sigma(x)$ since G is vertex-transitive. $\sigma(A)$ is a positive atom (the exercise 7.2.4). Since $y \in \sigma(A) \cap A$, we have $\sigma(A) = A$ by Theorem 7.6. Let

$$\Sigma = \{\sigma \in \mathrm{Aut}\,(G) : \ \sigma(A) = A\}.$$

Clearly, Σ is a subgroup of $\mathrm{Aut}\,(G)$ and the constituent of Σ acts on A transitively. Let

$$\Pi = \{\pi \in \Sigma : \ \pi(x) = x, x \in A\}.$$

Then Π is a stable subgroup of Σ, and is normal. Thus there is an injective homomorphism from the quotient Σ/Π to $\mathrm{Aut}\,(G[A])$ whereby each coset of Π is associated with the restriction to A of any representative. This proves that $G[A]$ is vertex-transitive.

(b) Since G is vertex-transitive, for a fixed $y \in A$ and any $x \in V(G)$, there is $\pi \in \mathrm{Aut}\,(G)$ such that $x = \pi(y)$. Thus $\pi(A)$ is a positive atom containing x by the exercise 7.2.4, and $G[\pi(A)] \cong G[A]$ by the exercise 7.2.1. If $x \notin A$, then $\pi(A) \cap A = \emptyset$ by Theorem 7.6. Thus G has at least two atoms. Hence, for any $x \in V(G)$, there exists a positive atom A_x containing x and $G[\pi(A_x)] \cong G[A]$.

Furthermore, for any $y \in V(G)$ other than x, either $A_x = A_y$, or $A_x \cap A_y = \emptyset$. Thus these positive atoms $A_1, A_2, \cdots, A_m, m \geq 2$ form a partition of $V(G)$, and $G[A_i] \cong G[A]$ for each $i = 1, 2, \cdots, m$. ∎

Using the concept and the decomposition theorem of atoms, we can obtain some results on connectivity of a transitive graph. The results given here are due to Watkins [171] and Mader [123] for undirected case, and Hamidoune [86, 87] for directed case, respectively.

Theorem 7.8 Let G be a strongly connected edge-transitive digraph. Then $\kappa(G) = \lambda(G) = \delta(G)$.

Proof We need only show that $a(G) = 1$ (by the exercise 7.2.5). Suppose to the contrary that $a(G) > 1$. Let A be an atom of G. Without loss of generality, suppose that A is a positive atom. Consider two vertices x and y in A, and a vertex z in $N_G^+(A)$ such that $(x, y), (y, z) \in E(G)$. Because of edge-transitiveness of G, there is $\sigma \in \mathrm{Aut}\,(G)$ such that $y = \sigma(x)$ and $z = \sigma(y)$. Thus $\sigma(A)$ is a positive atom of G by the exercise 7.2.4. Noting $z \in \sigma(A)$ but $z \notin A$, we have $\sigma(A) \neq A$. On the other hand, $\sigma(A) = A$ by

Theorem 7.6 since $y \in A \cap \sigma(A)$. This contradiction shows that we should have $a(G) = 1$. ∎

Theorem 7.9 Let G be a connected vertex-transitive graph, then $\lambda(G) = \delta(G)$.

Proof Since G is a connected vertex-transitive graph, G is δ-regular (by the exercise 7.2.2), where $\delta = \delta(G) > 0$. Thus for any nonempty proper subset X of $V(G)$, $d^+_G(X) = d^-_G(X)$ by Example 1.4.1.

To prove this theorem, we need to only prove $\lambda = \lambda(G) \geq \delta$ by Whitney's inequality. Choose $X \subset V(G)$ such that $d^+_G(X) = \lambda = d^-_G(X)$ and $|X|$ is as small as possible. Then $|X| \leq \frac{1}{2}v$ and $G[X]$ is a strongly connected component of G.

If $|X| = 1$, then $\lambda = d^+_G(X) = d^-_G(X) = \delta$. We now suppose that $|X| \geq 2$.

We first prove that $G[X]$ is vertex-transitive. Since G is vertex-transitive, for any $x, y \in X$, there is $\sigma \in \text{Aut}\,(G)$ such that $y = \sigma(x)$. Let $Y = \sigma(X)$, then $X \cap Y \neq \emptyset$. Moreover, $|X| = |Y|$ and $d^+_G(X) = \lambda = d^+_G(Y)$ since $G[X] \cong G[Y]$ by the exercise 7.2.1. Since

$$|X \cup Y| = |X| + |Y| - |X \cap Y| \leq \tfrac{1}{2}\nu + \tfrac{1}{2}\nu - 1 = \nu - 1,$$

we have $X \cup Y \subset V(G)$. Since $d^+_G(X \cup Y) \geq \lambda$ and $d^+_G(X \cap Y) \geq \lambda$, we have, from the exercise 1.3.7, that

$$d^+_G(X \cup Y) + d^+_G(X \cap Y) \leq d^+_G(X) + d^+_G(Y) = 2\lambda,$$

which means $d^+_G(X \cap Y) = \lambda$. Because $X \cap Y \subseteq X$ and the minimality of X, we have $X \cap Y = X$. Noting $|X| = |Y|$, we have $X = Y$. Let

$$\Sigma = \{\sigma \in \text{Aut}\,(G) :\ \sigma(X) = X\}.$$

Then Σ is a subgroup of $\text{Aut}\,(G)$ and acts on X transitively. This means that $G[X]$ is vertex-transitive.

Let the degree of $G[X]$ be k. Then $0 < k \leq \delta - 1$ and $k \leq |X| - 1$. Since $\lambda = d^+_G(X) = (\delta - k)|X|$, we have that

$$\lambda = (\delta - k)|X| \geq (\delta - k)(k + 1) = \delta + k(\delta - k - 1) \geq \delta,$$

as required. ∎

Determining the connectivity $\kappa(G)$ of a vertex-transitive graph G seems difficult. In the following theorem, we present some basic results on the connectivity of a vertex-transitive graph.

Theorem 7.10 Let G be a connected graph, $\delta = \delta(G)$. If G is vertex-transitive, then

(a) $\kappa(G) \geq a(G)$;

(b) $\kappa(G) \geq \left\lceil \frac{1}{2}(\delta+1) \right\rceil$;
(c) $\kappa(G) \geq \left\lceil \frac{1}{3}(2\delta+1) \right\rceil$ if G is a digraph without symmetric edges;
(d) $\kappa(G) = \lambda(G) = \delta$ if G has a prime number order;
(e) $\kappa(G) = na(G)$ if G is an undirected graph, where $n \geq 2$ is an integer;
(f) $\kappa(G) \geq \left\lceil \frac{2}{3}(\delta+1) \right\rceil$ if G is an undirected graph;
(g) $\kappa(G) = \delta$ if G is an undirected graph and $\delta = 2, 3, 4, 6$.

Proof Without loss of generality, suppose that G is a non-complete graph. Assume that A is an atom of G, and without loss of generality, A is a positive atom if G is a digraph. Then $|A| = a(G)$.

(a) Let $G_A = G[A]$. Then G_A is vertex-transitive by Theorem 7.7 (a). Thus G_A is r-regular and $r < \delta$. Let $T = N_G^+(A)$. Then

$$d_G^+(A) = |A|(\delta - r) = \sum_{x \in T} |N_G^-(x) \cap A|.$$

Choose $y \in T$ such that $|N_G^-(y) \cap A|$ is as large as possible. Thus

$$|T||N_G^-(y) \cap A| \geq |A|(\delta - r). \tag{7.5}$$

On the other hand, there is an atom A' containing y by Theorem 7.7. Since $y \notin A$, $A' \cap A = \emptyset$ by Theorem 7.6. Let $G_{A'} = G[A']$. Then

$$|N_G^-(y) \cap A| \leq \delta^-(G) - \delta^-(G_{A'}) = \delta - r. \tag{7.6}$$

Combining (7.5) with (7.6), we have $|T| \geq |A|$, and, hence

$$\kappa(G) = |N_G^+(A)| = |T| \geq |A| = a(G).$$

(b) By (a) we have

$$\delta \leq \delta(G[A]) + \kappa(G) \leq |A| - 1 + \kappa(G) \leq 2\kappa(G) - 1.$$

This implies $\kappa(G) \geq \lceil \frac{1}{2}(\delta+1) \rceil$.

(c) If G has no symmetric edge, then from (a), we have

$$\delta \leq \delta(G[A]) + \kappa(G) \leq \tfrac{1}{2}(|A| - 1) + \kappa(G) \leq \tfrac{1}{2}(\kappa(G) - 1) + \kappa(G).$$

This implies $\kappa(G) \geq \lceil \frac{1}{3}(2\delta+1) \rceil$.

(d) By Theorem 7.7, there is a partition $\{A_1, A_2, \cdots, A_m\}$ of $V(G)$ such that $G[A_i] \cong G[A]$ for each $i = 1, 2, \cdots, m$, $m \geq 2$. Thus $v(G) = m|A|$. Since $v(G)$ is prime, $a(G) = |A| = 1$. By Theorem 7.9, the result (d) follows.

(e) Let G be an undirected graph and let $S = N_G(A)$. Then S is a κ-separating set of G. By Theorem 7.7, there is a partition of atoms $\{A_1, A_2, \cdots, A_m\}$ of $V(G)$ such that $G[A_i] \cong G[A]$ for each $i = 1, 2, \cdots, m$, $m \geq 2$. By Corollary 7.5, either $S \subseteq A_i$ or $S \cap A_i = \emptyset$ for each $i = 1, 2, \cdots, m$. This fact implies there is a positive integer n such that $\kappa = |S| = n|A|$. If $|A| = 1$, then $\kappa(G) = \delta \geq 2$ by Theorem 7.9, and so $n = \delta \geq 2$. Suppose $|A| > 1$ below. If $n = 1$, then S is also an atom of G and $G[S] \cong G[A]$. Since G is vertex-transitive, A is also a κ-separating set of G. Let F_1 and F_2 be two fragments of G. Then both F_1 and F_2 contain elements of S, which implies $G[S]$ is disconnected, a contradiction. Thus $n \geq 2$.

(f) If G is an undirected graph, then from (a) and (e), we have

$$\delta \leq \delta(G[A]) + \kappa(G) \leq |A| - 1 + \kappa(G) \leq \tfrac{3}{2}\kappa(G) - 1.$$

This implies $\kappa(G) \geq \lceil \frac{2}{3}(\delta + 1) \rceil$.

(g) If G is undirected, then for $\delta = 2, 3, 4$, $\kappa(G) = 2, 3, 4$ by (f), respectively. For $\delta = 6$, $5 \leq \kappa(G) \leq 6$ by (f) and Whitney's inequality. If $\kappa(G) = 5$, then $a(G) \geq 2$ by the exercise 7.2.5. Thus there is an integer $n \geq 2$ such that $5 = n\,a(G)$ by (e). This is impossible and so $\kappa(G) = 6$ if $\delta(G) = 6$. ∎

Exercises

7.2.1 Prove that for any $\sigma \in \mathrm{Aut}\,(G)$, its restriction to X is an isomorphism between $G[X]$ and $G[\sigma(X)]$ for any nonempty $X \subseteq V(G)$, where $\sigma(X) = \{y \in V(G) : \ y = \sigma(x), x \in X\}$.

7.2.2 Let G be a vertex-transitive graph of order n (≥ 2). Prove that

(a) G is regular, and

(b) all subgraphs of order $n - 1$ of G are isomorphic.

7.2.3 Prove that Corollary 7.4.1, Corollary 7.4.2 and Corollary 7.4.3.

7.2.4 Let A be a positive (resp. negative) atom of G. Prove that $\sigma(A)$ is a positive (resp. negative) atom of G for any $\sigma \in \mathrm{Aut}\,(G)$.

7.2.5 Let G be a strongly connected digraph. Prove that $a(G) = 1 \Leftrightarrow \kappa(G) = \lambda(G) = \delta(G)$.

7.2.6 Prove that n-cube Q_n is vertex-transitive.

7.3 Graphic Presentation of Groups

We have, in Section 7.1, seen that every graph G is associated with a group $\mathrm{Aut}(G)$. A natural question was asked by König [113]: for a given abstract group Γ, whether there exists a simple undirected graph G such that $\mathrm{Aut}(G) \cong \Gamma$? An affirmative answer to this question was given constructively by Frucht [69] by making use of the Cayley graph of the group Γ, which will be define as follows.

Let Γ be a nontrivial finite group, S be a nonempty subset of Γ that does not contain the identity element of Γ. Define a graph G as follows.

$$V(G) = \Gamma; \quad (x, y) \in E(G) \Leftrightarrow x^{-1}y \in S, \text{ for any } x, y \in \Gamma.$$

The graph so defined, proposed by A. Cayley [27], is called a *Cayley graph* of the group Γ with respect to S, denoted by $C_\Gamma(S)$. Clearly, $C_\Gamma(S)$ contains no loops and parallel edges, and so is simple since S does not contain the identity element of Γ. Moreover, $C_\Gamma(S)$ can be thought of as an undirected graph if $S^{-1} = S$, and a digraph otherwise.

Example 7.3.1 Consider the additive group $Z_n\,(n \geq 2)$ of residue classes modulo n, that is, the ring group of order n, zero is the identity element, the inverse of i is $n - i$. If $S = \{1\}$, then $S^{-1} = S$ for $n = 2$; and $S^{-1} \neq S$ otherwise. Thus the Cayley graph $C_{Z_2}(\{1\}) = K_2$, the Cayley graph $C_{Z_n}(\{1\})$ is a directed cycle C_n of length n if $n \geq 3$, while the Cayley graph $C_{Z_n}(\{1, n-1\})$ is an undirected cycle C_n of length n if $n \geq 3$.

Generally, let $S \subseteq \{1, 2, \cdots, n-1\}, n \geq 3$. Then the Cayley graph $C_{Z_n}(S)$ is the circulant digraph $G(n; S)$ if $S^{-1} \neq S$; the circulant undirected graph $G(n; \pm S)$ if $S^{-1} = S$. We show this fact for the directed case. Arbitrarily choose $i, j \in Z_n = V(G(n; S))$. From definitions of the Cayley graph and the circulant digraph,

$$\begin{aligned}
(i, j) \in E(C_{Z_n}(S)) &\Leftrightarrow i^{-1} + j \pmod n \in S \\
&\Leftrightarrow (n - i) + j \pmod n \in S \\
&\Leftrightarrow \text{ there is } s \in S \text{ such that } j - i \equiv s \pmod n \\
&\Leftrightarrow (i, j) \in E(G(n; S)).
\end{aligned}$$

∎

Example 7.3.2 Let S_3 be the symmetric group on the set $X = \{1, 2, 3\}$, that is, $S_3 = \{g_0, g_1, g_2, g_3, g_4, g_5\}$, where

$$g_0 = (1), g_1 = (12), g_2 = (123), g_3 = (132), g_4 = (23), g_5 = (13).$$

The graphs shown in Figure 7.5 are Cayley graphs $C_{S_3}(\{g_1, g_2\})$ and $C_{S_3}(\{g_2\})$, where the digit k nearby the edge (g_i, g_j) implies that $g_i^{-1} g_j = g_k$. ∎

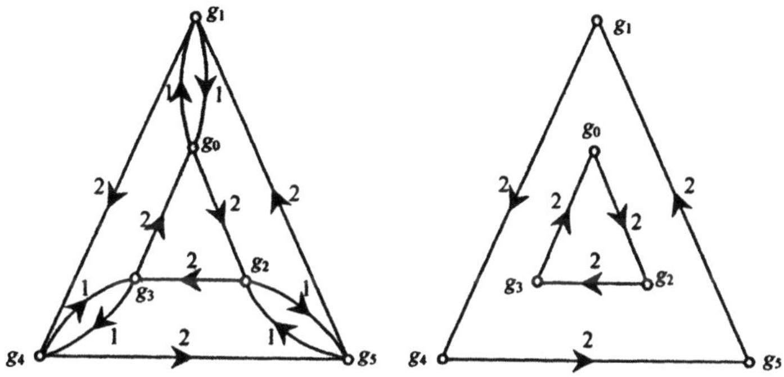

Figure 7.5: Two Cayley graphs $C_{S_3}(\{g_1, g_2\})$ and $C_{S_3}(\{g_2\})$

Since $G = C_\Gamma(S)$ is regular, by the exercise 1.5.4 (c), it is strongly connected if and only if it is connected. The following theorem gives another sufficient and necessary condition for $G = C_\Gamma(S)$ to be strongly connected.

Theorem 7.11 $G = C_\Gamma(S)$ is strongly connected if and only if $\Gamma = < S >$, i.e., S is a generating set of Γ.

Proof ($\Rightarrow$) It suffices to show that $\Gamma \subseteq < S >$. Suppose that y is any vertex of G. Since $G = C_\Gamma(S)$ is strongly connected, for any $x \in S$, there is an (x, y)-path $P = (x_0, x_1, \cdots, x_m)$ in G, where $x_0 = x$ and $y = x_m$. Since $(x_{i-1}, x_i) \in E(G)$ for each $i = 1, 2, \cdots, m$, there is $s_i \in S$ such that $x_i = x_{i-1} s_i$. Thus y can be expressed as $y = x s_1 s_2 \cdots s_m \in S$. This shows that $\Gamma \subseteq < S >$.

($\Leftarrow$) In order to show that G is strongly connected, by the above statement, it is sufficient to show that G is connected.

Assume that x and y are two distinct vertices of G and g_0 is the identity element of Γ. Since $\Gamma = < S >$, there exist $g_1, g_2, \cdots, g_m \in S$ such that $y = g_1 g_2 \cdots g_m$. Since $g_i = (g_0 g_1 \cdots g_{i-1})^{-1}(g_0 g_1 \cdots g_{i-1} g_i) \in S$, $(g_0 g_1 \cdots g_{i-1}, g_0 g_1 \cdots g_i) \in E(G)$ for each $i = 1, 2, \cdots, m$. Thus there is a (g_0, y)-path $P = (g_0, g_0 g_1, g_0 g_1 g_2, \cdots, g_0 g_1 g_2 \cdots g_{m-1}, y)$, which means that g_0 and y are connected. Similarly, g_0 and x are connected. Thus x and y are connected, and so is G because of the arbitrariness of x and y. ∎

We will prove that Cayley graphs are vertex-transitive. To this aim, we first introduce the concept of the color-preserving automorphism of a Cayley graph.

Let $\Gamma = \{g_0, g_1, \cdots, g_{n-1}\}$ be a nontrivial finite group of order n whose identity element is g_0, and let S be a subset of Γ that does not contain g_0. The *Cayley color-graph* $C_\Gamma(S)$ has an edge (g_i, g_j) that is assigned the color k if and only if there is $g_k \in S$ such that $g_k = g_i^{-1} g_j$.

For instance, see the Cayley graphs shown in Figure 7.5, the digit k nearby an edge is the color assigned on the edge.

An element $\phi \in \mathrm{Aut}(C_\Gamma(S))$ is said to be a *color-preserving automorphism* of $C_\Gamma(S)$ if for any $(g_i, g_j) \in E(C_\Gamma(S))$, the edges (g_i, g_j) and $(\phi(g_i), \phi(g_j))$ have the same color.

For instance, for the Cayley graph $C_{S_3}(\{g_1, g_2\})$ shown in Figure 7.5, it is easy to verify the permutation defined by

$$g_i \mapsto g_1 g_i, \quad i = 0, 1, 2, 3, 4, 5$$

is a color-preserving automorphism of $C_{S_3}(\{g_1, g_2\})$.

It is a routine exercise to prove that the set of all color-preserving automorphisms of $C_\Gamma(S)$ forms a subgroup of $\mathrm{Aut}\,(C_\Gamma(S))$ (the exercise 7.3.3). This subgroup is called the *color-preserving automorphism group* of $C_\Gamma(S)$, denoted by $\mathrm{Aut}^c(C_\Gamma(S))$.

For each $i = 0, 1, \cdots, n-1$, define a mapping

$$\begin{aligned} \phi_i : \Gamma &\to \Gamma \\ g_j &\mapsto \phi_i(g_j) = g_i g_j, \quad j = 0, 1, 2, \cdots, n-1. \end{aligned}$$

It is not difficult to verify that ϕ_i is a permutation on Γ, and

$$\phi(\Gamma) = \{\phi_i : \ i = 0, 1, 2, \cdots, n-1\}$$

is a permutation group (the exercise 7.3.3).

The major significance of the color-preserving automorphism group of a Cayley graph is contained in the following theorem, due to Frucht [69].

Theorem 7.12 Let $\Gamma = \{g_0, g_1, \cdots, g_{n-1}\}$ be a nontrivial finite group with generating set S that does not contain the identity element g_0. Then

$$\mathrm{Aut}^c(C_\Gamma(S)) = \phi(\Gamma) \cong \Gamma.$$

Proof Arbitrarily choose $\phi_m \in \phi(\Gamma)$, then ϕ_m is a permutation on Γ. For any $g_i, g_j \in \Gamma$, the edge (g_i, g_j) in $C_\Gamma(S)$ is assigned the color

$$\begin{aligned} k &\Leftrightarrow g_i^{-1} g_j = g_k \in S \\ &\Leftrightarrow (g_m g_i)^{-1}(g_m g_j) = g_k \in S \\ &\Leftrightarrow \phi_m^{-1}(g_i)\phi_m(g_j) = g_k \in S \\ &\Leftrightarrow (\phi_m(g_i),\ \phi_m(g_j)) \in E(C_\Gamma(S)). \end{aligned}$$

Therefore ϕ_m is a color-preserving automorphism of $C_\Gamma(S)$, and so $\phi_m \in \mathrm{Aut}^c(C_\Gamma(S))$. This shows

$$\phi(\Gamma) \subseteq \mathrm{Aut}^c(C_\Gamma(S)). \tag{7.7}$$

Conversely, Arbitrarily take $\sigma \in \mathrm{Aut}^c(C_\Gamma(S))$. Assume $\sigma(g_0) = g_m$, where g_0 is the identity element of Γ. We need prove $\sigma = \phi_m$.

Arbitrarily choose $g_j \in \Gamma$. Since S is a generating set of Γ, the Cayley graph $C_\Gamma(S)$ is strongly connected by Theorem 7.11. Thus there exists a (g_0, g_j)-path $P = (g_{j_0}, g_{j_1}, \cdots, g_{j_p})$ in $C_\Gamma(S)$, where $g_0 = g_{j_0}$ and $g_j = g_{j_p}$. Suppose that the edge $(g_{j_{i-1}}, g_{j_i})$ has been assigned the color k_i for any $i = 1, 2, \cdots, p$. Then g_j can be expressed as $g_j = g_0 g_{k_1} g_{k_2} \cdots g_{k_p}$, where $g_{k_i} \in S$ for each $i = 1, 2, \cdots, p$. Observe that σ is a color-preserving automorphism if and only if $\sigma(gh) = \sigma(g)h$ for every $g \in \Gamma$ and $h \in S$. By successive applications of such a simple observation, we have that

$$\sigma(g_j) = \sigma(g_0) g_{k_1} g_{k_2} \cdots g_{k_p} = \sigma(g_0) g_j = g_m g_j = \phi_m(g_j),$$

which shows that $\sigma = \phi_m \in \phi(\Gamma)$. Because of the arbitrariness of $\sigma \in \mathrm{Aut}^c(C_\Gamma(S))$, we have that

$$\mathrm{Aut}^c(C_\Gamma(S)) \subseteq \phi(\Gamma). \tag{7.8}$$

Combining (7.7) with (7.8), we have that $\mathrm{Aut}^c(C_\Gamma(S)) = \phi(\Gamma)$.

We now want to prove $\Gamma \cong \phi(\Gamma)$. To this aim, define a mapping

$$\begin{aligned} \Phi :\ \Gamma &\to \phi(\Gamma) \\ g_k &\mapsto \Phi(g_k) = \phi_k,\ k = 0, 1, 2, \cdots, n-1. \end{aligned}$$

It is easy to verify that $\Phi : \Gamma \to \phi(\Gamma)$ is bijective. In order to prove that Φ is an isomorphism, it suffices to prove that Φ is operation-preserving. For any $g_i, g_j \in \Gamma$, let $g_i g_j = g_k$. Since for any $g_h \in \Gamma$, we have that

$$\begin{aligned}\phi_k(g_h) &= g_k g_h = (g_i g_j) g_h = g_i(g_j g_h) = \phi_i(g_j g_h) \\ &= \phi_i(\phi_j(g_h)) = \phi_i \phi_j(g_h).\end{aligned}$$

Thus $\phi_k = \phi_i \phi_j$, and so

$$\Phi(g_i g_j) = \Phi(g_k) = \phi_k = \phi_i \phi_j = \Phi(g_i)\Phi(g_j),$$

which means that Φ is operation-preserving. Thus we show that $\Gamma \cong \phi(\Gamma)$. ∎

Theorem 7.13 Any Cayley graph $C_\Gamma(S)$ is vertex-transitive.

Proof Suppose that g_i, g_j are any two elements of $V(C_\Gamma(S)) = \Gamma$, and let $g_k = g_j g_i^{-1}$. Then

$$\phi_k(g_i) = g_k g_i = (g_j g_i^{-1}) g_i = g_j.$$

This shows that there exists $\phi_k \in \mathrm{Aut}^c(C_\Gamma(S)) \subseteq \mathrm{Aut}(C_\Gamma(S))$ such that $\phi_k(g_i) = g_j$. This means that $C_\Gamma(S)$ is vertex-transitive. ∎

Theorem 7.12 shows that for a given finite group Γ, there exists a simple digraph G and a subgroup H of Aut (G) such that $H \cong \Gamma$. The following theorem, known as Frucht's theorem, affirmatively answers to König's question mentioned at the beginning this section.

Theorem 7.14 For any given finite group Γ, there exists a simple undirected graph G such that Aut $(G) \cong \Gamma$.

Proof Let $\Gamma = \{g_0, g_2, \cdots, g_{n-1}\}$. If $n = 1$, let $G = K_1$, then the theorem holds clearly. Suppose $n \geq 2$, and let S be a generating set S of Γ that does not contain the identity element g_0. Then $\mathrm{Aut}^c(C_\Gamma(S)) \cong \Gamma$ by theorem 7.12. We now construct a simple undirected simple graph $G_\Gamma(S)$, called the *Frucht graph*, as follows.

Replace each directed edge (g_i, g_j) of $C_\Gamma(S)$ that is of the same color k by an undirected path $(g_i, u_{ij}, u'_{ij}, g_j)$ of length 3, and then join a new undirected path P_{ij} of length $2k-2$ with u_{ij} and join a new undirected path P'_{ij} of length $2k-1$ with u'_{ij}. See Figure 7.6, which shows the Frucht graph $G_{S_3}(\{g_1, g_2\})$ of the Cayley graph $C_{S_3}(\{g_1, g_2\})$ shown in Figure 7.5.

It is easy to see that the Frucht graph $G_\Gamma(S)$ of the Cayley graph $C_\Gamma(S)$ has $n(2s^2+s+1)$ vertices and $2ns(s+1)$ edges, where $s = |S|$ (the exercise 7.3.4). We now show Aut $(G_\Gamma(S)) \cong \mathrm{Aut}^c(C_\Gamma(S))$.

Noting that every $\alpha \in \mathrm{Aut}^c(C_\Gamma(S))$ uniquely decides one element in Aut $(G_\Gamma(S))$, we need to only prove that every $\beta \in$ Aut $(G_\Gamma(S))$ is decided by some element $\alpha \in \text{Aut}^c(C_\Gamma(S))$.

Since $C_\Gamma(S)$ is strongly connected, each vertex of $C_\Gamma(S)$ is neither cut-vertex of $G_\Gamma(S)$ nor 1-degree vertex of $G_\Gamma(S)$. Thus $\beta(g) \in V(C_\Gamma(S))$ for every $g \in V(C_\Gamma(S))$, that is, $V(C_\Gamma(S))$ is fixed under Aut $(G_\Gamma(S))$.

Since the subgraph G_{ij} used to replace the edge (g_i, g_j) of $C_\Gamma(S)$ to obtained the Frucht graph according to the above construction is a connected component of $G_\Gamma(S) - V(C_\Gamma(S)$, it follows that $\beta \in \text{Aut}\,(G_\Gamma(S))$ maps one connected component to other one; a path of length $2k-2$ to other one of the same length; a path of length $2k-1$ to other one of the same length. Thus if let $\overline{\beta} = \beta\,|V(C_\Gamma(S))$ then $(g_i, g_j) \in E(C_\Gamma(S))$ with the color k if and only if $(\overline{\beta}(g_i), \overline{\beta}(g_j)) \in E(C_\Gamma(S))$ with the color k. This implies $\overline{\beta} \in \text{Aut}^c(C_\Gamma(S))$. Thus we have Aut $(G_\Gamma(S)) \cong \text{Aut}^c(C_\Gamma(S)) \cong \Gamma$ as desired. ∎

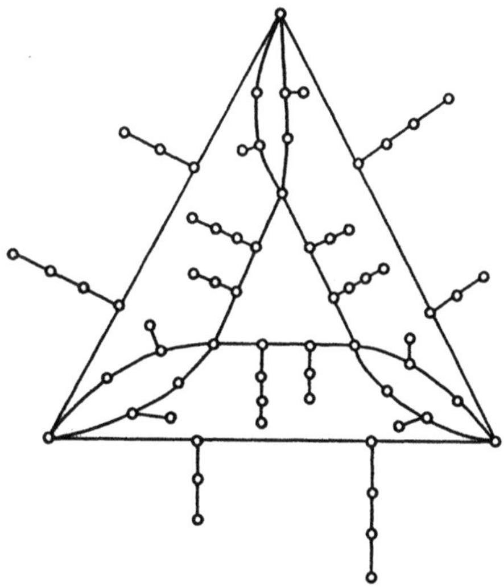

Figure 7.6: Frucht graphs $G_{S_3}(\{g_1, g_2\})$

Graph theory has become an important analytical tool in group theory. In particular, in 1968, Higman and Sims [94] discovered the well-known Higman-Sims group of order 44,352,000 by graph theoretical method. Since that time several other sporadic simple groups have been constructed as automorphism group of graphs. There are many interesting results concerning the relationship between graphs and groups. The interested reader is referred to a book by Biggs and White[11].

Exercises

7.3.1 Let $C_\Gamma(S)$ be a Cayley graph. Prove that

(a) the converse $\overleftarrow{C_\Gamma(S)}$ of $C_\Gamma(S)$ is also a Cayley graph and $\overleftarrow{C_\Gamma(S)} = C_\Gamma(S^{-1})$;

(b) if $S^{-1} = \{s^{-1} : s \in S\} = S$, then $C_\Gamma(S)$ is a symmetric digraph, hence an undirected graph;

(c) the edges of $C_\Gamma(S)$ can also be defined as $(x, y) \in E(G) \Leftrightarrow xy^{-1} \in S$.

7.3.2 Let S_4 be the symmetric group S_4 on the set $X = \{1, 2, 3, 4\}$, $a = (12), b = (134) \in S_4$, $S = \{a, b, ab\}$ and $F = \{e, a\}$, where e is the identity element of S_4.

(a) Construct the Cayley $C_{S_4}(S)$ and the Frucht graph $C_{S_4}(S)$.

(b) Prove that F is a positive atom of $C_{S_4}(S)$, and $C_{S_4}(S)$ has no negative atom. (G. Zemor, 1989)

7.3.3 Prove that the set of all color-preserving automorphisms of $C_\Gamma(S)$ forms a subgroup of Aut $(C_\Gamma(S))$.

7.3.4 Prove that the Frucht graph $G_\Gamma(S)$ of the Cayley graph $C_\Gamma(S)$ has $n(2s^2 + s + 1)$ vertices and $2ns(s + 1)$ edges, where $s = |S|$.

7.3.5 Prove that Petersen graph is a vertex-transitive graph, but not a Cayley graph.

7.3.6 Let $G = C_\Gamma(S)$ be a Cayley graph, and A be an atom of G that contains the identity element of Γ. Prove that A is a subgroup of Γ. Furthermore, A is generated by $S \cap A$ if and only if $|A| \geq 2$.

7.3.7 Proved that any Cayley graph of abelian group must contain both positive and negative atoms. (Y. O. Hamidoune, 1985)

7.3.8 Prove that if S is a minimal generating set of a group Γ that does not contain the identity element, then the connectivity $\kappa(C_\Gamma(S)) = |S|$. (C. D. Godsil (1981) for the undirected case and Y. O. Hamidoune (1984) for the directed case.)

Applications

7.4 Design of Interconnection Networks

Modern science, engineering and technology have put forward a great number of important problems whose solutions require an enormous amount of computational or processing power, and in some cases the solutions must be obtained within a small time period. For example, image processing, radar signal processing, weather tracking and prediction, control of fast and complex nuclear or chemical reactions, aerodynamic simulation and many more. It is imperative that one need have a very large-scale multiple processor system consisting of high speed processors capable of executing parallel algorithms.

The advent of very large scale integrated (VLSI) circuit and the development with great speed of the modern communication technology such as fiber optics materials have enabled us to design more complicatedly, more conveniently and economically high performance supercomputers and massive interconnection networks. The Connection Machine (see Hillis [95]), consisting of 2^{12} processors, has been demonstrated the viability of constructing massively parallel supercomputers using current hardware technology. A natural question is how to interconnect such a large number of processors that the system is of high performance and low cost as far as possible.

A connection pattern of the components in a system is called an *interconnection network* of the system. It quite natural that an interconnection network may be modelled by a simple graph whose vertices represent components of the network and whose edges represent physical communication links, where directed edges represent one-way communication links and undirected edges represent two-way communication links. Such a graph is called the *topological structure* of the interconnection network. For example, the Connection Machine adopts the 12-cube Q_{12} as its topological structure.

Clearly, a topological structure of an interconnection network is important for design of a supersystems. When network requirements can be expressed in terms of graph-theoretic parameters, the problem of an interconnection network design becomes:

Find a graph G satisfying some specified requirements.

Following Hillis [95], some of these specified requirements are as follows.

1. Small and fixed degree. The degree of a graph corresponds the number of connections to each component. This number is bounded by the number of the interfaces available for I/O devices attached to each component in the network. An excess of any physical connection will result in replacement of the components in the network to increase the number of the interfaces. The larger degree, the more wiring. More wiring not only costs much money, and also is disadvantageous to implementation of VLSI layout. Thus a small or fixed maximum degree is desirable.

2. Small diameter or average distance. Since transmission delay or signal degradation for sending a message from one vertex to another is approximately proportional to the number of times that a message has to be stored and forwarded by intermediate vertices. Thus, a small average distance or diameter is desired to obtain a highly efficient interconnection network. In particular, diameter should be bounded by a given value for a real-time processing system.

3. Maximum connectivity. The network must continue to work in case of vertex or edge failures. The maximum connectivity is desirable since it corresponds to not only the maximum fault tolerance of the network but also the maximum number of internally (or edge-) disjoint paths between any two distinct vertices.

4. Symmetry. We can desire that all components behave in the same manner and that they communicate in similar ways, or at least that there is some balance in the communication patterns. This implies at least some regularity and some symmetric properties on the graph. A highly symmetry network is desirable since it is advantage to construction and simulation of some algorithms.

5. Extendability. It should be possible to build a network of any given size, or at least to build arbitrarily large versions of the network. Furthermore, it would be easy to construct large networks from small ones. When a small network is extended, some desirable properties should be remained and some useful parameters should be calculated easily.

Note that the wish list contains contradictions. For example, a complete graph satisfies all of the above-mentioned requirements, except no efficient layout of VLSI circuits, but the cost is too big, making it impractical for large system. On the other hand, a tree has a simple structure and easy extendability, but its has very poor fault tolerance. Similarly, we can not have both a small degree and a big connectivity. We can, therefore, say that any decision will be a compromise or tradeoff among many factors when a topological structure is chosen for an interconnection network.

For a very large scale computer system, it is necessary that we should have a systematic method for constructing a topological structure of the system. The cartesian product method is a very effective method for constructing a larger graph from several specified small graphs. We have seen in the preceding chapters that a number of important graph-theoretic parameters of a cartesian product graph, such as degree, diameter and connectivity, can be easily calculated from its factor graphs according to the exercise 1.4.6, Theorem 1.3 and Theorem 4.6, respectively.

For example, as we have known, the n-cube Q_n is the cartesian products of n identical complete graph K_2, that is, $Q_n = K_2 \times K_2 \times \cdots \times K_2$. It is a n-regular graph with connectivity n and diameter n.

We can also easily show that the cartesian product preserves Hamilton property, Euler property and bipartiteness of factor graphs, that is, the cartesian product of hamiltonian (resp. eulerian, bipartite) graphs is a hamiltonian (resp. eulerian, bipartite) graph, the proofs are left to the reader as exercises for details (the exercise 7.4.2). As an immediate consequence of these results, the n-cube Q_n is bipartite, hamiltonian if $n \geq 2$ and eulerian if n is even.

In this section, we will further introduce some of other useful properties of cartesian product graphs.

Theorem 7.15 The cartesian product graph $G_1 \times G_2 \times \cdots \times G_n$ is vertex-transitive if G_i is vertex-transitive for each $i = 1, 2, \cdots, n$.

Proof Let $G = G_1 \times G_2 \times \cdots \times G_n$ and let $x = x_1x_2\cdots x_n$ and $y = y_1y_2\cdots y_n$ be two distinct vertices in G, where $x_i, y_i \in V(G_i)$ for each $i = 1, 2\cdots, n$. Since G_i is vertex-transitive there exists $\sigma_i \in \text{Aut}\,(G_i)$ such that $\sigma_i(x_i) = y_i$ for each $i = 1, 2, \cdots, n$. It is easy to verify that the mapping ϕ defined by

$$\phi(x_1x_2\cdots x_n) = \sigma_1(x_1)\sigma_2(x_2)\cdots\sigma_n(x_n)$$

is an element in $\text{Aut}\,(G)$ and $\phi(x) = y$. Thus, G is vertex-transitive. ∎

As an application of Theorem 7.15, it is straightforward that the n-cube Q_n is vertex-transitive since $Q_n = K_2 \times \cdots \times K_2$, and K_2 is vertex-transitive.

As we have known from Theorem 7.13, Cayley graphs are vertex-transitive. And Theorem 7.15 asserts that the cartesian product of vertex-transitive graphs is also vertex-transitive. A natural question is whether or not the cartesian product of Cayley graphs is a Cayley graph. The answer to the question is the affirmative.

To state this result, let us first recall the cartesian product of groups. Let $\Gamma = \Gamma_1 \times \Gamma_2 \times \cdots \times \Gamma_n = (X, \circ)$ be the cartesian product of n finite groups $\Gamma_i = (X_i, \circ_i)$ $(i = 1, 2, \cdots, n)$, where $X = X_1 \times X_2 \times \cdots \times X_n$. The operation "$\circ$" is defined by

$$(x_1 x_2 \cdots x_n) \circ (y_1 y_2 \cdots y_n) = (x_1 \circ_1 y_1)(x_2 \circ_2 y_2) \cdots (x_n \circ_n y_n),$$

where $x_i, y_i \in X_i$ $(i = 1, 2, \cdots, n)$. The inverse of an element $x_1 x_2 \cdots x_n$ in Γ is defined by

$$(x_1 x_2 \cdots x_n)^{-1} = x_1^{-1} x_2^{-1} \cdots x_n^{-1},$$

where x_i^{-1} is the inverse of the element x_i in the group Γ_i for each $i = 1, 2, \cdots, n$. The identity element of Γ is $e = e_1 e_2 \cdots e_n$, where e_i is the identity element of the group Γ_i for each $i = 1, 2, \cdots, n$.

Theorem 7.16 The cartesian product of Cayley graphs is a Cayley graph. More precisely speaking, let $G_i = C_{\Gamma_i}(S_i)$ be a Cayley graph of a finite group $\Gamma_i = (X_i, \circ_i)$ with respect to a subset S_i, then $G = G_1 \times G_2 \times \cdots \times G_n$ is a Cayley graph $C_\Gamma(S)$ of the group $\Gamma = \Gamma_1 \times \Gamma_2 \times \cdots \times \Gamma_n$ with respect to the subset

$$S = \bigcup_{i=1}^{n} \{e_1 \cdots e_{i-1}\} \times S_i \times \{e_{i+1} \cdots e_n\},$$

where e_i is the identity element of Γ_i for each $i = 1, 2, \cdots, n$.

Proof It suffices to prove that the assertion holds for $n = 2$. Under our assumption, $G = G_1 \times G_2$, $\Gamma = \Gamma_1 \times \Gamma_2$, and $S = (\{e_1\} \times S_2) \cup (S_1 \times \{e_1\})$. Arbitrarily choose two elements $x_1 x_2$ and $y_1 y_2$ in $X_1 \times X_2$, where $x_i, y_i \in X_i$ for $i = 1, 2$. We need only show that

$$(x_1 x_2, y_1 y_2) \in E(G) \Leftrightarrow (x_1 x_2)^{-1} \circ (y_1 y_2) \in S.$$

By definition of the cartesian product digraph, we have

$$(x_1 x_2, y_1 y_2) \in E(G) \Leftrightarrow \begin{cases} x_1 = y_1, \ (x_2, y_2) \in E(G_2), & \text{or,} \\ x_2 = y_2, \ (x_1, y_1) \in E(G_1) \end{cases}$$

Since $G_i = C_{\Gamma_i}(S_i)$, we have $(x_i, y_i) \in E(G_i) \Leftrightarrow x_i^{-1} \circ_i y_i \in S_i$ for each $i = 1, 2$. It follows that

$$\begin{aligned} x_1 = y_1,\ (x_2, y_2) \in E(G_2) &\Leftrightarrow \\ (x_1x_2)^{-1} \circ (y_1y_2) &= (x_1^{-1}x_2^{-1}) \circ (y_1y_2) \\ &= (x_1^{-1} \circ_1 y_1)(x_2^{-1} \circ_2 y_2) \\ &= (x_1^{-1} \circ_1 x_1)(x_2^{-1} \circ_2 y_2) \\ &= e_1(x_2^{-1} \circ_2 y_2) \in \{e_1\} \times S_2 \subseteq S. \end{aligned}$$

Similarly, we have

$$\begin{aligned} x_2 = y_2,\ (x_1, y_1) \in E(G_1) &\Leftrightarrow \\ (x_1x_2)^{-1} \circ (y_1y_2) &= (x_1^{-1}x_2^{-1}) \circ (y_1y_2) \\ &= (x_1^{-1} \circ_1 y_1)(x_2^{-1} \circ_2 y_2) \\ &= (x_1^{-1} \circ_1 y_1)(x_2^{-1} \circ_2 x_2) \\ &= (x_1^{-1} \circ_1 y_1)e_2 \in S_1 \times \{e_2\} \subseteq S. \end{aligned}$$

These imply that $G = G_1 \times G_2$ is a Cayley graph $C_\Gamma(S)$ of the group $\Gamma = \Gamma_1 \times \Gamma_2$ with respect to the subset $S = (S_1 \times \{e_2\}) \cup (\{e_1\} \times S_2)$. Thus, the theorem follows. ∎

Theorem 7.16 is very useful since it provides a method for constructing a larger Cayley graph from some small Cayley graphs such that it contains them as its subgraphs.

As applications, we give some examples. Consider the additive group Z_n of residue classes modulo n. The identity element $e = 0$ and the inverse $i^{-1} = n - i$. The Cayley graph $C_{Z_2}(\{1\})$ is K_2. For $n \geq 3$, the Cayley graph $C_{Z_n}(\{1, n-1\})$ is an undirected cycle C_n; the Cayley graph $C_{Z_n}(\{1\})$ is a directed cycle C_n.

Example 7.4.1 Consider the group $Z_2 \times Z_n$ ($n \geq 3$). The operation is defined by

$$(x_1x_2) \circ (y_1y_2) = (x_1 + y_1)(\text{mod } 2)(x_2 + y_2)(\text{mod } n),$$

where $x_1, y_1 \in \{0, 1\}$ and $x_2, y_2 \in \{0, 1, \cdots, n-1\}$. If we set

$$S = (\{1\} \times \{0\}) \cup (\{0\} \times \{1, n-1\}) = \{10, 01, 0(n-1)\},$$

then $S = S^{-1}$, where the first element is self-inverse and the others are mutually inverse. So the Cayley graph $C_{Z_2 \times Z_n}(S)$ is an undirected graph, which is $K_2 \times C_n$ by Theorem 7.16.

The graph shown in Figure 7.7 (a) is Cayley graph $C_{Z_2\times Z_5}(\{10,01,04\})$, which is $K_2 \times C_5$. If we set

$$S = (\{1\} \times \{0\}) \cup (\{0\} \times \{1\}) = \{10, 01\},$$

then $S^{-1} \neq S$ and so the Cayley graph $C_{Z_2\times Z_n}(S)$ is a digraph, which is $K_2 \times C_n$ by Theorem 7.16, where C_n is a directed cycle.

The digraph shown in Figure 7.8 (b) is $C_{Z_2\times Z_5}(\{10,01\})$, which is $K_2 \times C_5$, where C_5 is a directed cycle and an undirected edge denotes two symmetric directed edges. This example is of interest, which shows that a Cayley digraph may not be hamiltonian. However, a problem whether or not a Cayley undirected graph is hamiltonian has not been solved as far. ∎

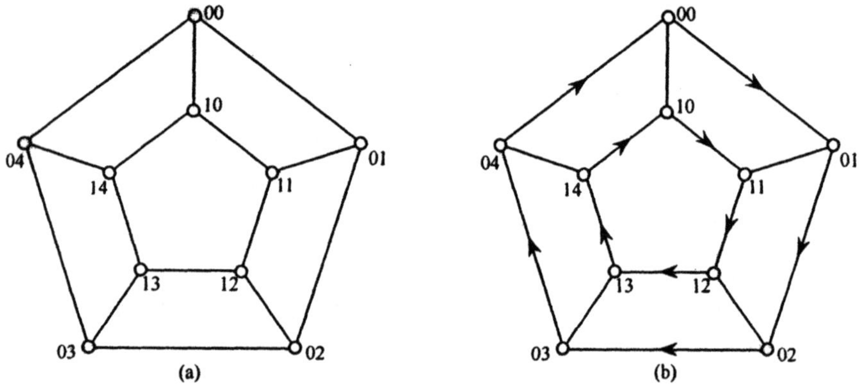

Figure 7.7: Cayley graphs $C_{Z_2\times Z_5}(\{10,01,04\})$ and $C_{Z_2\times Z_5}(\{10,01\})$

Example 7.4.2 Consider the group $\Gamma = Z_2 \times Z_2 \times \cdots \times Z_2$. The operation is defined by

$$(x_1 \cdots x_n) \circ (y_1 \cdots y_n) = (x_1 + y_1)(\text{mod}2) \cdots (x_n + y_n)(\text{mod}2)$$

where $x_i, y_i \in \{0, 1\}$ and $e_i = 0$ for $i = 1, 2, \cdots, n$. Let

$$S = \bigcup_{i=1}^{n} \{e_1 \cdots e_{i-1}\} \times S_i \times \{e_{i+1} \cdots e_n\}$$
$$= \{100\cdots 00, 010\cdots 00, \cdots, 000\cdots 01\},$$

where $S_i = \{1\}$ for $i = 1, 2, \cdots, n$. It is easy to see that all elements in S are self-inverse. Thus $S^{-1} = S$ and so the Cayley graph $C_\Gamma(S)$ is an undirected graph, which is $K_2 \times K_2 \times \cdots \times K_2 = Q_n$ by Theorem 7.16. ∎

Example 7.4.3 Consider a little complex example due to Carlesson *et al* [24]. Use $(Z_2)^n$ to denote $Z_2 \times Z_2 \times \cdots \times Z_2$, which is the cartesian product of n sets $Z_2 = \{0, 1\}$. Let

$$M = \begin{pmatrix} 0 & 0 & 0 & \cdots & 0 & 1 \\ 1 & 0 & 0 & \cdots & 0 & 0 \\ 0 & 1 & 0 & \cdots & 0 & 0 \\ \vdots & \vdots & \vdots & \ddots & \vdots & \vdots \\ 0 & 0 & 0 & \cdots & 0 & 0 \\ 0 & 0 & 0 & \cdots & 1 & 0 \end{pmatrix}$$

be an n-square matrix. For an element $\mathbf{v}$ in $(Z_2)^n$, thinking of $\mathbf{v}$ as a column vector, we let M act on $\mathbf{v}$ in the normal manner, except that all additions are computed modulo 2. Then $M\mathbf{v}$ is also an element of $(Z_2)^n$.

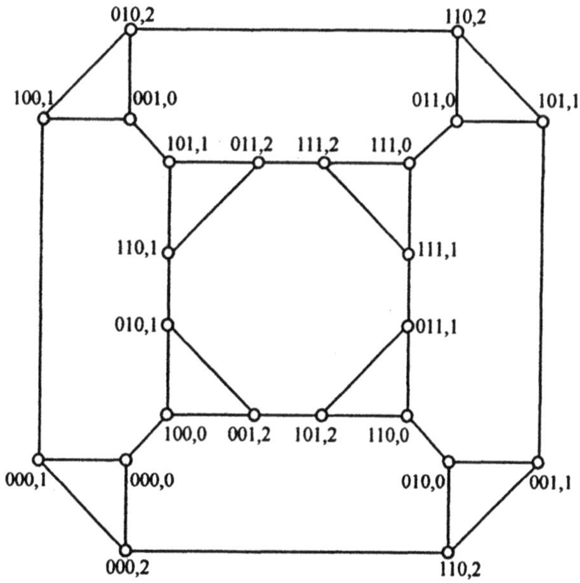

Figure 7.8: Cayley graph $C_{(Z_2)^3 \times Z_3}(\{100, 001, 002\})$

We can define a new group $\Gamma = (Z_2)^n \times Z_n$. For $(\mathbf{x}; i), (\mathbf{y}; j) \in (Z_2)^n \times Z_n$, the operation "$\circ$" of Γ is defined as follows.

$$(\mathbf{x}; i) \circ (\mathbf{y}; j) = (M^j\mathbf{x} + \mathbf{y}; i + j),$$

where the first addition is componentwise modulo 2 (in $(Z_2)^n$) and the second is modulo n (in Z_n). It is a simple exercise to check that this new operation makes $\Gamma = (Z_2)^n \times Z_n$ a group. Its identity element is $(0,0)$ and the inverse $(\mathbf{x}; i)^{-1} = (-M^{n-i}\mathbf{x}; n-i)$. Let

$$S = \{(10\cdots 0; 0),\ (00\cdots 0; 1),\ (00\cdots 0; n-1)\},$$

where the first two elements are mutually inverse and the last is self-inverse. Thus $S^{-1} = S$ and the Cayley graph $C_\Gamma(S)$ is an undirected graph.

This Cayley graph $C_\Gamma(S)$ is the well-known n-dimensional *cube-connected cycle* $CCC(n)$. The graph shown in Figure 7.8 is $C_{(Z_2)^3\times Z_3}(\{100, 001, 002\})$, which is $CCC(3)$. ∎

We have seen that n-cube Q_n possesses many desirable excellent graph-theoretic properties. Thus it goes without saying that the n-cube becomes the first choice as the most popular and efficient topological structures of interconnection networks for parallel processing and computing systems.

There are many factors that affect performance of interconnection networks, also there are a number of graph-theoretic parameters that can be used to measure performance of interconnection networks. The interested reader can be referred to a book by Xu [181].

Exercises

7.4.1 Let $G = G_1 \times G_2 \times \cdots \times G_n$ and $G' = G'_1 \times G'_2 \cdots \times G'_n$. Prove that $G \subseteq G'$ if $G_i \subseteq G'_i$ for each $i = 1, 2, \cdots, n$.

7.4.2 Let G_1 and G_2 be undirected graphs. Prove that

(a) $G_1 \times G_2$ is hamiltonian if one factor is hamiltonian and the other contains Hamilton paths;

(b) $G_1 \times G_2$ is eulerian if both G_1 and G_2 are eulerian;

(c) $G_1 \times G_2$ is bipartite if both G_1 and G_2 are bipartite.

7.4.3 Let Q_n be an n-cube. Prove that

(a) the mean distance $m(Q_n) = \dfrac{n2^{n-1}}{2^n - 1}$;

(b) there are n internally disjoint xy-paths such that all are of length n if $d = n$, and d of them are of of length d, otherwise $d+2$ if $d \leq n-1$, where $d = d_{Q_n}(x, y)$ for any two vertices x and y in Q_n.

Bibliography

[1] Ahuja, R. K., Magnanti, T. L., and Orlin, J. B., *Networks Flows: Theory, Algorithms, and Apllications.* Prentice-Hall, Englewood Cliffs, NJ, 1993

[2] Aigner, M., On the line graph of a directed graph. *Math. Z.*, **102** (1967), 56-61

[3] Appel, K., and Haken, W., Every planar map is four colorable. *Bull. Amer. Math. Soc.*, **82** (1976), 711-712

[4] Appel, K., and Haken, W., Every planar map is four colorable. Part I, Discharging. *Illinois J. Math.*, **21** (1977), 429-490

[5] Appel, K., Haken, W., and J. Koch., Every planar map is four colorable. Part II, Reducibility. *Illinois J. Math.*, **21** (1977), 491-567

[6] Appel, K., and Haken, W., Every planar map is four colorable. With the collaboration of J. Koch. *Contemporary Mathematics*, **98** (1989), 1-741

[7] Behzad, M., Chartrand, G., and Wall, C., On minimal regular digraphs with given girth. *Fund. Math.*, **65** (1970), 225-231

[8] Bellmore, M., and Nemhauser, G. L., The travelling salesman problem: a survey. *Operations Reaserch*, **16** (1968), 538-558

[9] Berge, C., Two theorems in graph theory. *Proc. Nat. Acad. Sci. USA*, **43** (1957), 842-844

[10] Biggs, N. L., Lloyd, E. K., and Wilson, R. J., *Graph Theory, 1736-1936.* Clarendon Press, Oxford, 1976

[11] Biggs, N. L., and White, A., *Permutation Group and Combinatoral Structures*. Cambridge University Press, Cambridge, 1979

[12] Birkhoff, G.D., The reducibility of maps. *Amer. Math.*, **35**(1913), 115-128

[13] Boesch, F. T., The strongest monotone degree condition for n-connectedness of a graph. *J. Combin. Theory* (B), **16** (1974), 162-165

[14] Bondy, J. A., Properties of graphs with constraints on degree. *Studia Sci. Math. Hungar.*, **4** (1969), 473-475

[15] Bondy, J. A., Induced subsets. *J. Combin. Theory* (B), **12** (1972), 201-202

[16] Bondy, J. A., A remark on two sufficient conditions for Hamilton cycles. *Discrete Math.*, **22** (1978), 191-193

[17] Bondy, J. A., and Chvatal, V., A method in graph theory. *Discrete Math.*, **15**(1976), 111-136

[18] Bondy, J. A., and Murty, U. S. R., *Graph Theory with Applications.* London and Basingstoke, MacMillan Press LTD, 1976

[19] Bondy, J. A., and Thomassen, C., A short proof of Meyniel's theorem. *Discrete Math.*, **19** (1977), 195-197

[20] Bridges, W. G., and Toueg, S., On the impossibility of directed Moore graphs. *J. Combin. Theory* (B), **29** (1980), 319-341

[21] Brooks, R. L., On colouring the nodes of a network. *Proc. Cambridge Philos. Soc.*, **37** (1941), 194-197

[22] Brooks, R. L. Smith, C. A. B., Stone, A. H., and Tuttle,W. T., The dissection of rectangles into squares. *Duke Math. J.*, **7** (1940), 312-340

[23] Camion, P., Chemins et cricuits hamiltoniens des graphes complets. *C. R. Acad. Sci. Paris*, **249**(1959), 2151-2152

[24] Carlsson, G. E., Cruthirds, J. E., Sexton, H. B., and Wright, C. G., Interconnection networks based on a generalization of cube-connected cycles. *IEEE Trans. Computers*, **34** (8) (1985), 769-772

[25] Cayley, A., On the colouring of maps. *Proc. Roy. Geog. Soc. (New Ser.)*, **1** (1879), 259-261

[26] Cayley, A., A theorem on trees. *Quart. J. Pure Appl. Math.*, **23** (1889), 376-378

[27] Cayley, A., The theory of graphs, graphical representation. *Mathematical Papers*, Cambridge, **10** (1895), 26-28

[28] Chan, Shu-Park, *Graph Theory and its Applications in Electrical Networks.* Academic Press, Beijing, 1982

[29] Chartrand, G., and Harary, F., Graphs with prescribed connectivities. In *Theory of Graphs* (P. Erdös and G. Katona, eds). Akadémiai Kiadó, Budapest, (1968), 61-63

[30] Chartrand, G., and Lesniak, L., *Graphs and Digraphs.* Wadsworth. Inc., Belmont. California, (1986), 172-176

[31] Chaty, G., On critically and minimally k vertex (arc) stronly connected digraphs. *Proceedings of the Keszthely*, (1976), 193-203.

[32] Christofides, N., Worst-case analysis of a new heuristic for the travelling salesman problem. *Technical Report*, Graduate School of Industrial Administration, Carnegie-Melon University, Pittsburgh, PA, 1976

[33] Chvàtal, V., and Erdös, P., A note on hamiltonian circuits. *Discrete Math.*, **2** (1972), 111-113

[34] Dantzing, G. B., On the shortest route through a network. *Management Sci.*, **6** (1960), 187-190

[35] De Bruijn, N. G., A combinatorial problem. *Koninklije Nedderlandse van Wetenshappen Proc.*, **49A**(1946), 758-764

[36] Demoucron, G., Malgrange, Y., and Pertuiset, R., Graphes planaires: recommaissance et construction de représentation planaires topologiques. *Rev. Francaise Recherche Opérationnelle*, **8** (1964), 33-47

[37] Dijkstra, E. W., A note on two problems in connection with graphs. *Numer. Math.*, **1** (1959), 269-271

[38] Dirac, J. A., Some theorems on abstract graphs. *Proc. London Math. Soc.*, **2**(1952), 69-81

[39] Dirac, J. A., A property of 4-chromatic graphs and some remarks on critical graphs. *J. London Math. Soc.*, **27**(1952), 85-92

[40] Dirac, G. A., Généralisations du théoréme de Menger. *C. R. Acad. Sci. Paris*, **250** (1960), 4252-4253

[41] Dirac, G. A., Short proof of Menger's graph theorem. *Mathematika*, **13** (1966), 42-44

[42] Dirac, G. A., and Schuster, S., A theorem of Kuratowski. *Indagationes Math.*, **16** (1954), 343-348

[43] Du, D. Z., Hsu, D. F., and Lyuu, Y. D., On the diameter vulnerability of Kautz digraphs. *Discrete Math.*, **151** (1996), 81-85

[44] Duijvestijn, A. J. W., Simple perfect squared of lowest order. *J. Combin. Theory* (B), **25** (1978), 240-243

[45] Duijvestijn, A. J. W. Simple perfect squared squares and 2 × 1 squared rectangles of order 26. *Mathematics of Computation.* **65**(215) (1996), 1359–1364

[46] Dulmage, A. L., and Mendelsohn, N. S., Graph and matrices. *Graph Theory and Theoretical Physics* (edited by Harary, F.). London, Academic Press, 1967, 167-227

[47] Edmonds, J., Paths, tree an flows. *Canad. J. Math.*, **17** (1965), 449-467

[48] Edmonds, J., and Johnson, E. L., Matching: A well-solved class of integer linear programs. *Combinatorial Structures and Their Applications*, Gordon and Breach, New York, (1970), 89-92

[49] Edmonds, J., and Johnson, E. L., Matching, Euler tours and the Chinese postman. *Math. Programming*, **5** (1973), 88-124

[50] Edmonds, J., and Karp, R. M., Theoretic improvements in algorithmic efficiency for network flow problems. *J. Assoc. Comput. Mach.*, **19** (1972), 248-264

[51] Egerváry, J., Matrixok kombinatorikus tulajonságairól. *Mat. és Fiz. Lapok*, **38** (1931), 16-28

[52] Elians, P., Feinstein, A, and Shannon, C. F., A note on the maximum flow through a networks. *IRE Trans. Inform. Theory*, **2** (1956), 117-119

[53] Elspas, B., Topological constraints on interconnection-limited logic. In *Proceedings of the 5th Annual Symposium on Switching Circuit Theory and Logic Design*, **5** (1964), 133-147

[54] Erdös, P., On the graph-theorem of Turán (in Hungaian). *Mat. Lapok*, **21** (1970), 249-251 264-286

[55] Erdös, P., Fajtlowicz, S. and Hoffman, A. J., Maximum degree in graphs of diameter 2. *Networks*, **10** (1980), 87-90

[56] Erdös, P., and Wilson, R. J., On the chromatic index of almost all graphs. *J. Combin. Theory* (B), **26** (1977), 255-257

[57] Euler, L., Solutio problematis ad geometriam situs pertinentis. *Comment. Academiae Sci. I. Petropolitanae*, **8** (1736), 128-140

[58] Euler, L., Demonstratio nommullarum insignium proprietatum quibus solida hedris planis inclusa sunt praedita. *Novi Comm. Acad. Sci. Imp. Petropol*, **4** (1752-3), 140-160

[59] Fan Genghua, New sufficient conditions for cycles in graphs, *J. Combin. Theory* (B), **37** (1984), 221-227

[60] Fáry, I., On straight line representation of planar graphs. *Acta Sci. Math. Szeged.*, **11** (1948), 229-233

[61] Federico, P. J., Squaring rectangles and squares. *Graph Theory and Related Topics*, Proceedings of the Conference held in honour of Professor W. T. Tuttle on the occasion of his sixieth birthday (edited by J. A. Bondy and U. S. R. Murty), Academic Press, New York, 1979, 173-196

[62] Fiol, M. A., Yebra, J. L. A., and Alegre, L., Line digraph iterations and (d, k) digraph problem. *IEEE Trans. Computers*, **33** (5) (1984), 400-403

[63] Folkman, J. H., Regular line-symmetric graphs. *J. Combin. Theory*, **3** (1967), 215-232

[64] Forbenius, G., Über zeriegbare Dertminaten, Sitzugsber, König. *Preuss. Akad. Wiss.*, XVIII (1917), 274-277

[65] Ford, L. R. Jr., and Fulkerson, D. R., Maximal flow through a network. *Canad. J. Math.*, **8** (1956), 399-404

[66] Ford, L. R. Jr., and Fulkerson, D. R., A simple algorithm for fiding maximal network flows and an application to Hitchcock problem. *Canad. J. Math.*, **9** (1957), 210-218

[67] Ford, L. R. Jr., and Fulkerson, D. R., *Flows in Networks.* Princeton University Press, Princeton, 1962

[68] Frink, O., and Smith, P. A., Irreducible non-planar graphs. *Bull. Amer. Math. Sci.*, **36** (1930), 214

[69] Frucht, R., Herstellung von Graphen mit vorgegebener abstrakten Gruppe. *Compositio Math.*, **6** (1938),239-250

[70] Gallai, T., Über extreme Punkt-und Kantenmengen. *Ann. Univ. Sci. Budapest, Eötvös Sect. Math.*, **2** (1959), 133-138

[71] Gallai, T., On directed paths and circuits. in *Theory of Graphs* (edited by P. Erdös and G. Katona), Academic Press, New York, 1968, 115-118

[72] Geller, D., and Harary, F., Connectivity in digraphs. In *Recent Trends in Graph Theory* (edited by M. Capobuanco, J. B. Frechen and M. Krolik), Lecture Notes in Mathematics, **186**, Springer-Verlag, 1971, 105-115

[73] Ghouila-Houri, A., Une Condition suffisance déxistece dún circuit hamiltonien. *C. R. Acad. Sci. Paris*, **251** (1960), 495-497

[74] Gibbons, A., *Algorithmic Graph Theory.* Cambrisge Unversity Press, Cambridge, Loandon, 1985

[75] Good, I. J., Normal recurring decimals. *J. London Math. Soc.*, **21** (1946), 167-169

[76] Grarey, M. R., and Johnson, D. S., *Computers and Intractability, A guide to the theory of NP-completeness.* W. H. Freeman, 1979

[77] Graver, J. E. and Yackel, J., Some graph theoretic results associated with Ramsey's theorem of Turán (in Hungaian). *J. Combin. Theory*, **4** (1968), 125-175

[78] Grinberg, È. Ja., Plane homogeneous graphs of degree three without Hamiltonian circuits (in Russian). *Latvian Math. Yearbook*, **4**(1968), 51-58

[79] Grinstead, C. M. and Roberts, M., On the Ramsey numbers $R(3,8)$ and $R(3,9)$. *J. Combin. Theory* (B), **33**(1) (1982), 27-51

[80] Guan, M. K., Graphic programming using odd or even points (in Chinese). *Chinese Math.*, **10** (1960), 263-266

[81] Guy, R. K., Crossing number of graphs. *Graph Theory and Applications*, Springer-Verlag, New York, (1972), 111-124

[82] Hall, P., On representatives of subsets. *J. London Math. Soc.*, **10** (1935), 26-30

[83] Halmos, P. R., and Vaughan, H. E., The marriage problem. *Amer. J. Math.*, **72** (1950), 214-215

[84] Hamidoune, Y. O., Sur les atomes d'un graph oriente. *C. R. Acad. Sci. Paris* (A), **284** (1977), 1253-1256

[85] Hamidoune, Y. O., Connectivity of transitive digraphs and a combinatorial property of finite groups. *Annals of Discrete Math.*, **8** (1980), 61-64

[86] Hamidoune, Y. O., An application of connectivity theory in graphs to factorizations of elements in groups. *Europ. J. Combin.*, **2** (1981), 349-355

[87] Hamidoune, Y. O., Quelques problèmes de connexit e dans les graphes orientès. *J. Combin. Theory* (B), **30** (1981), 1-10

[88] Harary, F., *Graph Theory.* Addison-Wesly Publishing Company, 1969

[89] Haynes, T.W., Hedetniemi, S.t., and Slater, P.J., *Fundaments of Domination in Graphs.* Marcel Dekker, New York, 1998

[90] Haynes, T.W., Hedetniemi, S.t., and Slater, P.J., *Domination in Graphs: Advanced Topics.* Marcel Dekker, New York, 1998

[91] Heawood, P. J., Map colour theorem. *Quart. J. Pure. Appl. Math.*, **24** (1890), 332-338

[92] Heesch, H., Untersuchungen zum Vierfarbenproblem. Hochschulskriptum, **810**/a/b, Bibliographisches Institut, Mannheim, 1969

[93] Hieholzer, C., Über die Möglichkeit, einen Linienzug ohne Wiederholung und ohne Unterbrechnung zu umfahren. *Math. Ann.*, **6** (1873), 30-32

[94] Higman, D. G., and Sims, C. C., A simple group of order 44,352,000. *Math. Z.*, **105** (1968), 110-113

[95] Hillis, W. D., *The Connection Machine.* Cambridge, Mass.: MIT Press, 1985

[96] Hochbaum, D. S., *Approximation Algorithms for NP-hard Problems.* PWS Publishing Company, 1997

[97] Hoffman, A. J., and Singleton, R. R., On Moore graphs with diameters 2 and 3. *IBM J. Research and Development*, **4** (1960), 497-504.

[98] Holyer, I., The NP-completeness of edge-colouring. *SIAM J. Comput.*, **10** (1981), 718-720

[99] Hopcroft, J., and Tarjan, R. E., Efficient palarity testing. *JACM*, **21** (4) (1974), 549-568

[100] Imase, M., Soneoka, T., and Okada, K., Fault-tolerant processor interconnection networks. *Systems and Computers in Japan*, **17** (8) (1986), 21-30

[101] Jolivet, J. L., Sur la connexite des graphes. *C. R. Acad. Sc. Paris* (A), **274** (1972), 148-150

[102] Kalbfleisch, J. G., Chromatic graph and Ramsey's theorem. Ph. D. thesis, University of Waterloo, January, 1966

[103] Kautz, W. H., Design of optimal interconnection networks for multiprocessor. In *Architecture and Design of Digital Computers*, Nato Advanced Summer Institute, 1969, 249-272

[104] Kempe, A. B., On the geographical problem of four colours. *Amer. J. Math.*, **2** (1879), 193-200

[105] Kennedy, J. W., Quintas, L. V., and Syslo, M. M., The theorem on planar graphs. *Historia Math.*, **12** (1985), 356-368

[106] Kirchhoff, R. G., Über die Auflösung der Gleichungen, auf welche man bei der Untersuchung der linearen Vertheilung galvanischer Ströme geführt wird. *Ann. Phys. Chem.*, **72** (1847), 497-508. English translation, *IRE Trans. Circuit Theory*, **5**(1) (1958), 4-7

[107] Kirkman, T. P., On the representation of polyedra. *Phil. Trans. Roy. Soc. London*, **146** (1856), 413-418

[108] Klein, M., A primal method for minimal cost flows. *Man. Sci.*, **14** (1967), 205-220

[109] Kleitman, D. J., The crossing number of $K_{5,n}$. *J. Combin. Theory*, **9** (1970), 315-323

[110] Klotz, W., A constructive proof of Kuratowski's theorem. *Ars Combin.*, **28** (1989), 51-54

[111] König, D., Über Graphen und ihre Anwendung auf Determinatentheorie und Mengenlehre. *Math. Ann.*, **77** (1916), 453-465

[112] König, D., Graphs and matrices (in Hungarian). *Mat. Fiz. Lapok*, **38** (1931), 116-119

[113] König, D., *Theorie der endlichen und unendlichen Graphen.* Leipzig: Ajademische Verlagsgesell schaft, 1936; Reprinted Chelsea, New York, 1950

[114] Kruskal, J. B. Jr., On the shortest spanning subtree of a graph and the saleman problem. *Proc. Amer. Math. Soc.*, **7** (1956), 48-50

[115] Kuhn, H. W., The Hungarian method for assignment problem. *Naval. Res. Logist. Quart.*, **2** (1955), 83-97

[116] Kuratowski, C., Sur le problème des courbes gauches en topologie. *Fund. Math.*, **15** (1930), 271-283

[117] Lawler, E.L., Lenstra, J.K., Rinnooy Kan, A.H.G., and Shmoys, D.B., *The Travelling Salesman Problem.* John Wiley, 1985

[118] Liu, Bolian and Lai, Hong-Jian, *Matrices in Combinatorics and Graph Theory.* Kluwer Academic Publishers, Dordrecht/Boston/London, 2000.

[119] Liu, Yanpei, Boolean planarity characterization of graphs. *Acta Mathematica Sinica*, **4**(4) (1988), 316-329

[120] Lovász, L., Connectivity in digraphs. *J. Combin. Theory* (B), **15** (1973), 174-177

[121] Lovász, L., Three short proofs in graph theorey. *J. Combin. Theory* (B), **19** (1975), 111-113

[122] Lovász, L., and Plummer, M.D., *Matching Theory.* Annals of North-Holland Mathematics Studies, **121**, Discrete Mathematics (29), North-Holland, Amsterdam/New York/Oxford/Tokyo, 1986

[123] Mader, W., Eine Eigenschft der Atome andlicher Graphen. *Archives of Mathematics* (Basel), **22** (1971), 333-336

[124] Mader, W., Grad und lokaler Zusammenhang in endlichen Graphen. *Math. Ann.*, **205** (1973), 9-11

[125] Makarychev, Y., A short proof of Kuratowski's graph pplanarity criterion. *J. Graph Theory*, **25** (1997), 129-131

[126] Mckay, B. D., and Zhang Ke Min, The value of the Ramsey number $R(3,8)$. *J. Graph Theory*, **16**(1) (1992), 99-105

[127] Mckee, T. A., Recharacterizing Eulerian: intimations of duality. *Discreter Math.*, **51** (1984), 237-242

[128] McCuaig, W., A simple proof of Menger's theorem. *J. Graph Theory*, **8** (1984), 427-429

[129] McLane, S., A combinatorial condition for planar graphs. *Fund. Math.*, **28** (1937), 22-32

[130] Menger, K., Zur allgemeinen Kurventheorie. *Fundamenta Mathematicae*, **10** (1927), 96-115

[131] Menger, K., On the origin of the n-are therorem. *J. Graph Theory*, **5** (1981), 341-350

[132] Meyniel, H., Une condition suffisante d'existence d'un circuit hamiltionien dans an graphe oriente. *J. Combin. Theory* (B), **14** (1973), 137-147

[133] Moon, J. W., On subtournaments of tournament. *Canad. Math. Bull.*, **9** (1966), 297-301

[134] Moon, J. W., and Pallman, N. J., On the power of tournament matrices. *J. Combin. Theory*, **3** (1967), 1-9

[135] Moore, E. F., The shortest path through a maze. In *Proc. International Symopium on the Theory of Switching*, Part II, Cambridge Mass: Harward Univ. Press, 1957, 285-292

[136] Moroń, Z., On the dissection of a rectangle into squares. *Przeglad Mati Fiz.*, **3** (1925), 152-153

[137] Munkres, J., Algorithms for assignment and transportation problems. *J. Soc. Indust. Appl. Math.*, **5** (1957), 32-38

[138] Nash-Williams, C. St. J. A., Orientation, connectivity and odd-vertex pairing in finite graphs. *Canad. J. Math.*, **12** (1960), 555-567

[139] Nash-Williams, C. St. J. A., Hamiltonian circuits in graphs and digraphs. *The Many Facets of Graph Theory, Lecture Notes in Mathematica*, Springer-Verlag, **110** (1969), 237-243

[140] Nash-Williams, C. St. J. A., and Tutte, W. T., More proof of Menger's theorem. *J. Graph Theory*, **1** (1977), 13-14

[141] Ore, O., Note on Hamilton circuits. *Amer. Math. Month.*, **67**(1960), 55

[142] Papadimitriou, C., and Yannakakis, M., The travelling salesman problem with distances one and two. *Math. Oper. Research*, **18** (1993), 1-11

[143] Petersen, J., Die Theorie der regulären Graphs. *Acta Math.*, **15** (1891), 193-220

[144] Peyrat, C., Diameter vulnerability of graphs. *Discrete Applied Math.*, **9** (1984), 245-250

[145] Plesnik, J., Critical graphs of given diameter. *Acta Facultatis Rerum Naturalium Universitatis Comenianae -Mathematics*, **30** (1975), 71-93

[146] Plesnik, J., and Znam, S., Strongly geodetic directed graphs. *Acta Fac. Rerum Natur. Univ. Comenian. Math.*, **29** (1974), 29-34

[147] Prim, R. C., Shortest connection networks and some generations. *Bell System Tech. J.*, **36** (1957), 1389-1401

[148] Rado, R., Note on the transfinite case of Hall's theorem on representatives. *J. London Math. Soc.*, **42** (1967), 321-324

[149] Ramsey, F. P., On a problem of formal logic. *Proc. London Math. Soc.*, **30** (1930), 264-286

[150] Rédei, L., Ein kombinatorischer Satz. *Acta. Litt. Sci. Szeged*, **7** (1934), 39-43

[151] Rizzi, R., A short proof of Köning's matching theory. *J. Graph Theory*, **33** (2000), 138-139

[152] Robbins, H. E., A theorem on graphs, with an application to a problem of traffic control. *Amer. Math. Monthly*, **46** (1939), 281-283

[153] Robertson, N., Sanders, D., Seymour, P., and Thomas, R., The four-color theorem. *J. Combin. Theory* (B), **70** (1997), 2-44

[154] Rosenblat, D., On the graphs and asymptotic forms of finite boolean relation matrices. *Naval. Res. Quart.*, **4** (1957), 151-167

[155] Roy, B., Nombre chromatique et plus longs chemins d'un graphe. *Rev. Francaise Automat. Informat. Recherche Opérationelle Sér. Rouge*, **1** (1967), 127-132

[156] Sahni, S., and Gonzalez, T. *P*-complete approximation problems. *JACM*, **23** (1976), 555-565

[157] Sprague, R., Beispiel einer Zerlegung des Quadrats in lauter verschiedene Quadrate. *Math. Z.*, **45** (1939), 607-608

[158] Tait, P. G., On the colouring of maps. *Proc. Royal Soc. Edinburgh*, **10** (1878-1880), 501-503

[159] Thomassen, C., Kuratowski's theorem. *J. Graph Theory*, **5** (1981), 225-241.

[160] Toida, S., Properties of an Euler graph. *J. Franklin Inst.*, **295** (1973), 343-345

[161] Turán, P., An extremal problem in graph theory (in Hungarian). *Mat. Fiz. Lapok*, **48** (1941), 436-452

[162] Tutte, W. T., On Hamiltonian circuits. *J. London Math. Soc.*, **21** (1946), 98-101

[163] Tutte, W. T., The factoeization of linear graphs. *J. London Math. Soc.*, **22** (1947), 107-111

[164] Tutte, W. T., *Introduction to Matroid Theory*. Elsevier, New York, 1971.

[165] Tverberg, H., A prrof of Kuratowski's theorem. *Ann. Discrete Math.*, **41** (1989), 417-419

[166] Vizing, V. G., On an estimate of the chromatic class of a P-graph. *Diskret. Analiz.*, **3** (1964), 25-30

[167] Vizing, V. G., Critical graphs with a given chromatic class. *Diskret. Analiz.*, **5** (1965), 9-17

[168] Vizing, V. G., Some unsolved problems in graph theory. *Uspekhi Mat. Nauk*, **23** (1968), 117-134 = *Russian Math. Surveys*, **23** (1968), 125-142

[169] Wagner, K., Bemerkunman zum Vierfarbenproblem. *Jahresber. Deut. Math.-Verein.* **46** (1936), 26-32

[170] Wagner, K., Über eine Eigenschaft der ebenen Komplexe. *Math. Ann.*, **114** (1937), 57-590

[171] Watkins, M., Connectivity of transitive graphs. *J. Combin. Theory*, **8** (1970), 23-29

[172] Wernicke, P., Über den kartographischen Viefarbensatz. *Math. Ann.*, **58** (1904),413-426

[173] Whitney, H., Non-separable and planar graphs. *Proc. Nat. Acad. Sci. U. S. A.*, **17** (1931), 125-127

[174] Whitney, H., Non-separable and planar graphs. *Trans. Amer. Math. Soc.*, **34** (1932), 339-362

[175] Whiting, P. D., and Hillier, J. A., A method for finding the shortest route through a road network. *Operational Res. Quart.*, **4** (1960), 37-40

[176] Whitney, H., Congruent graphs and the connectivity of graphs. *Amer. J. Math.*, **54** (1932), 150-168

[177] Wielandt, H., Unzerlegbare, nicht negative matrizem. *Mathematische Zeitschrift*, **52** (1950), 642-648

[178] Woodall, D. R., Sufficient conditions for circuits in graphs. *Proc. London Math. Soc.*, **24**(1972), 739-755

[179] Xu, Junming, Local Strongly arc-connectivity in regular bipartite digraphs. *J. Combin. Theory* (B), **59**(2) (1993), 188-190

[180] Xu, Junming, A very short proof of Vizing's theorem. *J. China Univ. Sci. Tech.*, **29**(2) (1999), 199-201

[181] Xu, Junming, *Topological Structure and Analysis of Interconnection Networks.* Kluwer Academic Publishers. Dordrecht/Boston/Loandon, 2001

[182] Xu, Junming, Tao, Yingfeng and Xu, Keli, Note on bounded length paths of de Bruijn digraphs. *J. China Univ. Sci. Tech.*, **32** (2002), 19-21

[183] Yamada, T., and Kinoshita, H., Finding all the negative cycles in a directed graph. *Discrete Applied Math.*, **118** (2002), 279-291

[184] Zhang, Limin, Every planar graph with maximum degree 7 is of class 1. *Graphs and Combinatorics*, **16** (4) (2000), 467-495

List of Symbols

This list gives the symbols used in the book, but not complete.

General Mathematical Symbols

$\setminus$: set-theoretic difference
$\overline{S}$: $= V \setminus S$, $S \subset V$
$|S|$: cardinality of the set S
$\times$: cartesian product
$\lceil r \rceil$: smallest integer not less than the real number r
$\lfloor r \rfloor$: greatest integer not exceeding the real number r
$\cup$: union
$\cap$: intersection
$\subseteq$: subset
$\subset$: proper subset
$\binom{k}{n} := \dfrac{n(n-1)\cdots(n-k+1)}{k!}$, $k \le n$
g.c.d. $(n_1, n_2, \cdots, n_l)$: the greatest common divisor of $n_1, n_2, \cdots, n_l$

Graph-Theoretic Symbols with the English Alphabet

Graph-Theoretic Symbols with the Greek Alphabet

Graph-Theory Operations

Index